W9-AFT-653

COLORIMETRIC DETERMINATION OF ELEMENTS

Original title

DOSAGES COLORIMÉTRIQUES DES ÉLÉMENTS MINÉRAUX. PRINCIPES ET MÉTHODES

DEUXIÈME EDITION
ENTIÈREMENT REFONDUE
MASSON ET CIE., PARIS, 1961

SOME SECTIONS HAVE BEEN REVISED BY THE AUTHOR
FOR THIS FIRST ENGLISH EDITION

TRANSLATED BY
EXPRESS TRANSLATION SERVICE,
28 ALEXANDRA ROAD,
WIMBLEDON, LONDON

COLORIMETRIC DETERMINATION OF ELEMENTS

PRINCIPLES AND METHODS

by

G. CHARLOT

Professor of Analytical Chemistry, Faculté des Sciences, École Supérieure de Physique et de Chimie Industrielles, Paris

ELSEVIER PUBLISHING COMPANY

AMSTERDAM - LONDON - NEW YORK

1964

ELSEVIER PUBLISHING COMPANY
335 JAN VAN GALENSTRAAT, P.O. BOX 211, AMSTERDAM

AMERICAN ELSEVIER PUBLISHING COMPANY, INC.
52 VANDERBILT AVENUE, NEW YORK 17, N.Y.

ELSEVIER PUBLISHING COMPANY LIMITED
12B, RIPPLESIDE COMMERCIAL ESTATE
RIPPLE ROAD, BARKING, ESSEX

QD
113
.C513

LIBRARY OF CONGRESS CATALOG CARD NUMBER 63-19819

WITH 72 ILLUSTRATIONS AND 12 TABLES

ALL RIGHTS RESERVED. THIS BOOK OR ANY PART THEREOF
MAY NOT BE REPRODUCED IN ANY FORM,
INCLUDING PHOTOSTATIC OR MICROFILM FORM,
WITHOUT WRITTEN PERMISSION FROM THE PUBLISHERS

PREFACE

This new edition was undertaken because of the rapid strides which have been made in colorimetric determinations since the first edition was published. As before, it is intended for use by the practising chemist.

In the first part will be found an account of the principles necessary for full understanding of the method. Also included are accounts of the newer separation methods, particularly those of extraction and ion-exchange.

Emphasis has been given throughout to the most accurate estimations which are now possible through the development of differential methods using double-beam spectrophotometers. The growing importance of spectrophotometric titrations has also been stressed.

The second part of the book deals with the actual estimation of the more important elements. Methods of separation and selected estimation procedures are given for each one. This section of the work is not simply a compilation of well-known methods; indeed, it incorporates not only new methods selected and developed in our own laboratories, but also experience gained in the following organisations with which the author is connected: Commission des Méthodes d'Analyse du Commissariat à l'Energie Atomique, Association Française de Normalisation, Sections de Chimie Analytique de la Société Chimique de France, Société de Chimie Industrielle, etc. Numerous discussions with industrial laboratories either directly or through former pupils have also played their part in the selection of methods, although not all those quoted have been subjected to this treatment. On the other hand, many elements of particular interest at the present time [uranium, thorium, titanium, zirconium, niobium, tantalum, boron, beryllium etc.] have been especially studied, both in a general way and in the determination of small traces.

I am conscious of the fact that many imperfections must be present in both sections of the book, and will greatly welcome any comments or criticisms which the reader may send to me.

GASTON CHARLOT

Paris
December 1963

54346

CONTENTS

PART I. THEORETICAL BACKGROUND

PART II. ESTIMATION OF THE PRINCIPAL ELEMENTS

Part I

Theoretical Background

Characteristics of the colorimetric method

The names absorptiometry or absorption spectrophotometry refer to a very widely applicable method, in which a beam of light of a given wavelength is analyzed after passing through the substance to be determined. The latter is generally a solution, but may also be a gas or even a solid. The concentration of the absorbing substance is deduced from the proportion of the luminous intensity absorbed by the solution.

The spectral ranges covered by various radiations are given below:

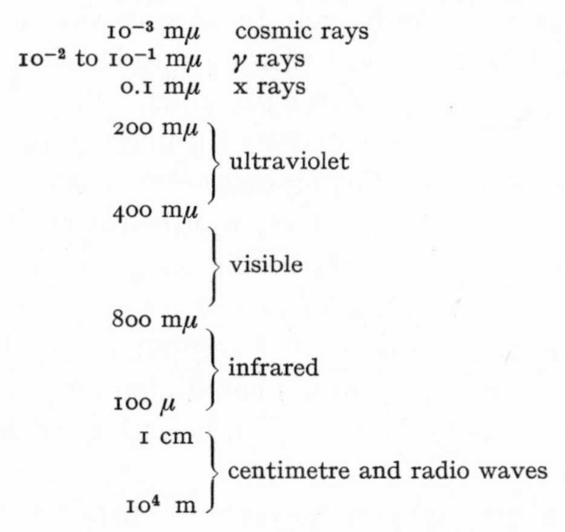

We shall here particularly concern ourselves with the methods relating to that part of the spectrum which extends from ultraviolet (220 mμ) to the beginning of infrared (1 μ), as the above regions are of interest in the case of inorganic compounds. The same principles are applicable, however, with some variations, to other parts of the spectrum, and particularly to the infrared region.

The method is called absorptiometry, absorption photometry, spectrophotometry, and more usually, although improperly, colorimetry. (In actual fact the term 'colorimetry' denotes the methods of analysis, specification, and description of colours). The term 'colorimetric methods of analysis' can however be used without ambiguity.

Colorimetric methods of analysis are at present undergoing considerable development and may, assuming a judicious choice of conditions, be applied to a great number of estimations with adequate accuracy. At the same time, a wide and growing variety of superior quality instruments is now becoming available commercially.

The field of application and advantages of the method. Formerly, 'colorimetry' was almost exclusively confined to the estimation of trace amounts, owing to its poor degree of accuracy. At the present time, however, improved methods of instrumentation and the use of the differential method sometimes permit an accuracy to be obtained which is comparable with, or better, than that of current volumetric methods (0.2–0.5%). Absorptiometry has acquired an even greater importance than volumetric estimation, and is today the most important method of analysis. Its main advantages are the following:

(a) The method is capable of very general application; thus if the substance to be determined exhibits only low absorption, suitable reagents may be added to form compounds of higher absorbing power.

(b) The method may be made extremely sensitive, particularly if organic reagents are used. The estimation of trace amounts is now carried out by this method more frequently than by any other.

(c) The method can be extremely rapid since direct measurements may be carried out without the necessity of using standardized solutions and owing to the simplicity of the experimental procedure. Moreover, separations can be simplified or avoided, since to eliminate the effect of interfering ions, all factors which are used in the chemical manipulation of solutions are also available here: selection of reagent, oxidation-reduction, adjustment of pH, formation of complexes, and the use of organic solvents, in addition to a supplementary factor, *i.e.* selection of the wavelength used.

(d) Finally, the method lays open a very important possibility. We shall see that, by comparison with a solution containing only the interfering materials ('blank test'), a material can very often be estimated in the presence of other absorbing substances.

In addition, 'colorimetry' permits a reaction to be followed and its equivalence point to be determined (see *Colorimetric Titrations*, discussed in Chapter 7).

Chapter 1

Beer's Law. Additivity of optical densities

(1) The Lambert-Beer law

Consider a beam of monochromatic light passing through a thickness l of a solution of an absorbing material (Fig. 1) and let I_0 and I be the respective

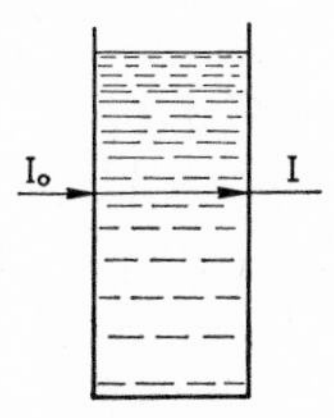

Fig. 1.

intensities of the beam before entering and after leaving the solution. If the concentration of the absorbing material is c, then according to the Lambert-Beer law

$$\log \frac{I_0}{I} = \varepsilon l c.$$

This law states that the proportion of light absorbed does not depend on the incident intensity, and that it varies as the total number of absorbing molecules or ions encountered by the light; *i.e.* that l and c play the same role.

(i) Definitions. The 'transmittance' T is defined as the ratio of the two luminous intensities $T = I/I_0$. The optical density D (or extinction E, or absorbance A) is the logarithm of the inverse ratio $D = \log I_0/I$. The length l is always expressed in centimetres. ε is called the molar extinction coefficient if the concentration is expressed in gram-ions or moles per litre.

NOTE. Beer's law remains valid only if the luminous energy absorbed is converted into thermal energy. Deviations from the law are observed if any fluorescence occurs, or if the solution is colloidal (loss of light by reflection and diffusion).

(ii) Properties of the extinction coefficient. According to Beer's law, ε is independent of the concentration. In the case of solutions, it depends on the dis-

solved material, the wavelength and the temperature, being in principle independent of the solvent.

As an example, the molar extinction coefficient ε of the permanganate ion is plotted as a function of the wavelength λ in Fig. 2, showing that permanganate possesses an absorption band (region of maximum absorption) at about 500–570 mμ.

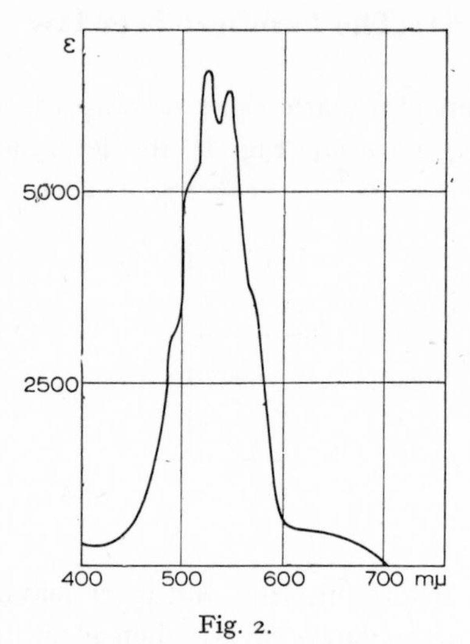

Fig. 2.

The extinction coefficient can vary considerably, the highest values known being in the region of 100,000 (for rhodamine B chloroantimonate).

The shape of absorption curves, such as that illustrated in Fig. 2, depends on the temperature, the absorption bands being as a rule displaced towards longer wavelengths at higher temperatures. The influence of temperature on ε is therefore particularly marked when the slope $d\varepsilon/d\lambda$ of the curve $\varepsilon = f(\lambda)$ is large, *i.e.* at the edges of the absorption bands.

For example, the extinction coefficient of the *o*-dinitrophenoxide ion, which has an absorption maximum at about 336 mμ, varies by -0.3% per degree at 334 mμ, $+1\%$ per degree at 436 mμ, and is temperature independent at 366 mμ (G. Kortüm, *Z. Phys. Chem.*, 170 (1934) 212).

The variations are generally of this order of magnitude.

(2) Additivity of optical densities

Consider a monochromatic beam of light passing successively through two different solutions of the same thickness l, and let ε_1 and c_1, and ε_2 and c_2, be their respective extinction coefficients and concentrations (Fig. 3).

The total optical density is then

$$D = \log \frac{I_0}{I_2} = \log \frac{I_0}{I_1} \cdot \frac{I_1}{I_2} = \log \frac{I_0}{I_1} + \log \frac{I_1}{I_2} = D_1 + D_2.$$

This value depends only on the number of absorbing particles encountered and would be the same if the two absorbing compounds were dissolved, in the same concentrations, in a single solution of thickness l (Fig. 4).

This may, in general, be written:

$$D = \varepsilon_1 l c_1 + \varepsilon_2 l c_2 + \ldots . \varepsilon_n l c_n.$$

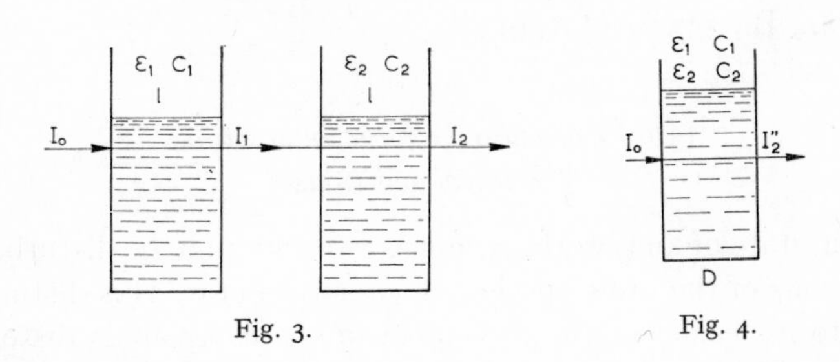

Fig. 3. Fig. 4.

This relationship is important being, as will be shown, the basis of most determinations and, in particular, permitting the estimation of absorbing materials in the presence of one another.

Importance of Beer's law. The extent to which Beer's law holds true is an important factor, since it makes available a measurable magnitude D which is proportional to the concentration to be determined. Beer's law need of course only hold true to the extent that the experimental error involved in applying the formula is negligible, *i.e.* the approximation can be of the order of 0.2%, 1%, etc., according to circumstances.

When this law is not obeyed, the calibration of the system prior to measurement is complicated (see p. 51), and mixtures of coloured materials and also the performance of a blank test present problems which are relatively difficult to solve (see p. 63).

It is therefore important to know what are the reasons for failures of Beer's law and to determine the limits of its application.

(3) Validity of Beer's law

Beer's law is a limiting law which requires ideal conditions both for the beam of light, which must be monochromatic, and for the solution, in which the absence of any action on the absorbing material not due to the beam of light is assumed. We shall consider in turn several diffent causes of any deviations which may arise.

(3a) *Refractive index of the solution*

The coefficient ε depends on the refractive index n of the solution. According to theory of dispersion it is not ε, but the function $\frac{\varepsilon n}{(n^2 + 2)^2}$ which is independent of the concentration. Generally n, and consequently ε, increase with the concentration of the dissolved material (since $n > 1$).

In the majority of measurements the relative error rarely reaches 0.1% and is therefore negligible if the concentration does not exceed 10^{-2} *M*. This is no longer the case at higher concentrations and in certain particular cases. *e.g.* eosin in 0.07 *M* solution gives a deviation of 1.8% at 527 mμ (G. KORTÜM, *Z. Phys. Chem.*, B., 33 (1936) 243).

(3b) *Molecular or ionic interactions*
Chemical reactions

The electronic state of an absorbing molecule or ion may be disturbed by other molecules or ions of the same species which surround it. This disturbance may also be due to molecules of the solvent or of the compounds dissolved in the same solution. These effects are very variable, ranging from simple mutual 'orientation' and 'association' between polar molecules to chemical reactions in the fullest sense.

This explains the (sometimes very slight) variations which can be observed on passing from one solvent to another. For example, the absorption maximum of benzeneazophenol changes from 348 to 351 mμ and the corresponding extinction coefficient from 26,100 to 26,300 when normal butyl alcohol is used as a solvent instead of methanol (according to W. R. BRODE, *J. Phys. Chem.*, 30 (1926) 56). This interaction is also the cause of the slight variations in ε which may be observed on addition of a large excess of a nonabsorbing compound.

The interactions between absorbing particles increase with their concentration and consequently lead to a variation of ε with c. These interactions can only be determined experimentally. For example, the molar coefficient of the chromate ion CrO_4^{2-} in the presence of 5×10^{-3} *M* sodium hydroxide varies by 0.76% at 436 mμ when the concentration changes from 6.8×10^{-4} to 3.7×10^{-2} *M*, and by only 0.02% at 303 mμ when the concentration changes from 9.8×10^{-4} to 1.9×10^{-2} *M* although no known chemical reaction takes place (G. KORTÜM, *Z. Phys. Chem.*, B, 33 (1936) 243).

A further example of disturbance due to foreign salts is also given by KORTÜM (*Z. Phys. Chem.*, B, 30 (1935) 317); the molar coefficient of 10^{-4} *M* dinitrophenoxide ion in a 10^{-3} *M* solution of potassium hydroxide at 436 mμ varies by 0.5% when $10^{-1.3}$ *M* sodium perchlorate or $10^{-2.6}$ *M* lanthanum nitrate is added.

(i) Chemical equilibria. In contrast to these small variations, the deviations

caused by the recurrence of chemical reactions are often very important. Thus, the dimerization of methylene blue causes its molar extinction coefficient (at pH 3.4 and 656.5 mμ) to change from 38,800 to 16,700 when the concentration is increased from 2×10^{-6} to 2×10^{-3} M (E. RABINOVITCH AND L. F. EPSTEIN, *J. Amer. Chem. Soc.*, 63 (1941) 69).

In this case, the deviation is predictable and can sometimes be reduced by certain operative procedures.

Consider, for example, a complex AB which dissociates according to $AB \leftrightarrows A + B$, with $\frac{[A]\,[B]}{[AB]} = K$. Let α be the degree of dissociation (the fraction of [total AB] dissociated) and c_0 the total concentration. Then:

$$\frac{\alpha^2 c^2_0}{c_0(1 - \alpha)} = K, \quad \text{or} \quad \frac{\alpha^2}{1 - \alpha} = \frac{K}{c_0}$$

It is thus clear that further dissociation of the complex may be promoted by lowering c_0. Values of [AB] as a function of the total concentration for the cases $K = 10^{-4}$ and $K = 10^{-10}$ are tabulated below to illustrate this point (Table 1).

TABLE 1

c_0	*[AB] for* $K = 10^{-4}$	*[AB] for* $K = 10^{-10}$
10^{-1}	0.997×10^{-1}	0.999×10^{-1}
10^{-2}	0.905×10^{-2}	0.999×10^{-2}
10^{-3}	0.78×10^{-3}	0.997×10^{-3}
10^{-4}	0.40×10^{-4}	0.990×10^{-4}

The same argument applies to the dissociation of acids.

When the complex is considerably dissociated, Beer's law is *apparently* no longer obeyed.

If α is to remain less than 1%, it is sufficient that $K/c_0 \leqslant 10^{-4}$, so that the deviation remains negligible, to within 1% or less, as long as $c \geqslant 10^4$ K. In other words, the more stable complexes can be diluted to a greater extent, without disturbances being observed (see G. CHARLOT AND R. GAUGUIN, *Les methodes d'analyse des reactions en solution*, Masson, Paris, 1951).

If the complex is not too unstable, α can still be kept below 1%, in spite of dilution, by adding a large excess of complexing ions.

If the complex is very unstable, the excess concentration of the complexing agent must be kept constant during dilution.

Other expedients may be used in particular cases. For example, let us consider the determination of the concentration of a compound such as bromophenol blue R, which takes part in the equilibrium $R \rightleftarrows R^- + H^+$, where R and R^- are of different colour. The optical density of this system is given by:

$$D = \varepsilon_R lc(1 - \alpha) + \varepsilon_{R^-} \cdot l\alpha c = lc[\varepsilon_R(1 - \alpha) + \varepsilon_{R^-} \cdot \alpha],$$

where c is the total concentration of bromophenol blue. D is proportional to c if α is constant, or if $\varepsilon_R = \varepsilon_{R^-}$. The first condition is realized in a medium buffered with respect to the pH; the second is achieved by carrying out the measurements at the wavelength of the isobestic point ($\varepsilon_R = \varepsilon_{R^-}$).

NOTE. In the case of chemical side reactions, additional deviations from Beer's law are caused by the foreign salts present and by the temperature, apart from their action mentioned above. Thus the dissociation of complexes is increased at higher temperatures and is affected by changes in the ionic strength I, consequent upon the addition of foreign salts to the solution.

For example, in the case of the ferrithiocyanate equilibrium:

$$\frac{c_{Fe^{3+}} \cdot c_{SCN^-}}{c_{FeSCN^{2+}}} = K \frac{f_{FeSCN^{2+}}}{f_{Fe^{3+}} \cdot f_{SCN^-}} = K',$$

where f is the activity coefficient. To a first approximation, K′ is increased by a factor of 5 when the ionic strength I changes from 10^{-2} to 10^{-1}. Consequently, if an excess of SCN^- is added, such that $c_{SCN^-} = K'$ for $I = 10^{-2}$, for $I = 10^{-1}$, we have $c_{SCN^-} = K'/5$; thus α changes from 1/2 to 5/6. If we had taken $c_{SCN^-} = 100\ K'$ for $I = 10^{-2}$, α would have changed from 0.01 to 0.05. The effect of the ionic strength becomes negligible in the case of very stable complexes.

(3c) *Deviations due to the polychromaticity of the light used*

In principle, Beer's law is only valid for ideally monochromatic light. In practice, this is never available.

Consider two beams with different wavelengths λ and λ' and intensities I_0 and I'_0. We have:

$$\log \frac{I_0}{I} = \varepsilon_\lambda lc \text{ and } \log \frac{I'_0}{I'} = \varepsilon_{\lambda'} lc.$$

If Beer's law were obeyed for all of the light used, we should have:

$$\log \frac{I_0 + I'_0}{I + I'} = \varepsilon lc,$$

which obviously cannot be deduced from the previous relation which, on the contrary, gives:

$$\log I_0 + \log I'_0 - (\log I + \log I') = (\varepsilon_\lambda + \varepsilon_{\lambda'})lc.$$

Measuring devices (eye or photoelectric cell) are sensitive to mI + nI′ and not to log I + log I′, and their readings will therefore no longer be proportional to c.

NOTE. It is only in the case where $\varepsilon_\lambda = \varepsilon_{\lambda'}$ that we can write:

$$\frac{I_0}{I} = \frac{I_0}{I'} = \frac{I_0 + I'_0}{I + I'} \text{ and } \log \frac{I_0 + I'_0}{I + I'} = \varepsilon_\lambda lc.$$

More precisely, let us suppose that the total intensity $I_0 + I'_0$ is kept constant. If a greater proportion of the less absorbed radiation is present, the total emergent intensity increases and the apparent optical density is decreased.

It should also be noted that the emergent light is richer in the less absorbed radiations than the incident light. If then a second solution identical with the first is placed in the path of the beam emerging from the latter, by applying the results obtained with a single solution, we see that the second solution is illuminated with a light richer in the less absorbed radiation and that, consequently, its apparent optical density is less than that of the first solution: $D_2 < D_1$. For each monochromatic part of the beam, only the number of absorbing particles encountered comes into consideration, *i.e.* the product lc. As the same argument obviously applies to the superposition of different beams, it may be concluded that when it is no longer l but c which is multiplied by 2, an optical density equal to less than twice the initial value is obtained. A more detailed study would show that the apparent extinction coefficient lies between the extreme values of those of the different incident beams and that, when c increases, it tends towards the minimum of these coefficients. Furthermore, this tendency is the more rapid the richer is the incident light in the less absorbed radiation.

EXAMPLE. Two identical cells 1 cm thick were filled with 0.1 N dichromate. Using a Wratten No. 38 filter (light blue, with a transmission maximum at 460 mμ and a wide transmission band), the optical density of each of them was established as 0.284. When the two cells were placed simultaneously in the path of the light beam, the optical density of the combination was found to be 0.381 instead of 0.568. According to this final result, the apparent optical density of the second cell, is 0.097 instead of 0.284.
In practice, the range of wavelengths used is never zero and ε_λ is not generally constant. The ideal conditions are approached by reducing the range of wavelenghts or by selecting it so that it straddles a maximum (or a minimum) of the $\varepsilon = f(\lambda)$ curve of the material under consideration. Under these conditions, the variations of ε_λ for the beam used are reduced and Beer's law is obeyed, to a certain degree of approximation, over a wider range of concentrations.

EXAMPLE. A permanganate solution has two absorption maxima, at 524 and

534 mμ (Fig. 2). With a filter which allows a 50 mμ band of wavelength to pass with a transmission maximum at about 520 mμ, Beer's law is obeyed to within at least 0.2%, as long as the concentration of the permanganate is less than 2.7×10^{-4} *M* (S. LACROIX AND M. LABALADE, *Anal. Chim. Acta*, 3 (1949) 262).

On the other hand, the absorption maximum of silicomolybdate is in the ultraviolet, and a filter with a transmission maximum in the region of 440 mμ and with a band width of 50 mμ does not therefore yield results consistent with Beer's law. Thus, when the concentration changes from 1 to 5 mg of Si per litre, the molar extinction coefficient changes from 280 to 186, while the optical density remains low (D $\leqslant$ 0.08). It is moreover probable that the complex dissociates slightly on dilution in spite of the presence of an excess of molybdate. (S. LACROIX AND M. LABALADE, *Anal. Chim. Acta*, 3 (1949) 383).

(4) Validity of the principle of additivity of optical densities

The problem to be considered in this section involves cases in which several absorbing substances exist together in a solution.

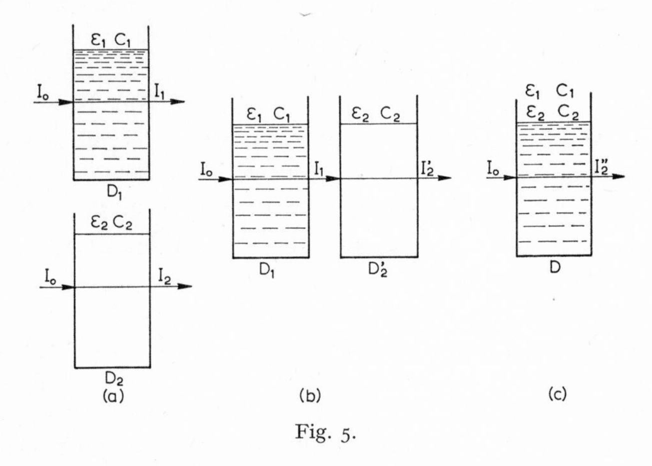

Fig. 5.

Let us consider two different absorbing solutions of known characteristics I_0, ε_1, c_1, l, I_1, and I_0, ε_2, c_2, l, I_2 (Fig. 5,*a*). We have:

$$D_1 = \log \frac{I_0}{I_1} \text{ and } D_2 = \log \frac{I_0}{I_2}.$$

Let us now suppose that solutions 1 and 2 are placed in series in the path of the same beam of light (Fig. 5,*b*); the successive optical densities are then

$$D_1 = \log \frac{I_0}{I_1} \text{ and } D'_2 = \log \frac{I_1}{I'_2}.$$

Finally, let us suppose that we have a solution of thickness l containing the two substances simultaneously in concentrations c_1 and c_2 (Fig. 5,*c*). The optical density of this solution is given by:

$$D = \log \frac{I_0}{I_2''}.$$

If the optical densities may be treated additively, we may write

$$\underset{(1)}{D_1 + D_2 = D_1} + \underset{(2)}{D'_2 = D}.$$

The above additive treatment of the optical densities permits the overall optical density, *i.e.* the optical density measured experimentally, to be related to the individual optical densities. This is very useful, as will be shown later.

1. In monochromatic light, the relationship between the apparent optical density measured and the individual optical densities is always additive. In fact, even if Beer's law does not apply, the deviations can only be due to the properties of the solution (refractive index, intermolecular interactions, or chemical equilibria) and are independent of the intensity of the incident beam.

Equation (1) is therefore valid. Equation (2) will also be valid except in cases where mixing the compounds in question alters the intermolecular forces, causes chemical reactions, or displaces equilibria. Each individual case should be checked experimentally.

2. The deviations due to the polychromatic character of the beam do not depend on the spatial distribution of the molecules encountered, and equation (2) consequently remains valid.

On the other hand, equation (1) no longer applies in this case. We have seen, in fact, that the apparent optical density of a solution depends on the spectral composition of the incident beam if ε is not constant. Consequently, if solution 2 does not satisfy these conditions, the beam I_1 should have the same composition as I_0, *i.e.* ε_1 should be constant if D'_2 is to be equal to D_2.

In this case therefore, one of the two solutions should always have an extinction coefficient independent of the wavelength in the range used.

If this condition is not satisfied, the deviation observed depends on the spectral composition of the incident beam and on the respective shapes and positions of the absorption bands of each of the absorbing materials.

EXAMPLE. Two cells, 1 cm thick, were respectively filled with a 0.5 *M* nickel salt solution and with 0.1 *N* potassium dichromate. Using a Wratten No. 38 filter (light blue, transmission maximum at 460 mμ) the optical density of the nickel solution was found to be 0.295, and that of the dichromate solution 0.280. The total combined optical density of the two cells has been found to be 0.627 instead of 0.575 corresponding to a deviation of +9%.

On the other hand, with a Wratten No. 47 filter (deep blue, transmission maximum at 440 mμ) experimental values of 0.495 for the dichromate (0.01 N), 0.220 for the nickel (0.1 M), and 0.677 for the combination were obtained; since the theoretical value was 0.715, the deviation is of the order of -5%.

Generally speaking, if the condition that ε is constant is only true for k out of the n absorbing substances present in the solution, the additivity relation reduces to:

$$D = \sum^{k} \varepsilon_i l c_i + D',$$

where D′ is the optical density due to the materials not possessing a constant coefficient ε.

The practical consequences of these considerations are very important. In fact, the optical density due to a coloured compound which is to be estimated is often obtained in the following way. A first determination is made of the optical density of the combination of cell, solvent, and all the absorbing compounds present. In a second experiment, the optical density is again measured of the above combination minus the compound whose concentration is to be determined. The difference between the two values obtained corresponds specifically to the optical density of the material to be estimated, measured in a pure solution and using the same incident beam, provided the light used is monochromatic or, if this is not the case, if it and all the other materials obey Beer's law.

It is clearly sufficient that these conditions are satisfied to within 0.2 or 1%, according to the degree of accuracy required.

Chapter 2

Principles of operation and equipment

The present discussion will be confined to the principles of the methods and the equipment used in colorimetry, with particular reference to subject matter of direct interest to the analyst. The characteristics of the various types of apparatus will only be indicated in relation to their use, as the optical or electrical arrangements cannot receive a detailed treatment in a work of this scope.

Determinations of optical density always involve the comparison of two luminous intensities. The methods used can be classified into two groups, differing according to whether the eye or a photoelectric cell is used as the receiver sensitive to luminous intensity.

No mention will be made of photographic methods, which are today of little importance in absorption spectrophotometry.

(1) Methods of visual comparison

The use of the eye is limited to assessing whether two luminous intensities are equal. Moreover, as may be seen from Fig. 6, the sensitivity of a human eye to differences in intensity depends on the wavelength of the light concerned.

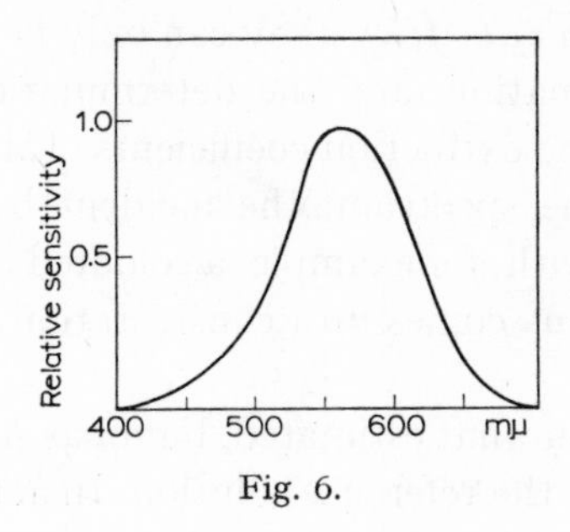

Fig. 6.

In addition, the two beams which are to be compared should enter the eye simultaneously. In general, therefore, visual methods rely on the equalization of the intensities of two light beams, of which one has passed through a reference solution and the other through the solution to be analyzed. The two solutions can thus be brought to the same optical density or, alternatively, the difference in their optical densities can be determined by compensating for the difference in the intensities of the emergent beams with the aid of a calibrated device.

(1a) *Methods of equalizing the optical densities*

The two solutions are illuminated by identical light beams. If the light is monochromatic, solutions which allow the same luminous intensity to pass, *i.e.* which possess the same optical density, are selected; for such solutions

$$\varepsilon lc = \varepsilon' l' c'.$$

Provided the reference solution and the solution to be analyzed contain the same material, the two extinction coefficients in the above equation cancel out, so that

$$lc = l'c',$$

from which c can be calculated if l, l' and c' are known. The use of monochromatic light is not essential and white light is often used, but in such cases the solutions are treated so as to 'equalize' their colours. In fact, if two solutions illuminated by identical beams have the same colour, they must both absorb each individual component of the incident beam to the same extent. The identity

$$\varepsilon_\lambda lc \equiv \varepsilon'_\lambda l'c'$$

must thus be obeyed for every wavelength λ. It is thus clear that the only possible way of achieving this is to make

$$\varepsilon_\lambda \equiv \varepsilon'_\lambda \quad (3) \qquad \text{and} \qquad lc = l'c' \quad (4)$$

If the reference solution also contains the material to be estimated, equation (3) is satisfied automatically, without reference to the spectral composition of the incident light. Equation (4) is used in the form given above.

If the colours of two solutions are not strictly identical, *i.e.* the latter do not possess the same function $\varepsilon_\lambda = f(\lambda)$, they can only be equalized to a greater or lesser degree of approximation, and the determination of one of the concentrations is imprecise. If the extinction coefficients of the two solutions are equal in a certain region of the spectrum, the incident beam must be reduced to radiations in this region with, for example, a coloured filter. The spectral composition of the radiation thus comes into consideration when equation (3) is not satisfied.

A material different from that estimated, but of as similar a colour as possible, is thus sometimes used as the reference solution. In actual fact, however, colour equalization can only be approximate in such cases as it is impossible to achieve a complete identity of the values of ε_λ for all wavelengths. It is therefore advantageous to interpose a coloured filter which eliminates radiations for which the extinction coefficients of the two solutions differ too markedly.

It should be noted that coloured filters may also serve to increase the sensitivity of the measurement by eliminating the wavelengths which are only slightly absorbed.

These principles may be applied in various ways.

(i) Comparison with a series of standards. In this procedure the unknown solution is compared with a series of solutions of the same material, of various concentrations but prepared in an identical manner. In particular, if the coloured material is obtained by the action of a reagent on the material to be estimated, each solution must contain the same excess of the reagent.

Equal volumes of each solution are placed in identical containers, for example, Nessler tubes or graduated test-tubes (Fig. 7), and the standard solution which possesses the colour closest to that of the unknown is then found.

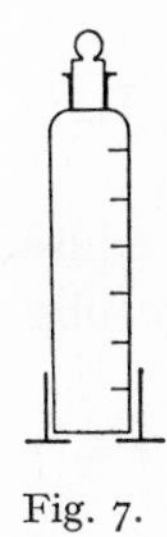

Fig. 7.

Whatever the method of observation, (lateral, vertical, etc.), deviations from Beer's law do not falsify the results, since identical conditions prevail in each case, and accuracy is only limited by the lack of sensitivity of the eye to small differences in luminous intensities. For the latter reason the variations in the concentration of different standard solutions should not be made too slight.

A fixed scale of permanently standardized coloured solutions of organic or inorganic materials is frequently used, particularly when the material to be estimated is unstable.

EXAMPLE. Soluble silica can be estimated colorimetrically in the form of the yellow silicomolybdate complex with the aid of permanently stable, standard chromate solutions.

(ii) The dilution method. This may be exemplified as follows: A known volume of the coloured solution to be analyzed is placed in a test tube, and the solvent in another, identical one; a measured volume of a standardized solution of the coloured material to be estimated is then added to the tube containing the solvent until the two colours become equal. If the volumes of the two solutions were also equal, this would be identical with the preceding case. Whatever method of observation is used, the two test tubes contain the same amount (and the same concentration) of the coloured material.

EXAMPLE. Manganese may be estimated colorimetrically after conversion into permanganate. A standard solution of the same concentration is made by diluting a standardized permanganate solution until the two solutions are of the same colour.

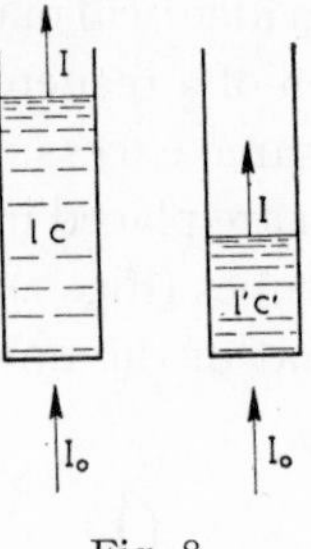

Fig. 8

To simplify the experimental procedure, it is frequent practice to neglect the volume of the solutions, which are observed vertically (Fig. 8). When the colours are identical,

$$lc = l'c'.$$

The two solutions are identical, and the columns have the same cross-section S, so that:

$$Slc = Sl'c', \quad \text{or} \quad Vc = V'c'.$$

The two solutions therefore contain the same amount of coloured material.

This simplification can only be used if the substance to be estimated does not take part in any chemical equilibrium and change its apparent extinction coefficient on dilution. This difficulty may however be avoided in certain cases. Thus, for example, when the coloured compound formed by the addition of the reagent to the material to be estimated dissociates on dilution, the initial solvent may be replaced by a solution of the reagent of the same concentration as in the unknown solution. In accordance with what has already been said, the apparent extinction coefficient can be made independent of dilution if the concentration of the reagent is sufficiently large.

Due to the relatively approximate nature of such methods, other sources of error—refractive index and molecular interactions—may generally be neglected, particularly since the concentrations c and c' are in any case never very different.

NOTE. An interesting variation consists of the so-called 'sensitive shade' method. The principle is as follows: when two differently coloured substances are present in a single solution, if the ratio of their concentrations is changed, it is found that a certain range of compositions exists in which a small variation of this ratio is accompanied by a pronounced change of colour; the latter is called the 'sensitive shade'.

For example, dithizone is green in a solution of chloroform whilst the metallic complexes which it forms are often red; in consequence mixtures of dithizone and such complexes exhibit colours varying from green to red, through a sensitive shade—grey. In practice, an excess of dithizone is added to the solution to be analyzed until the sensitive shade is obtained. A comparison solution containing the same total quantity of dithizone is then made up, and a standardized solution of the metal ion to be estimated is added until the same shade is obtained. The amounts of metal in the two solutions are then equal.

Methods employing dilution or comparison against a series of standards are interesting because they do not require any special apparatus; they are however, of low precision. The relative error is often in the neighbourhood of 5%, but may in certain unfavourable cases reach 10 or even 20%. Under the most favourable conditions, this error may be reduced to as low as 2%. Such methods are generally suitable for the estimation of trace amounts, where extreme accuracy is in general not essential.

(1b) *Methods of determining the optical densities*

In principle, the intensities of the two beams to be compared are diminished in one case by absorption during its passage through the unknown solution, and in the other by a calibrated device which can be adjusted until the luminous intensities are equal. This gives the optical density of the solution directly.

The concentration of the solution may then be calculated from the equation $D = \varepsilon lc$. The extinction coefficient ε is determined by a previous measurement made on a solution of known concentration and under strictly identical conditions; this value may be used repeatedly.

If Beer's law is to be obeyed, it is essential that the light used should be as monochromatic as possible, unless the extinction coefficient ε is reasonably constant in the wavelength range employed. It is therefore generally necessary to use a coloured filter which transmits radiation close to the maximum of the absorption band of the material under investigation. Alternatively, better results may be obtained with a monochromatic prism or grating, which further reduce the polychromaticity. Strict monochromaticity is also necessary for another reason: the eye may only be relied on to appreciate the equality of luminous intensities of light of the same colour. If filtration of the light is omitted, or is inadequate, the two beams received by the eye are not of the same colour and comparison becomes difficult if not impossible.

The device which permits the intensity of the reference beam to be diminished may be a diaphragm, a neutral grey wedge, a system of nicol prisms, or an arrangement consisting of a rotating blade which allows light to pass only in definite pulses.

These devices allow an accuracy of 1% to be obtained under the most favourable conditions. Their accuracy is generally limited by the sensitivity of the eye. The advantage is that there is no longer any need to provide standard solutions, which is particularly useful when the coloured compound is not sufficiently stable.

(2) Photoelectric methods

Outline of the apparatus. In contrast to the eye, photoelectric cells possess the advantage of allowing a comparison of unequal luminous intensities, which are individually observed at different moments and, in addition, their use is not limited to visible radiation.

Precise colorimetric determinations, in which the error is reduced to $\pm$ 0.1%, may only be achieved if apparatus of the above type is employed.

Such accuracy may be lowered by a number of causes, in particular with apparatus which does not comprise a monochromator, where errors caused by the non-additivity of optical densities are often observed.

(2a) *Photoelectric cells*

Two types of cells are used: photovoltaic cells and photoemissive cells.

(i) Photovoltaic or barrier layer cells. These usually consist of a layer of selenium sandwiched between a sheet of iron or copper and a transparent layer of lead, platinum, copper or gold.

The current delivered when the cell is illuminated is to all intents and purposes proportional to the luminous intensity received, although deviations appear when the external resistance is increased. The mean proportionality constant between the current delivered and the luminous intensity received is a function of the wavelength, as may be seen from Fig. 9.

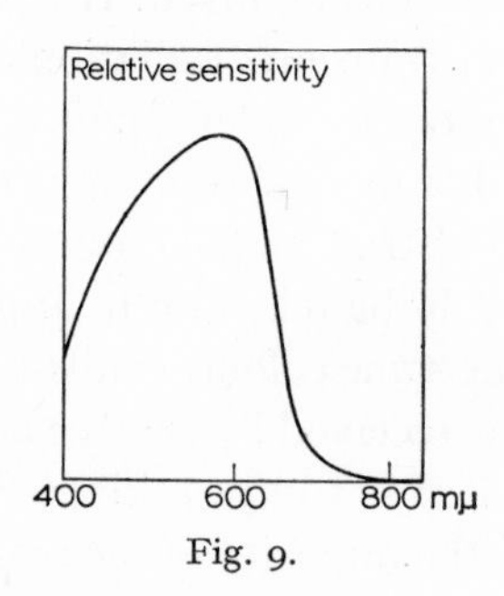

Fig. 9.

The output of such cells can reach considerable values, of *e.g.* 120 microamperes per lumen.

Apart from the systematic error due to the non-proportionality between the luminous intensity and the current, certain additional errors may arise if the following precautions are neglected:

(a) The temperature of the cell must remain constant throughout the experiment, since the current delivered may otherwise vary by as much as 1.5% per degree in certain cases.

(b) The relative positions of the optical accessories (including the luminous source) must be fixed, to ensure that the beam always illuminates the same portion of the cell.

(c) The cell should be checked for the (unavoidable) effect of ageing, *i.e.* a very slow and irreversible reduction in the current delivered for a constant illumination.

(d) If the current delivered by the cell for a given illumination is observed to fall off with time, stabilization must be awaited. This may take some minutes, and is particularly important with higher luminous intensities.

Finally, we may note that with certain types of apparatus, in which the beam of light strikes the cell intermittently, the inertia of the cell becomes apparent for a frequency as high as 50 c/s and causes additional deviations from the proportionality between the luminous intensity and the current.

(ii) Photoemissive cells. These make use of the fact that certain metals brought to a sufficiently high negative potential can liberate electrons under the action of light.

The sensitivity of such cells depends on the wavelength and on the metal used in the cell (see *e.g.* Fig. 10), and is generally of a lower order than that of selenium cells.

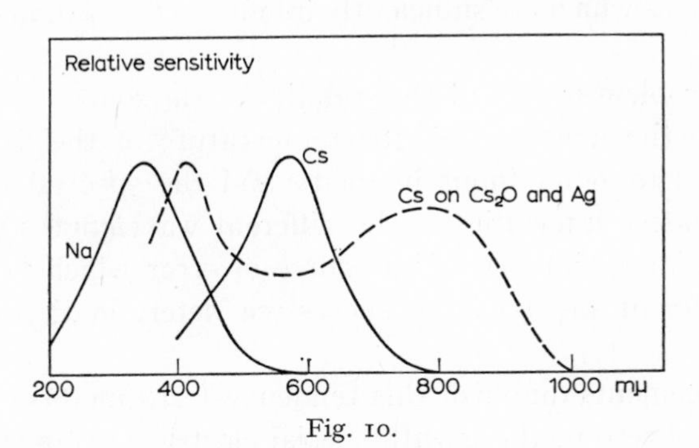

Fig. 10.

On the other hand, the other characteristics of such cells are generally more favourable: these comprise a low inertia and negligible cell fatigue. The proportionality between the current delivered and the intensity of incident light is reasonably consistent. Although the extent of proportionality depends on the wavelength and the applied potential difference, the deviations are less than 0.03%, at least for certain vacuum cells, in the normal range of luminous intensities.

(iii) Use of photoelectric cells. From the technological point of view, the selenium cell possesses the advantage of extreme simplicity of construction. Since however it cannot easily be used with a current-amplifying system, the intensity of the incident light should be high, which, as has already been stated, implies that the working range of wavelengths cannot be made too narrow; in practice it is not less than about 15–20 mμ. Selenium cells are thus suitable for use in inexpensive, non-monochromatic types of equipment.

In contrast to the above, the photoemissive cell requires a more complicated mounting but may be amplified; it is consequently used in cases where only low luminous intensities are available (monochromators).

(2b) *Other accessories of photoelectric equipment*

(i) The light source. This is selected according to the wavelengths required. Thus, from 350 to 1,300 mμ, incandescent lamps can generally be employed; their emission spectrum is continuous, and the desired radiations can therefore be separated out.

However, the absorption of the glass bulb substantially reduces the intensity of radiation below 350 mμ and cuts off wavelengths shorter than 300 mμ. In this region, hydrogen lamps with a quartz envelope are used. Under certain conditions, the hydrogen arc emits a reasonably continuous spectrum between 200 and 450 mμ.

In certain particular cases, spectra such as those of a mercury arc, containing certain wavelengths which are sufficiently intense to be used, may be employed.

An important problem is that of the stability of the source of light. When the voltage fed into the lamp varies, the temperature of the filament changes, altering both the total luminous intensity and the spectral distribution, *i.e.* changing the relative intensities of the different wavelengths present.

(a) The variation of intensity is a source of error which becomes relevant when an experiment involves the successive determination of two optical densities.

Certain arrangements diminish this tendency (zero methods) or suppress it (flicker method or zero methods with special electrical arrangements).

(b) Changes in the wavelengths of the light emitted are of no consequence if monochromatic light is used, but are a very important source of error if this is not the case.

Thus, G. KORTÜM AND J. GRAMBOW, *Z. Angew. Chem.*, 53 (1940) 183, studied the effect of variations in the voltage supplied to the lamp, on the calibration curve $D = f(c)$ of naphthol yellow, using a blue filter which allowed a 100 mμ band to pass. A variation of 10% in the feed voltage resulted, at certain points, in a change of 60% in the concentration required to maintain the apparent optical density at a constant value. Assuming that the errors are proportional to voltage variations, the voltage should be stabilized to within 0.02% if measurements are to be exact to ± 0.1%.

It is therefore necessary, in certain cases, to ensure that the supply voltage is perfectly stable. The lamp is fed by means of a battery of accumulators.

The lamp should periodically be checked for ageing, which produces similar disturbances, and the feed voltage adjusted accordingly.

(ii) Filters and monochromators. The desired spectral region is isolated from the spectrum emitted by the source, by means of a filter or a monochromator. Filters are absorbing systems which appreciably transmit only certain regions of the spectrum. The most widely used filters generally consist of coloured glass or gelatine sheets. The bands transmitted are generally more than 20 mμ wide, (measured between the points at which the transmission is equal to half the maximum value).

The absorption curve of a filter is displaced and deformed by variations in temperature, so that the spectral composition of light transmitted by a filter depends upon its temperature. The effect of this on the apparent optical density of the solution obviously depends on the respective positions of the band transmitted by the filter and the absorption band of the solution, and can exceed 0.1% per degree in unfavourable cases.

These variations must be taken into account when the filter is warmed by the source of light. In practice, heating by conduction may quite easily be avoided and it is only necessary to wait until the temperature of the filter, heated by radiant heat, has become stable. Radiant heat can moreover be diminished by placing before the filter a second filter which is opaque to infrared radiation.

Interference filters allow narrower bands of wavelengths to be obtained.

To obtain a very narrow band of wavelengths, the beam of radiation is dispersed by a prism or grating and the required radiation is selected by means of a slit. The higher the degree of dispersion of the system and the narrower the slit, the more nearly monochromatic is the radiation selected.

Quartz and constant-deviation glass prisms are respectively best suited for use with ultraviolet and visible light. The spectrum obtained from a parallel beam with the prism can be traversed across the slit of a collimator by rotating the prism. The transmission of quartz is adequate above 200 mμ.

Dispersion can also be achieved by means of a plane transmission grating or a concave reflection grating. Gratings can have a higher dispersive power than prisms, particularly in the high orders of the spectra, but there is an overlap of spectra of orders greater than 1; moreover, the beam obtained is less luminous than with a prism.

Light bands as narrow as 1 mμ may be obtained with some commercial instruments.

In addition, the beam obtained often contains radiations other than those isolated by the slit of the monochromator, owing to stray reflections; the apparatus must therefore be arranged so as to reduce these as much as possible.

Such equipment is generally calibrated directly in wavelengths. The precision of the regulation of the wavelength chosen must be related to the minimum width of the bands which can be used.

(iii) Amplification. Improvement in the monochromaticity is always accompanied by a reduction in the intensity of the incident beam. A compromise between monochromaticity and intensity must therefore be achieved. Equipment in which provision is made for amplifying the photocell current thus permits the use of more highly monochromatic light.

The degree of amplification is limited by the properties of the cells themselves.

Amplification can be carried out within the photoemissive cell itself if secondary electrodes ('electron multipliers') are added.

When a zero method is not used, the amplification must be perfectly linear as otherwise all the advantages of the method are lost.

(2c) *The principal types of apparatus*

Colorimetric measurements always involve the determination of the ratio of two luminous intensities, I_0 and I. In principle, I_0 is the incident luminous intensity, *i.e.* that which the photoelectric cell receives before the solution is placed in the path of the beam of light, and I is the emergent intensity—that received by the cell when the beam has passed through the solution. Various methods of determining this ratio will be considered below.

The various types of apparatus can be classified in the following way:

(i) The deflection instruments. (Single light beam). If the intensity of the current i produced by the cell is proportional to the luminous intensity I received, we have,

$$D = \log \frac{I_0}{I} = \log \frac{i_0}{i}.$$

In the simplest case, i_0 and i are determined by means of a galvanometer. For example, the deflection of the galvanometer is adjusted to the 100th division of the scale for i_0 and the reading corresponding to i then gives the transmission I/I_0 as a percentage. Alternatively, the graduation may be logarithmic so as to indicate optical densities. The measurement is only correct if i is accurately proportional to I, so that if the current is amplified, amplification must be perfectly linear. In addition, the source of light and the cell must remain identical during the measurement and the feed voltage of the source must therefore be maintained at a perfectly stable value.

Numerous devices of this type can still be found which are inexpensive and relatively imprecise (2–5%), and which simply measure the current produced by a selenium cell. Highly precise instruments are also available.

(ii) Zero instruments with electrical compensation. — 1. *Single beam method.* The galvanometer can be used as a null instrument. When I_0 is replaced by I, a change in the current produced by the cell, and consequently in the potential drop across a resistance, is observed; the galvanometer is then brought back to its initial position by means of an auxiliary calibrated potentiometer, the movement of which is directly recorded on a scale of relative intensities.

The same remarks apply as in the preceding case.

Numerous versions of this type of apparatus exist.

2. *Double beam method.* ('Electrical compensation by derived voltage'.) The two electric currents are balanced until the needle of a galvanometer is brought to zero; this offers all the advantages of double-beam equipment and some of those of instruments employing optical compensation (see below).

(iii) Instruments employing optical compensation. — 1. *Double beam method.* The two measurements may be carried out simultaneously to reduce the effect of any fluctuations in the source of light. The effect of variations in the total intensity of the incident beam can be eliminated in this way, but possible disturbances due to a change in the spectral composition of the beam between the moments of estimation and calibration may still occur if a monochromator is not used.

In principle, a single photocell cannot compare two luminous intensities simultaneously; a nearly equivalent effect may however be achieved by arranging a screen or mirror so that the cell is alternately illuminated by two light beams, of which one has passed through the solution to be estimated, and the other through a suitable reference system—for example, a second cell.

The two beams of light can be equalized, *i.e.* the pulsations of electric current indicated by a galvanometer can be nullified, by optical compensation, *e.g.* by reducing the intensity of one of the light beams.

This can be achieved by means of an absorbing wedge which absorbs equally at all wavelengths. The intensity of the light beam may be varied by interposing portions of the wedge of varying thickness in its path; since however such wedges are never perfect, the accuracy of this method is of the order of $\pm$ 1% (relative error).

The intensity of the light may also be varied, by, for example, venetian blind shutters placed in the path of the beam. The precision may then be better than 0.1%.

In the case of optical compensation and in methods employing monochromators, fluctuations of the light source are of little importance. Since the photoelectric cell is used only to establish the equality of the two beams, it is no longer vital that the intensity of the current should be proportional to that of the light. In addition, the amplification of current need no longer be linear. Finally, the two light beams fall on the same point of the photoelectric cell almost at the same instant, and the conditions for the comparison of the two luminous intensities are thus the best obtainable in practice.

2. *Single beam method.* A preliminary measurement is carried out with the solution to be analyzed, the galvanometer being zeroed by electrical compensation. The blank sample is then introduced, causing a deflection of the galvanometer, which is then cancelled out by varying the intensity of the light beam.

This method is in principle equivalent to the preceding one. The two measurements are carried out with the same luminous intensity I, but since they are made successively, absolute stability of the light source and the cell with time must be ensured.

In principle, double-beam instruments employing optical compensation possess the best characteristics for colorimetric estimations. It will be shown that with other types of equipment, it is necessary to employ a differential technique if a high degree of precision is required.

(3) Principles of the estimation procedures

(i) Direct method. If the optical density $D = \log I_0/I$ of the dissolved material is known, its concentration may be deduced from the equation $D = \varepsilon lc$. The value of ε (or rather εl) used is an established one, derived under strictly identical conditions, for a solution of known concentration placed in the same cell.

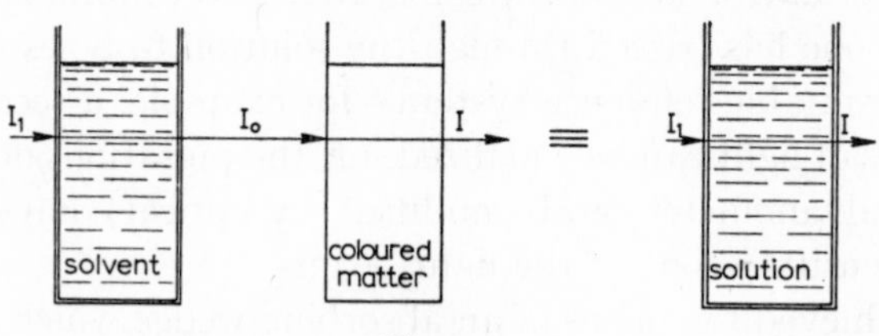

Fig. 11.

In practice, the incident intensity differs from the emergent intensity not only because the light has been partially absorbed by the material to be estimated, but also because some of it has been reflected by the surfaces of the cell or absorbed by the cell and by the solvent.

In order to take these extraneous factors into account, it is necessary to carry out a blank test. The intensity I_0 is thus not the incident intensity, but the intensity of the same light beam after having passed through the cell filled with solvent alone (Fig. 11).

NOTE. The question of whether the optical densities of the solution and those of the solvent and the cell are additive, is not relevant in this method, since the determination of εl is carried out under the same conditions in each case. The value of the coefficient εl measured during the blank test thus corresponds to a beam I_0 identical with that used in the actual measurement.

It is however possible that if the cell is changed, a given beam I_1 will give a beam I_0' of composition different from I_0 and, consequently, a different coefficient εl for the solution.

More generally, if it is desired to estimate a material A in the presence of other absorbing materials, including of course the solvent and the cell, the difference between the optical densities due to A may be determined by comparing the absorption due to all the materials present in the solution with that of all of them excluding A, if this is possible ('blank test').

This method is still used more frequently than any other, since it holds good for all types of instruments and possesses the advantage of being rapid. Only two measurements (I_0 and I), and one precise measurement of εl or, better, one calibration curve, are necessary to obtain the titre of a solution.

The precision which can be obtained may be excellent provided the apparatus itself is accurate.

(ii) Indirect method. It is sometimes possible to make use of the fact that the addition to a solution of the substance to be estimated causes the disappearance of the absorbing material, for instance by the formation of complexes. The concentration of the material to be estimated is then deducible from the reduction of optical density.

(iii) Differential colorimetry. Greater precision can be achieved by a method developed recently, in which the intensity I_0 is no longer that of the beam

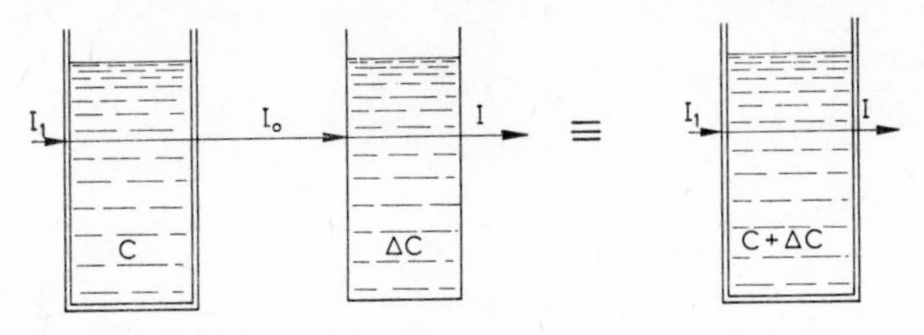

Fig. 12.

emerging from the cell filled with solvent, but with a solution of the coloured material at a known concentration which is slightly lower than the concentration to be determined. The method involves the determination of the increase in optical density $\log I_0/I$ due to the difference in concentrations, Δc, of the two solutions (Fig. 12, and *cf.* Fig. 11).

Δc can be deduced from this if ε (or εl) has been determined previously or if a calibration curve has been plotted.

The precision of this method may be improved by the use of more concentrated solutions. The optimum conditions for such estimations will be discussed under the section dealing with the precisions of colorimetric techniques.

(iv) Colorimetric titrations. It is possible to follow the reaction and to determine the point of equivalence in volumetric (or coulometric) titrations by observing the variations in the optical density of the solution (Chapter 7).

Chapter 3

Accuracy and precision of colorimetric estimations

The precision demanded of colorimetric methods, as indeed of all analytical methods, depends on the ultimate purpose of the analysis. In trace analysis it is often sufficient to determine merely the order of magnitude of the content of the element concerned; the result may then have a relative error of 10, or 20%, or sometimes even of 100%.

Visual colorimetry, involving comparison with standards or the dilution method, is often sufficient in such cases; occasionally, other very simple colorimetric techniques can also be useful. Except in the case of measurements in ultraviolet light, or of mixtures of coloured materials, this type of estimation does not require the use of expensive apparatus comprising amplification and monochromators. In all such cases it is unnecessary to take precautions which only become essential in precision colorimetry.

Special precautions must be taken when the colorimetric method is used for the estimation of large amounts of an element with a precision equivalent to that of volumetric or gravimetric methods, and the choice of apparatus becomes in consequence an important consideration.

In microanalysis, small amounts have to be estimated with high accuracy.

(1) Definitions

The following definitions shall be employed in order to avoid possible confusion.

(i) Precision. This characterises the reproducibility or the scatter of the results with respect to a mean value. This scatter is due to neglected or inadequately controlled factors which lead to chance variations (random errors).

Scattered results may thus be obtained with a photoelectric colorimeter in which, *e.g.* the temperature of the filters or of the solution is not kept sufficiently constant, or where the stabilization of the source is inadequate, if the same determination is repeated several times. For example, on a scale graduated in steps of 0.001 units of optical density, successive readings may give densities of 0.406, 0.400, 0.402, and so on.

It is obvious that in practice, causes of scatter associated with the operator are added to those inherent in the apparatus (as for example, the cleanliness of the cells, the rotation of the micrometer screws in the same direction, etc.).

The precision of an apparatus may then be characterized as the minimum scatter which is obtained with a careful experimenter.

The results may be treated statistically since the different causes of scatter operate independently of one another. The most probable value, which is used in practice, is the arithmetic mean $\bar{x}$ of the results obtained.

The standard deviation of this mean is given by

$$\sigma = \sqrt{\frac{\Sigma \varepsilon^2}{n(n-1)}},$$

where ε is the deviation of each measurement from the mean and n is the number of measurements. For example, assuming that there are no other errors, the true value of a measurement has a 95% chance of being between $\bar{x} - 3.2\sigma$ and $\bar{x} + 3.2\sigma$ if $n = 4$.

The precision of a chemical method is defined in an analogous manner. The causes of chance variations are, for example: variable adsorption, partial redissolution of the precipitate by washing, etc.

NOTE. Each component of a complicated device possesses its own precision; the worst of these sources of error determines the overall error of the whole apparatus.

(ii) Sensitivity. This is, in principle, the smallest detectable variation (ΔX) of the magnitude measured, X. It is usually expressed as the relative magnitude ΔX/X and often depends on X itself.

Suppose for example, that in a volumetric estimation, the sensitivity with which volume readings of the burette, ΔV, can be made, is 0.02 ml. The result could not then be known with an uncertainty smaller than this.

Consider also, a colorimetric determination made by means of photoelectric deflection apparatus. After adjustment, the optical density D is read directly on the scale of a galvanometer. The sensitivity of the electrical measuring apparatus is thus involved. There is an uncertainty ΔD in the measurement of the optical density, which is equal to the smallest variation in optical density corresponding to a perceptible deflection of the needle of the galvanometer.

In a complex apparatus, the uncertainty is determined by the least sensitive component and may depend not only on the apparatus considered, but also on the conditions of use.

Thus, let us consider a photoelectric apparatus with optical compensation. A single measurement involves two observations: the needle of the galvanometer must first be brought to zero, and the optical density then read off the dial connected to the compensating system. Two possibilities therefore arise, assuming that the apparatus is strictly reliable:

(a) The smallest perceptible deflection of the needle of the galvanometer from zero produces a clear deflection of the needle on the dial of optical densities. The sensitivity is the corresponding change of optical density.

(b) The smallest perceptible deflection of the needle on the optical density dial produces a large deflection of the galvanometer.

This time, the uncertainty is due to the sensitivity of the optical density scale.

(iii) Accuracy or correctness. This characterizes the systematic errors, *i.e.* the deviation between the true and the most probable values, taking the probable error and the sensitivity into account.

Thus, when Beer's law is used inappropriately, or if *e.g.* the compensating device of the colorimetric apparatus is badly calibrated, the result obtained is false, even though the equipment may be sensitive and reproducible. The same argument would apply to an estimation based on a reaction which is not truly quantitative.

This cause of error can be detected by carrying out a series of measurements on standard solutions. The deviation found is definite in magnitude and in sign for a given value of the magnitude measured. The results obtained in the course of subsequent measurements must then be corrected for this error. Certain estimations whose calculations incorporate an empirical coefficient, in fact involve this procedure.

The accuracy of a colorimetric apparatus must for instance be checked for given wavelengths, by means of a permanganate solution which is known to obey Beer's law over a large range of concentrations. If for one region of concentrations optical densities are found which are consistently too high or too low, a calibration curve $D_{true} = f(D_{measured})$ should be plotted, to allow the correction of this systematic error.

(iv) Reliability of a result. According to the above definitions the errors involved may be of three types:

1. Calculable or measurable systematic errors.

2. Errors due to the sensitivy of the apparatus or of the method used. These are, in principle, measurable or calculable from the characteristics of the least sensitive component.

3. Errors due to the lack of reproducibility of the apparatus or the method. After a certain number of measurements, the probable error can be calculated statistically.

When the result is corrected for systematic errors, a certain uncertainty due to the other two causes of error will still remain; the overall accuracy of the method is determined by the largest of these three errors.

Most conventional chemical estimations, whether gravimetric or volumetric, have been developed in such a way that the uncertainty due to reproducibility is negligible by comparison with that due to the sensitivity with which the measurements of weight or volume may be obtained. Thus, with a burette which can be read to 0.02 ml, the uncertainty of the result corresponds to this volume.

When the various uncertainties are of the same order of magnitude, the final uncertainty ΔD is given by the formula

$$\Delta D = \sqrt{\Sigma(\Delta_i D)^2},$$

where Δ_iD represents each particular uncertainty.

In practice, since in the above the uncertainties are squared before summation, only a slight difference between them is necessary for one of them to become preponderant. Consequently although the individual components of an apparatus may have sensitivities of the same order, in practice only the least sensitive component need be taken into consideration.

The precision or probable error of the result is expressed either in the form of an absolute error or, more generally, in the form of a relative error.

EXAMPLE. 50.0 ± 0.1% (absolute error) or 50.0% to within ± 0.2% rel. (relative error, %).

The aim of 'precision colorimetry' is to achieve relative errors less than or equal to 0.2%.

(2) Accuracy and precision in visual colorimetry

In general, sources of error (if any) arising out of the chemical method or the apparatus employed, are negligible in comparison with the error associated with the lack of sensitivity of the human eye to variations in luminous intensity.

As has already been mentioned, the eye can only judge whether two luminous intensities are equal; between two limits of intensity (particularly above 10^{-1} to 1 millilambert), it can carry out these comparisons with a relative constant error of dI/I (Weber-Fechner law).

The above constant depends on the colour of the light (Fig. 5), being least for green and yellow, and increasing in the red and blue ends of the visible spectrum. Under extremely favourable conditions it is of the order of ± 1%.

The error Δc which results from this in the determination of the concentration of coloured materials is obtained from the relation

$$\log \frac{I_0}{I} = \varepsilon l c,$$

which on differentiation becomes

$$\frac{dI}{I} \times \log_{10} e = -\varepsilon l dc,$$

so that

$$dc = -\frac{0.43}{\varepsilon l} \times \frac{dI}{I} \qquad (5)$$

and finally,

$$\frac{dc}{c} = \frac{-0.43}{\log\frac{I_0}{I}} \times \frac{dI}{I}. \tag{6}$$

According to equation (6), the error dc/c will become smaller as I_0/I and, consequently, D are increased. In practice, it is necessary to keep within the limits of validity of the Fechner-Weber law, since outside them dI/I increases rapidly and nullifies the gain achieved by increasing the optical density. The optimum conditions generally correspond to $0.7 < D < 1$. This would give $dc/c = 0.5\%$ (with $dI/I = 1\%$).

Such precision is rarely realized in practice. In any case, it is a limit which can be reached only when the eye is trained, when the colour of the light is that which allows maximum sensitivity, and when all possible precautions are taken, in for example the operation of the photometer. In the absence of screening, the illuminated surroundings adversely affect the sensitivity of the eye. Moreover, light of a colour other than that corresponding to the maximum sensitivity must often be used, so that the relative error due solely to the sensitivity of the eye may generally be reckoned as 5% and even 10 or 20% under unfavourable conditions.

NOTE. It is usual to employ white light, which is rich in poorly absorbed radiations; this causes a reduction of the apparent optical density, $\log I_0/I$, and produces an increase in the ratio dc/c.

The inherent limitations of the eye itself preclude the performance of precision colorimetry by means of visual methods.

Such methods may however be applied, with the advantage of simplicity, to the estimation of trace amounts of substances whose solutions absorb the radiations to which the eye is sensitive.

(3) Accuracy of photoelectric cell equipment

(3a) *Causes of error*

(i) Systematic errors. In principle, the following systematic errors are almost eliminated either by the manufacturer or by the operator:

(a) Errors of graduation of the reading scale.

(b) Errors due to the calibration of the optical compensation system. Where a neutral grey wedge is used, it is generally the imperfection of the latter which limits the experimental accuracy.

(c) Standardized amplifications—see 'multiple sensitivities' on p. 39.

(d) Errors in the reproducibility of the wavelength. The motion of the wavelength selection drum should be terminated in the same direction.

(e) Stray illumination. Whatever precautions are taken, a small proportion of light is always transmitted, of wavelengths outside the band selected. Although the intensity of this stray light is generally negligible in comparison with the total intensity of the light falling on the photoelectric cell, nevertheless, when high optical densities are involved, and particularly if the cell has a low sensitivity with respect to the wavelength chosen, this source of error becomes important; Beer's law is no longer obeyed, and the optical densities cease to be additive. In the limit, measurements become impossible with most devices.

However, the use of periodic illumination (as in the case of most double-beam instruments) minimizes the possibility of stray illumination originating from outside the monochromator.

(f) Errors relating to the chemical reaction.

(g) Errors relating to the determination of εl.

(h) Use of nonidentical cells in differential colorimetry.

(ii) Miscellaneous errors. The importance of cleanliness of the cells and of the solution is discussed on p. 56.

(a) Position of cells in the apparatus. The apparatus must be designed so that the cells may be placed in exactly the same position with great accuracy during each measurement. One face of each cell should be marked so that it is always oriented in the same direction.

(b) Errors in reading the scale. In general the error in reading, ΔI or ΔD, is less than the corresponding error introduced by the galvanometer–photoelectric cell arrangement.

For deflection apparatus, or devices with a single beam and electrical compensation:

(c) Nonreproducibility or nonlinearity of the response of the photoelectric cell.

(d) Variable or nonlinear amplification.

(e) Fluctuation of the light source, etc.

The majority of these errors can often be eliminated. At present, the most important error is in many cases, due to uncertainty of the current, Δi. This is extremely difficult to eliminate, and arises either as a consequence of the lack of sensitivity with which current can be measured or because the equipment is operated using wavelengths to which the photoelectric cell used is relatively unresponsive. In general, two cells are used to diminish this fault, the first of which, a caesium cell, operates satisfactorily above 600 mμ, while the second, an antimony cell, may be operated in the ultraviolet and in the visible range, below 600 mμ.

(3b) *Limitation of the accuracy by Δi*

With modern apparatus, the accuracy is often limited by the electrical measuring equipment. The most important error is then that which arises in connection with the evaluation of the current measured, or with the zeroing of the galvanometer. If the uncertainty in the evaluation of current is denoted by Δi, this ultimately corresponds to an uncertainty ΔI in the measurement of the intensity of the light, and its effects on the measurements of optical density D or concentration c may then be calculated.

(i) The error after a single measurement. The error ΔD after a single measurement may be deduced from the relationship

$$D = \log \frac{I_0}{I} = \varepsilon l c.$$

From which

$$\Delta D = \varepsilon l \, dc = 0.4 \frac{dI}{I} = 0.4 \frac{dI}{I_0} 10^D$$

whence

$$dc = \frac{0.4}{\varepsilon l} \frac{dI}{I} = \frac{0.4}{\varepsilon l} \frac{dI}{I_0} 10^D$$

and

$$\frac{dc}{c} = \frac{\Delta D}{D} = 0.4 \frac{dI}{I_0} \frac{10^D}{D}.$$

For a given value of dI/I_0, the error depends only on the measured optical density D. The variations $\Delta D = f(D)$ can be calculated from the table given below, and are shown in Fig. 13, from which it may be seen that ΔD tends towards a limit, and is least when the measured optical density tends towards zero. For optical densities lower than 0.1, $\Delta D \approx 0.4 - 0.5 \, dI/I_0$ and then increases with D, reaching a value 10 times as large as this when $D = 1$.

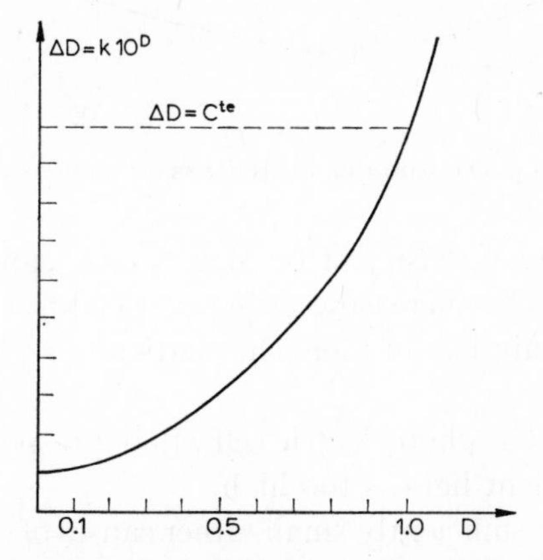

Fig. 13. Variations of ΔD as a function of D

On the other hand, $\Delta D/D$, the relative error of the optical density, changes in the manner shown by Table 2 and Fig. 14, *i.e.* it is very high at low optical densities, and varies from 3.4 dI/I_0 when $D = 0.2$ to 3.4 dI/I_0 when $D = 0.8$, passing through a minimum (at 2.7 dI/I_0) when $D \approx 0.5$.

$\Delta D/D$ finally rises at high optical densities.

For a given optical density D, the errors ΔD and $\Delta D/D$ can be minimized and the accuracy be consequently increased if the apparatus used permits an alteration of dI/I_0. This may be done in two ways:

(a) dI can be increased for a given value of Δi, by increasing the sensitivity of the galvanometer or augmenting the amplification. In particular, the latter may be improved considerably with the aid of electron photomultipliers.

dI is also a function of the sensitivity of the cell, which itself varies with the wavelength used, so that $dI = k_\lambda \Delta i$. The wavelength at which the cell is most sensitive should therefore be chosen if possible.

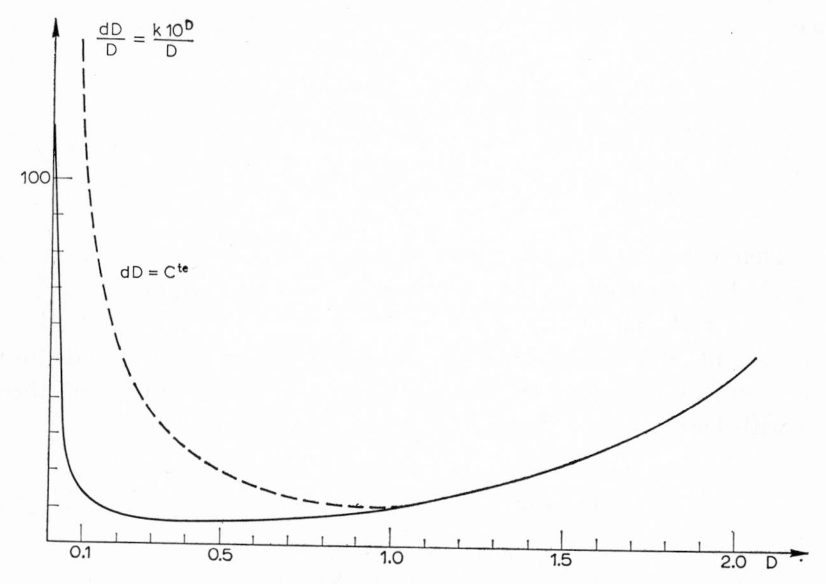

Fig. 14. The variation of $\Delta D/D$ as a function of D.

(b) To increase I_0, the intensity of the light source can be increased and the slit can be made larger by increasing either its height or its width. The latter course results in a certain loss of monochromaticity.

(ii) Limitations. — 1. The photoelectric cell will not respond to small variations if the intensity of incident light is too high.

2. When dI/I_0 becomes sufficiently small, other causes of error predominate and the above relations are no longer valid. To a first approximation, the uncertainty

TABLE 2

D	10^D	$\sqrt{1 + 10^{2D}}$	$\frac{10^D}{D}$	$\frac{\sqrt{1 + 10^{2D}}}{D}$	*Differential colorimetry (See p. 27)*
3	1000	1000	333	333	0.47
2	100	100	50	50	0.70
1	10	10	10	10	1.4
0.9	7.9_4	7.9	8.8	8.8	1.6
0.8	6.3_1	6.4	7.9	8.0	1.8
0.7	5.0_1	5.1	7.2	7.3	2.0
0.6	4.0 (3.98)	4.1	6.6	6.8	2.3
0.5	3.2 (3.16)	3.3	6.3	6.6	2.8
0.4	2.5_1	2.7	6.3	6.7	3.5
0.3	2.0_0	2.2	6.7	7.3	4.7
0.2	1.6 (1.58)	1.9	7.9	9.5	7.0
0.1	1.3 (1.26)	1.6	12.6	16	14.1
0.05	1.1_2	1.5	22	30	28.2
0.01	1.0_2	1.4	102	170	141
0	1.0	1.4	∞	∞	∞

due to the latter can in many cases be expressed as an error $d\mathrm{I}/\mathrm{I}$ = constant, or $\Delta\mathrm{D}$ = constant (Figs. 13 and 14).

NOTES. (1) The experimental accuracy will increase in proportion as the apparatus permits $d\mathrm{I}/\mathrm{I}_0$ to be as small as possible, for any given wavelength. The minimum value of $d\mathrm{I}/\mathrm{I}_0$ is therefore an important characteristic of such instruments. (2) With certain types of apparatus, the minimum possible value of $d\mathrm{I}/\mathrm{I}_0$, *i.e.* that for which the accuracy is highest, can only be attained using suitable methods of measurement—for example, differential colorimetry.
(3) The minimum value of $d\mathrm{I}/\mathrm{I}_0$ depends on the wavelength, since $d\mathrm{I} = k_\lambda \cdot \Delta i$.

(3c) *Possibilities offered by various types of apparatus*

(I) DEFLECTION INSTRUMENTS AND INSTRUMENTS EMPLOYING ELECTRICAL COMPENSATION

(i) The classical method. In this method, which is still most widely used today, the galvanometer is set at zero by means of a counter-current, corresponding to zero luminous intensity, and the latter, I′, is then adjusted so as to give a full deflection (100% transmission) by altering the amplification and the width (or, better, the height) of the spectrophotometer slit. The solution to be analyzed is then placed in the path of the light beam and the optical density corresponding to the new luminous intensity I, D = log I′/I, is read off directly. I′ is usually adjusted by placing in the path of the light beam an identical cell containing the pure solvent, or sometimes a solution of the interfering ab-

sorbing materials ('blank test'). The optical density of the blank may be denoted by D_1.

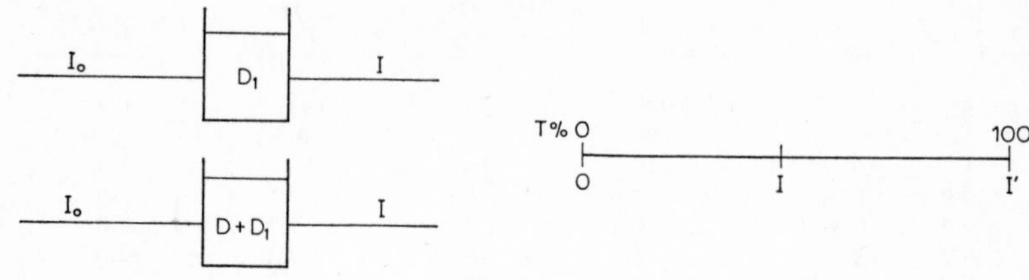

Fig. 15.

If the error in the adjustment of the scale zero is neglected, two separate errors occur, each in connection with one of the two measurements.

The uncertainty in the first optical density measured, D_1, is $\frac{dI}{I} = \frac{dI}{I_0} \cdot 10^{D_1}$. That of the second is $\frac{dI}{I} = \frac{dI}{I_0} \cdot 10^{D+D_1}$, dI being the same in the two cases. The uncertainty ΔD is given by the formula

$$\Delta D = 0.4 \sqrt{\left(\frac{dI}{I'}\right)^2 + \left(\frac{dI}{I}\right)^2} = 0.4 \frac{dI}{I_0} \sqrt{10^{2D_1} + 10^{2(D+D_1)}},$$

in which the following relationships have been taken into account:

$$\frac{I_0}{I'} = 10^{D_1} \text{ and } \frac{I_0}{I} = 10^{D+D_1}.$$

Hence,

$$\Delta D = 0.4 \frac{dI}{I} 10^{D_1} \sqrt{1 + 10^{2D}} \text{ and } \frac{\Delta D}{D} = \frac{dc}{c} = 0.4 \frac{dI}{I_0} 10^{D_1} \frac{\sqrt{1 + 10^{2D}}}{D}.$$

If D_1 (the blank test) is small, say < 0.1, then $10^{D_1} \approx 1$ and $\frac{dc}{c} = 0.4 \frac{dI}{I_0} \frac{\sqrt{1 + 10^{2D}}}{D}$. These variations are shown in the Table on p. 37 and are little different from those of $10^D/D$; thus in particular $\frac{\sqrt{1 + 10^{2D}}}{D} \approx \frac{10^D}{D}$ for $D > 0.5$.

It may be seen that for a given dI/I_0, the relative error dc/c is a minimum when the measured optical density is in the neighbourhood of 0.4–0.6. At both low and high optical densities, the error dc/c becomes very large (Fig. 14).

The measurements are preferably carried out at values of dI/I_0 as low as possible by amplifying the electric current, using a sensitive galvanometer, and increasing I_0 to the maximum. The errors in reading the scale could only be eliminated if a scale of considerable length were used. In practice, the scale length is limited in most instruments of this type by the maximum possible deflection of the galvanometer, and thus determines dI/I_0. It is thus not possible to increase I_0 or the sensitivity of the galvanometer indefinitely, or to amplify to any desired extent.

The scale is graduated linearly in luminous intensities, being arranged in such a way that the error in reading dI is negligible in comparison with the term dI = constant, which occurs due to the uncertainty in the current. The scale is also graduated logarithmically in optical densities.

NOTE. If the optical density of a solution is high, it can often be diluted down to give a value of D = 0.4–0.5, for which the error dc/c is a minimum.

Electrical compensation. It is also possible to zero the galvanometer in each of the two measurements described above, by means of current compensation, the compensating device being directly connected to give a scale reading. The errors involved are the same as in the preceding case but the galvanometer deflection no longer represents a limitation, and dI/I_0 can therefore be decreased to the point where the corresponding scale length is sufficiently long for reading errors to be negligible and, in addition, the electrical compensation may be made sufficiently sensitive.

In the case of both deflection instruments and instruments employing electrical compensation, a given limited reading scale can be used in various ways, as will be described below.

(ii) 'Multiple sensitivities'. In certain instruments the current may be amplified by an accurately known factor. This is a useful consideration in cases in which the optical density of the test solution is high.

As in the preceding case, the scale is first adjusted to read 0 and 100 with I = 0 and I = I_0 respectively. Since the optical density is high, I will be low and the error in reading the scale will be correspondingly significant, and will moreover depend on the uncertainty D in the very high optical density.

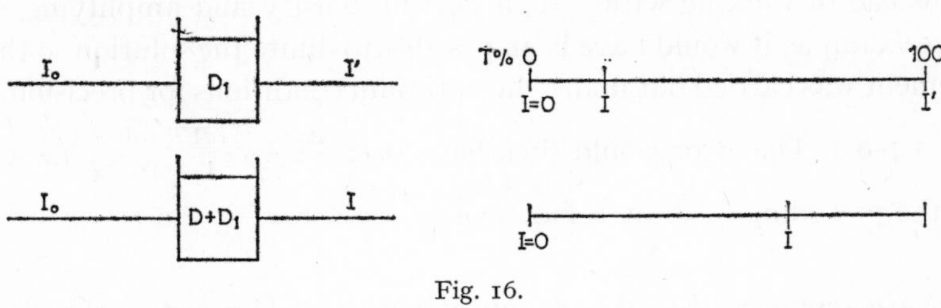

Fig. 16.

If however, the current is amplified by an accurately known factor 10^a, the optical density recorded is found to be diminished by the same factor, and the reading occurs on a portion of the scale where the error involved is low relative to the optical density in question. Furthermore, the error ΔD due to the uncertainty dI has in effect been diminished as well.

As in the preceding case, the error involved in the first measurement is given by:

$$\Delta D = 0.4 \frac{dI}{I_0} \cdot 10^{D_1}.$$

The second error is $\Delta D = 0.4 \frac{dI'}{I_0} \cdot 10^{D+D_1}$. In consequence of the 10^a amplification however, $dI' = dI/10^a$, and this error is therefore now equal to $0.4 \frac{dI}{I_0} \cdot 10^{D+D_1-a}$.

The overall uncertainty is therefore given by:

$$\Delta D = 0.4 \frac{dI}{I_0} \sqrt{10^{2D_1} + 10^{2(D+D_1-a)}} = 0.4 \frac{dI}{I_0} 10^{D_1} \sqrt{1 + 10^{2(D-a)}}.$$

If D_1 is small, $10^{D_1} \sim 1$, and

$$\frac{dc}{c} = \frac{\Delta D}{D} = 0.4 \frac{dI}{I_0} \frac{\sqrt{1 + 10^{2(D-a)}}}{D}.$$

If $(D - a)$ is large, the formula becomes:

$$\frac{dc}{c} = 0.4 \frac{dI}{I_0} \cdot \frac{10^{D-a}}{D}.$$

The relative error $\Delta D/D$ is divided by the amplification factor 10^a. The readings on the optical density scale of the galvanometer are simultaneously displaced by the factor a, so that the extremity of the scale, formerly corresponding to $D = 0$, now gives $D = a$.

The maximum precision will then occur at $(D - a) \sim 0.4$–0.5.

NOTE. Instead of working with a high optical density and amplifying i as in the above example, it would have been possible to dilute the solution so that the measurement was carried out under the optimum conditions for precision, given by $D \sim 0.4$–0.5. The error would then have been $\frac{dc}{c} = 0.4 \frac{dI}{I_0} \cdot \frac{10^{D-a}}{D-a}$, *i.e.* greater than in the preceding case, by a factor of $\frac{D}{D-a}$.

(iii) The differential method at high optical densities. The conventional differential procedure is undoubtedly the most important method used in practice in conjunction with deflection or electrically compensated instruments, since it allows all the advantages of the latter to be exploited. Thus, high values of I_0 may be used, and amplification can be carried out to the maximum.

The principle is as follows. A solution of known concentration and of optical density D_0, (apart from and in addition to D_1, the optical density of the cell, the solvent, and any foreign absorbing compounds), lower than that of the solution to be analyzed, is placed in the path of the light beam, and the luminous intensity I′ of the beam emerging from this solution is made to give a reading at the extremity of the scale (100% transmission), by using the amplifier and increasing I_0.

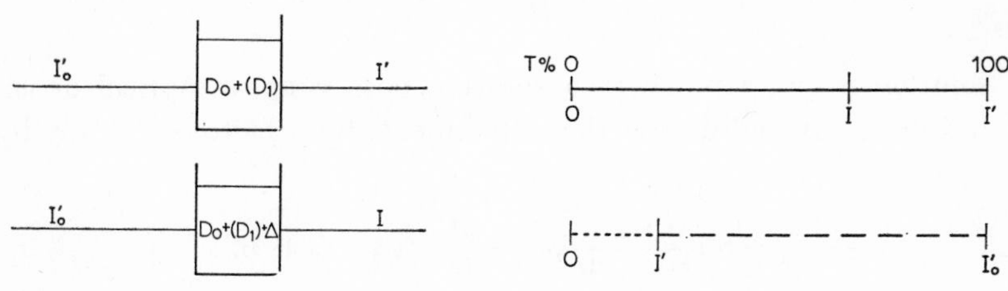

Fig. 17.

The measurement of I is then carried out with the solution to be analyzed, whose optical density $D = D_0 + \Delta(+D_1)$.

If the uncertainty in the luminous intensity under these new conditions is denoted by dI', then

$$\Delta D = 0.4 \sqrt{\left(\frac{dI'}{I}\right)^2 + \left(\frac{dI}{I'}\right)^2}, \text{ with } \frac{I'_0}{I} = 10^{D_0+D_1},$$

and

$$\frac{I'_0}{I} = 10^{D_0+\Delta(+D_1)}$$

whence

$$\Delta D = 0.4 \frac{dI'}{I_0'} 10^{D_0+D_1} \sqrt{1 + 10^{2\Delta}}.$$

In the direct method, dI/I was determined using a fixed scale. In this case, the scale has been altered by amplifying the current or increasing I_0 in a ratio such that its maximum reading corresponds to I′ instead of I_0'; thus, the ratio $I_0'/I_0 = 10^{D_0+D_1}$. Consequently, dI'/I_0' is $10^{D_0+D_1}$ times smaller than the ratio dI/I_0 obtained in the direct method, and $\Delta D = 0.4\,(dI/I_0)\sqrt{1+10^{2\Delta}}$, assuming $D_1 \approx 0$.

The error ΔD depends only on Δ, the difference between the optical densities of the two solutions, and reaches a minimum for $\Delta = 0$, or $D = D_0$. In this case, we have: $\frac{dc}{c} = \frac{dD}{D_0 + \Delta} = 0.6 \frac{dI}{I_0} \cdot \frac{I}{D_0}$.

The relative error with respect to the concentration becomes therefore smaller as the measured optical density is increased.

As D_0 increases, however, the value of I′ diminishes, and a point is reached at which no further manipulation of the apparatus can cause I′ to give a full-scale reading.

EXAMPLES. (1) Consider a solution with an optical density of 0.500. By the direct method,

$$\frac{dc}{c} = \frac{\Delta D}{D} = 0.4 \frac{dI}{I_0} \frac{\sqrt{1 + 10^{2D}}}{D} = \frac{dI}{I_0} \times 2.6 \text{ (Table 2)}.$$

If this solution is compared with a solution possessing an optical density of 0.4, dI/I_0 will be divided during the adjustment, by a factor $10^{0.4}$, *i.e.* by 2.5. Hence,

$$\frac{dc}{c} = 0.4 \frac{dI}{I_0} \frac{\sqrt{1 + 10^{2\Delta}}}{D} = \frac{dI}{I_0} \times 1.3. \text{ (Table 2)}.$$

The accuracy has thus been doubled.

(2) Consider a solution with an optical density of 2.0. The direct method would yield:

$$\frac{dc}{c} = 0.4 \frac{dI}{I_0} \frac{\sqrt{1 + 10^{2D}}}{D} = 20 \frac{dI}{I_0}.$$

If this is compared with a solution possessing an optical density of 1.9, we obtain:

$$\frac{dc}{c} = 0.4 \frac{dI}{I_0} \cdot \frac{\sqrt{1 + 10^{2\Delta}}}{D} = 0.3 \frac{dI}{I_0}.$$

In this case, the adjustment will diminish dI/I_0 by the factor $10^{1.9}$, or about 100. Changes of this order of magnitude may often be achieved by changing I_0 and amplifying the current.

The accuracy has been multiplied by about 70.

NOTE. If the same solution had been diluted to an optical density of 0.5, we should have had $dc/c = 2.5\, dI/I_0$, and the accuracy would have been reduced by a factor of 8.

N.B. The last column of Table 2 lists the values of $\sqrt{1 + 10^{2\Delta}}/D$ when Δ tends to zero. It can be seen, in particular, that for an optical density of 3.00, the accuracy is 333/0.47 = 700 times greater for the differential method than for the direct method.

Below an optical density of 0.2, there is practically no further gain.

(iv) The differential method at low optical densities. It has been shown that the

preceding differential method offers little improvement when the optical densities are low.

When this is the case, the following procedure may be adopted. Taking a known solution with an optical density D_1, the galvanometer is zeroed with respect to the intensity I_1 by suitably adjusting a counter-current, and I_0 is made to give a full scale reading by amplifying the current.

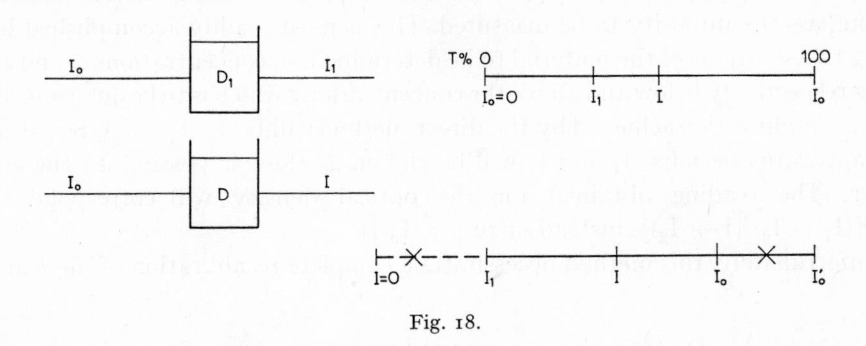

Fig. 18.

The measurement with the unknown solution is then carried out. The known optical density D_1 may be the optical density of a solution of the material to be estimated, of concentration higher than that which is to be determined.

The scale then corresponds to a variation in intensity $(I_0 - I_1)$ instead of I_0, which involves an amplification by a factor of $I_0/(I_0 - I_1)$.

The origin of the scale will be displaced from I_1, and the reading will be

$$D_{\text{reading}} = \log \frac{I_0 - I_1}{I_0 - I} = \log \frac{1 - 10^{-D_1}}{1 - 10^{-D}}.$$

The error dI/I will decrease at greater amplification, *i.e.* when $I_1 \sim I$ or $D_1 \sim D$. The amplification will then be $I_0/(I_0 - I)$.

The significant errors are those in respect to I, to I_0, and I_1.

If the three measurements were performed directly, then ΔD would be given by

$$\Delta D = 0.4 \frac{dI}{I_0} \sqrt{10^{2D_1} + 10^{2D} + 1}.$$

In this case, however,

$$\Delta D = 0.4 \frac{dI}{I_0} \times \frac{I_0 - I_1}{I_0} \sqrt{10^{2D_1} + 10^{2D} + 1}.$$

If D and D_1 are small, say < 0.1,

$$\Delta D = 0.4 \frac{dI}{I_0} \cdot \frac{I_0 - I_1}{I_0} \sqrt{3} = 0.7 \frac{dI}{I_0} \cdot \frac{I_0 - I_1}{I_0}; \quad \frac{dc}{c} = \frac{\Delta D}{D} = 0.7 \frac{dI}{I_0} \cdot \frac{I_0 - I_1}{I_0} \cdot \frac{1}{D}.$$

By contrast, in the direct method,

$$\frac{dc}{c} = 0.4 \frac{dI}{I_0} \frac{\sqrt{1 + 10^{2D}}}{D} \sim 0.4 \frac{dI}{I_0} \cdot \frac{1}{D}.$$

(v) Maximum amplification. Instead of amplifying so that the scale covers the range from I_1 to I_0, maximum amplification can be carried out so that the scale covers the range I_1 to I_2, these two luminous intensities being chosen so as to encompass the intensity to be measured. This is most readily accomplished by using two solutions of the material to be determined, of concentrations c_1 and c_2, lying respectively below and above the concentration c which is to be determined.

The amplification achieved by the direct method will be $I_0/(I_1 - I_2)$. So far as the apparatus permits, I_1 and I_2 will be chosen as close as possible to one another. The reading obtained for the optical density will correspond to $\log\ [(I_1 - I_2)/(I - I_2)]$, instead of to $\log\ (I_0/I)$.

Unfortunately, this method necessitates a complete recalibration of the scale.

(II) Double beam devices; optical compensation

The unknown solution $D + D_1$ is placed in the path of one of the beams and the blank sample D_1 (cell, solvent, and interfering coloured compounds identical with the first) is placed in the path of the other.

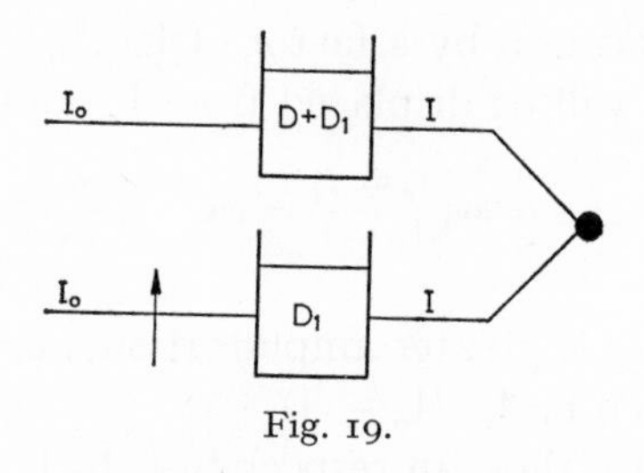

Fig. 19.

The two luminous beams are equalized by progressively diminishing the intensity of the latter beam by means of one of the devices described earlier (wedge, venetian blind shutters, etc.). The two beams are of equal intensity when the galvanometer registers a reading of zero.

This corresponds to the ideal case of differential colorimetry, because the comparison is made between two equal optical densities. Under these conditions, the optical density measured is zero, the error $\Delta D = 0.4\ dI/I_0$. In principle therefore, the relative error $dc/c = \Delta D/D$ will be the smaller, the greater the optical density measured.

Furthermore, there is in this case no theoretical limit to the extent to which dI/I_0 can be diminished, since there is nothing to limit the increase of I_0 (except the excessive illumination of the cell in the case of low optical desinties), the exploitation of the maximum sensitivity of the galvanometer, or the extent of amplification.

NOTES. (1) The apparatus incorporates the error due to the calibration of the optical compensating device. Such compensation can however be carried out with a very high accuracy by modern methods.
(2) In this method, the potential of the apparatus with respect to improving the accuracy is fully exploited; a limiting value of ΔD due to other causes of errors may be reached at certain wavelengths and at low and medium optical densities, the error in the electrical measurement having been almost eliminated. This represents a static value, *i.e.* $\Delta D =$ constant. Under these conditions, the scale of optical densities may be graduated linearly, to great advantage. ('Spectral' Jouan).

At high optical densities, the error ΔD due to the uncertainty in the measurement of the current again becomes predominant. In Figs. 13 and 14, the case where ΔD is constant has been represented by a broken line up to $D = 1$. In this particular case, $\Delta c/c$ is a minimum for $D = 1$.
(3) The scale of readings must naturally be sufficiently long so that the reading errors may be neglected.

Single beam instruments with optical compensation. The solution to be analyzed is placed in the path of the light beam and the galvanometer is zeroed by electrical compensation. The blank sample is then placed in position and the intensity of the light is reduced to bring the galavanometer back to zero.

The error ΔI therefore occurs twice.

This method is fundamentally the same as the preceding one, but the source of light and the photocells should be stable with time.

(3d) *Sensitivity*

The sensitivity is given by $\Delta D/dc$, which represents the increase in the magnitude read in relation to that of the magnitude to be determined.

Since $\Delta D/dc = \varepsilon l$, the sensitivity of the determination will be increased by altering ε (selection of the reagent and the wavelength), and the thickness of the cell (see p. 47).

Limit of sensitivity. The limit of sensitivity is defined as the smallest detectable value of the quantity to be measured—in this case, the optical density or the concentration—and is equal to the uncertainty in magnitude when the latter tends to zero.

It has already been mentioned on p. 35 that the uncertainty tends to a limiting value of ΔD_{lim} when D approaches zero. The limit of sensitivity with respect to the optical density D will therefore be $D = \Delta D_{lim}$. At this point D can no longer be distinguished from zero.

The limit of sensitivity decreases with decreasing dI/I_0, *i.e.* with increasing accuracy of the apparatus.

The limit of sensitivity with respect to the concentration, Δc_{lim}, is similarly $\Delta c_{lim} = \Delta D_{lim}/\varepsilon l$, and may therefore be decreased by increasing εl (see p. 47). To characterize the performance of the apparatus at a given wavelength the limit of sensitivity is regarded as an optical density, and is generally expressed in parts per million (ppm), *i.e.* in micrograms of the material to be estimated per millilitre of solution; this value is also a function of l.

Regardless of the apparatus used, the effect of any chemical reaction on the sensitivity is determined by ε.

NOTES. (1) For devices employing optical compensation, $\Delta D = 0.4\, dI/I_0 \cdot 10^{D+D_1}$, where D_1 represents the optical density of the blank sample.

When D tends to zero, ΔD tends to $0.4\, dI/I_0 \cdot 10^{D_1}$. If D_1 is itself small, for example less than 0.1, then $10^{D_1} \approx 1$ and

$$D_{lim} = \Delta D = 0.4 \frac{dI}{I_0},$$

from which the limiting concentration which can be detected is obtained as

$$dc = \frac{\Delta D}{\varepsilon l} = \frac{0.4}{\varepsilon l} \cdot \frac{dI}{I_0}.$$

If on the other hand D is large, owing to the presence of foreign absorbing ions, the limiting concentration can only increase by a factor of 10, as even if $D_1 = 1$, $10^{D_1} = 10$.

It may be seen that $c_{lim} = dc$ is smaller the larger the value of εl and the smaller that of dI/I_0.

(2) The differential method at low optical densities. The results are analogous, but dI/I_0 can be decreased to its minimum value.

Chapter 4

The experimental procedures in colorimetric estimations

(1) Selection of the conditions of measurement

In the performance of colorimetric estimation a certain amount of freedom exists in the selection of the wavelength, the extinction coefficient ε, the concentration of the solution, and sometimes the thickness of the solution traversed by the light beam. In addition, a choice of several reagents is occasionally available. The optimum selection of each condition follows directly from the considerations developed in the preceding chapters.

(i) Variations in ε. The molar extinction coefficient can vary greatly with wavelength (Fig. 2, p. 6). Knowledge of the form of this variation is important, and in consequence the optical density of a solution of known concentration is often determined at different wavelengths. At the present time, a number of devices exist which directly record absorption curves by plotting the variation in optical density with ε.

In addition, since ε varies from one substance to another, if an estimation may be performed using one of several alternative reagents, it is useful to know the curve $\varepsilon = f(\lambda)$ for each of them.

(ii) Variations in l. Many types of equipment employ cells of thicknesses, usually lying between 0.5 and 1 cm or between 5 and 10 cm. Consequently, the variation of l is, in practice, very limited in comparison with that of ε, which may change by a factor of as much as 10^4 with different wavelengths or reagents.

(iii) Variations in c. The control of c is governed by the circumstances of the extinction, regarding which there are two possibilities: (1) Only trace amounts are involved, and since the concentration is very low, there is no need for further dilution. (2) Appreciable amounts are to be estimated and the solution may be diluted if necessary.

(Ia) *Estimation of trace amounts*

In this case there is no control over c, the value of which approaches the limit of sensitivity of the estimation. In general, the accuracy is of little importance

and it is not essential for the conditions used to conform to Beer's law, the main requirement being to increase the sensitivity as far as possible. It has been seen that for this purpose, the maximum values of ε and l must be used.
(1) The coefficient ε may be increased in two ways. The wavelength, λ, can be selected to correspond to the absorption maximum of the substance to be estimated, provided this is permitted by the other prevailing conditions (*e.g.* other coloured substances present, etc.).

If the estimation is to proceed via the determination of the optical density of a reaction product of the substance to be estimated and an addition reagent, the latter is chosen so as to give the highest possible value of ε, even if the reaction is not quantitative. If, however, the other substances present may become involved in the reaction, the reagent should be chosen to be specific. Moreover, if other coloured compounds exist, the colour of the reactions product formed may influence the choice of the addition reagent.
(2) It is advantageous to increase l, although since the volume of the solution is often limited, it is sometimes necessary to use special apparatus. For example, the solution may be placed in a narrow tube along the axis of which the light is passed.

(3) It is advantageous to have dI/I_0 as small as possible (p. 46). This can be achieved with devices employing suitable optical compensation. If however, the minimum value of dI/I_0 cannot be used under the conditions of measurement, (*e.g.* deflection apparatus) the differential method may be used for low optical densities.

EXAMPLES. (1) With a certain wavelength, a given apparatus gives $dI/I_0 = 10^{-2}$; also, $\Delta D = 0.4 \times 10^{-2} \times 10^{D}$, and since D is small and $10^{D} \approx 1$, $\Delta D = 0.4 \times 10^{-2}$: thus, an optical density of 0.004 corresponds to the limit of sensitivity. The uncertainty of this reading is $\pm$ 0.004. The true figure may thus lie between 0 and 0.008.

If, as with many frequently used compounds, $\varepsilon = 10{,}000$, and if $l = 1$ cm, it follows that $c = 4 \times 10^{-7}\, M$, and if the molecular weight is 100, $c = 0.04$ ppm.
(2) A certain apparatus gives $dc/c = 0.5\%$ at an optical density of 0.5. Also,

$$\frac{\Delta D}{D} = 5 \times 10^{-3}, \text{ whence } \Delta D = 2.5 \times 10^{-3} = 0.43 \frac{dI}{I_0} 10^{D}.$$

Thus, $dI/I_0 = 2 \times 10^{-3}$. At low optical densities therefore, $\Delta D = 0.4\, dI/I_0 \approx 10^{-3}$.
The optical density corresponding to the limit of sensitivity will be 0.001. This is equal to the value chosen by E. B. SANDELL (*Colorimetric Determination of*

Traces of Metals, Interscience, 1959) to define the limit of sensitivity of colorimetric estimations.

(1b) *'Precision colorimetry'*

With many types of instruments and with a minimum of precautions it is frequently possible to decrease the relative error to 2% or even 1%. To limit the error to 0.2–0.5% is however much more difficult, since this may only be achieved when very rigorous precautions are taken.

In view of the continuously increasing importance of precision colorimetry in relation to other methods, the various requirements pertaining to maximum accuracy are outlined below. In particular, the use of monochromatic light considerably simplifies the problem.

(1) The conditions for 'precision colorimetry' with an apparatus which does not involve amplification most frequently necessitate the measurement of an optical density between 0.2 and 0.8, the optimum value being 0.4. The quantities ε, l, and c should thus be chosen so as to obtain this result. For a given absorbing substance therefore, if l is fixed, c and the wavelength must be chosen accordingly.

Selection of the wavelength is however occasionally governed by other, sometimes contradictory, considerations. Thus,

(a) It is advisable to select a band of radiation which forms a part of an absorption plateau, since the effect of the temperature of the solution and the influence of slight variations in the wavelength are then reduced to a minimum. Moreover, under these conditions Beer's law and the additivity of optical densities are most likely to remain valid.

(b) Selection of the wavelength may be governed by the presence of other coloured substances; the wavelength used must be only slightly absorbed by the latter.

(c) At certain wavelenghts, the apparatus is no longer sensitive. With selenium cell devices, this is the case in the red and the blue spectral regions.

These three conditions generally determine the choice of λ and consequently of ε. The solution is then diluted if necessary, in order to obtain εlc in the neighbourhood of 0.4.

(2) Determinations of an accuracy equivalent or superior to those carried out by gravimetric or volumetric methods are possible in the following cases:

(a) With suitable devices incorporating optical compensation;

(b) with suitable deflection equipment or devices employing electrical compensation, provided that differential methods are used.

This may be exemplified by estimation of pure metals. Usually, the percentage of a pure metal is obtained by subtracting from 100% the content of each of the impurities, estimated separately. In this case, it is often possible to compare a synthetic solution corresponding to the pure metal with the solution to be analyzed, with extreme accuracy. The difference between the two may be obtained with great accuracy.

In serial analyses, the direct comparison of a solution corresponding to the sample to be analyzed against standard solutions, often forms a very simple criterion of whether the sample is within the limits imposed.

In all cases however, the precautions appropriate to any precise determination are necessary.

(I) CHECK OF THE ACCURACY OF AN APPARATUS

Whenever a new apparatus is used, an acquaintance with its characteristics should first be acquired. Examples of two tests which can be made with this object in view are given below.

(1) A measurement is carried out with a solution of some perfectly stable colouring matter. The apparatus is left in operation for several hours and the measurement is repeated using various wavelengths, solutions of different colours and optical densities, and at various intervals of time. For a single solution and a single wavelength, the scatter of the results obtained is a measure of the accuracy (sensitivity + precision) which can be obtained with the apparatus.

(2) The precision and the systematic errors can be checked simultaneously by determining the extinction coefficient ε of a substance which obeys Beer's law, at different concentrations but at the same wavelength; ε (or εl) should be constant. The deviations observed between the values of εl obtained and the mean curve of $\varepsilon l = f(c)$ indicate the accuracy. If the deviation between $\varepsilon l = f(c)$ and a mean straight line εl = const. is greater than the scatter of the measured values of εl, the difference is attributed to a systematic error (Fig. 20). It is advisable to check whether this is reproduced in different series of measurements.

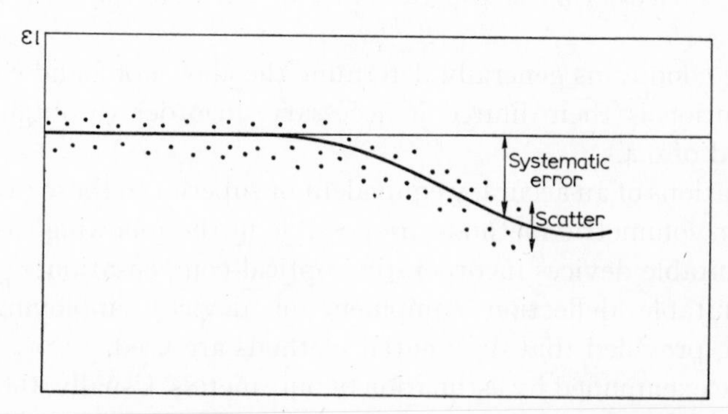

Fig. 20. Simultaneous determination of the systematic error and the scatter.

Permanganate constitutes a convenient reagent for the above check, since it obeys Beer's law even for quite wide bands of radiation, provided that they are situated in the neighbourhood of its absorption maximum (520–550 mμ). Thus, with a 50-mμ band obtained with a filter whose transmission maximum is at 520 mμ, Beer's law is obeyed to within 0.2% for concentrations lower than 15 mg of manganese per litre ($c < 3 \times 10^{-4}$ M, D < 0.5).

Before dilution, care must be taken to stabilize the permanganate with periodate in approximately 0.3 N sulphuric acid (200 mg of periodate per 100 ml of solution). Dilution is then carried out normally with 0.3 N sulphuric acid.

Similar determinations should be carried out in various regions of the spectrum.

Finally, and particularly in the case of devices employing deflection and electrical compensation, precision is governed by the relationship $d\text{I} =$ constant (see p. 34), at least for wavelengths with which the sensitivity is low and at low or high optical densities. The characteristics of the instrument may then be described in terms of the minimum value of $d\text{I}/\text{I}_0$ at different wavelengths.

(II) DETERMINATION OF ε (OR OF εl). CALIBRATION

(i) Confirmation of the validity of Beer's law. When a certain wavelength has been selected for use, an attempt is made to determine the value of ε or εl, and simultaneously to check the validity of Beer's law. Using solutions of known concentration, the curve $\log \text{I}_0/\text{I} = f(c)$ may be plotted. If Beer's law is obeyed, a straight line (Fig. 21) with slope εl is obtained.

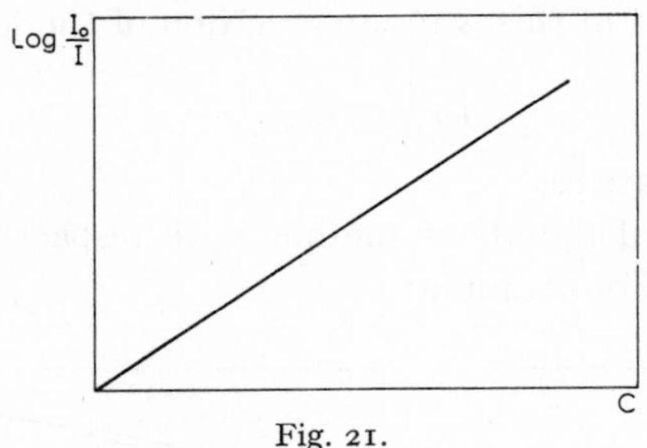

Fig. 21.

In practice, since rigorously monochromatic light is never available, the coefficient εl obtained is an apparent coefficient, the value of which varies with the type of apparatus used. This explains why the values quoted by different workers do not agree and cannot be used directly.

For example, for dichromate at 372 mμ, the following figures occur: 4,520; 4,719; 4,830 $\pm$ 30; and 4,800.

ε (or, more frequently εl) must therefore be determined independently in each case, and with each set of equipment.

The curve $\frac{\log I_0/I}{c} = f(c)$ may also be plotted. If Beer's law is obeyed, this should be a straight line parallel to the axis of c and with an ordinate intercept εl.

In practice, Beer's law only holds true to a greater or lesser degree of approximation and the plots $\log I_0/I = f(c)$ and $\frac{\log I_0/I}{c} = f(c)$ curve downwards as the concentration is increased (Figs. 22 and 23).

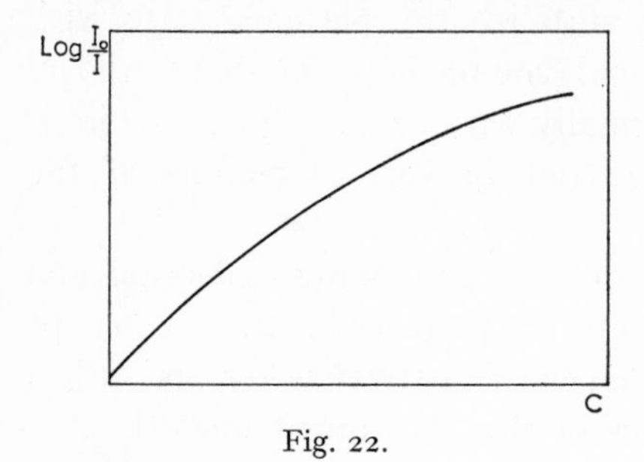

Fig. 22.

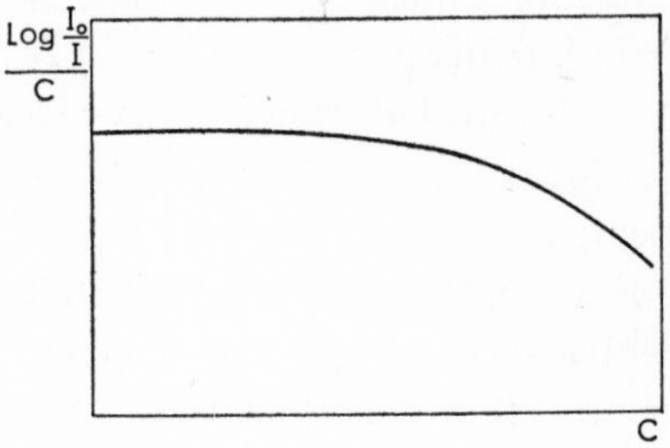

Fig. 23.

(ii) Calibration curve. If Beer's law is obeyed, it makes no difference whether the curve $\log I_0/I = f(c)$ is used or the concentration is calculated from the equation $D = \varepsilon lc$, after εl has been determined.

If however, this is not the case, the most convenient method consists of using the curve $D = f(c)$ as a calibration curve, since it directly relates concentrations with values of optical density. If the curvature is slight, the curve can be approximated by a series of straight lines OB, BC, etc., which represent the true optical density to within 0.2%, as in the example of Fig. 24. Within each of the intervals defined in this way, an equation of the form

$$\log \frac{I_0}{I} = \varepsilon l(c + a)$$

may be used to calculate c.

The disadvantages of the above method with respect to accurate measurements are numerous. For example:

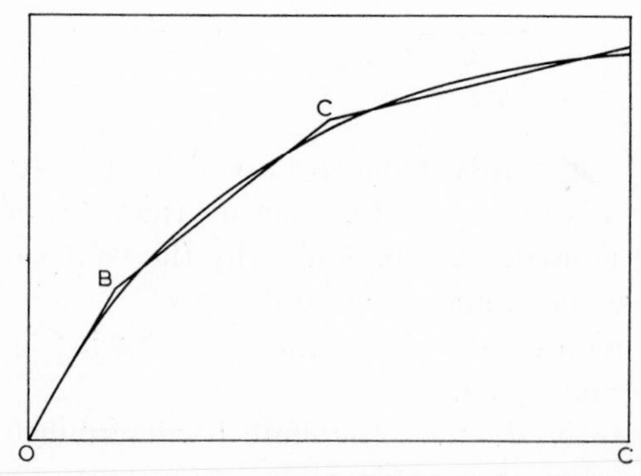

Fig. 24. Use of successive straight lines $d = l(c + a)$.

(1) In order to establish the curve with sufficient precision, a large number of points is required. This may entail a considerable amount of work, and it may be necessary to begin again if the conditions of measurement change (change of filters, temperature, etc.).
(2) Where the deviation from Beer's law is due to the non-monochromaticity of the light used, the optical densities may no longer be treated additively if the mixture of foreign substances present in addition to the material to be estimated does not have a constant coefficient ε in the band of wavelengths used. If this is the case, the calibration is rendered generally invalid in this form.

(iii) Other methods. 'Internal standard'. The use of an 'internal standard' has also been suggested. A substance which absorbs in a different region of the spectrum than the material to be estimated, is added to the solution in known concentration. Measurements are carried out using two wavelengths, one for the internal standard, such that:

$$D_1 = \varepsilon_1 l c_1;$$

and the other for the material to be estimated:

$$D_x = \varepsilon_x l c_x.$$

The ratio

$$\frac{D_x}{D_1} = \frac{\varepsilon_x}{\varepsilon_1} \frac{c_x}{c_1}$$

is then determined.

Reference may be made to a calibration curve. A certain number of causes of error are eliminated in this way.

(iv) 'Standard additions'. When a preliminary study has shown that, in spite of chemical reactions or molecular interactions involving impurities, the optical density remains proportional to the concentration of the material to be estimated, the method of increments may be used to good effect. This is equivalent to a calibration carried out under conditions strictly identical with those of the estimation.

The optical density is determined before and after the addition of a known amount of the material to be estimated. If a is the resulting increase in the concentration (assuming that errors of dilution have been eliminated), we may write:

$$D = \varepsilon l c \quad \text{and} \quad D' = \varepsilon l(c + a),$$

whence

$$\frac{D}{D'} = \frac{a}{c + a}.$$

From this, c may be deduced.

If other coloured substances are present, this is combined with a blank test to eliminate the parasitic absorption.

It is also possible, when the determination is not very accurate, to repeat the operation by successive 'standard additions' and to construct a $D' = f(a)$ curve from the experimental points, from which the value of D may be deduced with greater accuracy than before.

(III) CAUSES OF ERROR DUE TO THE APPARATUS

The most important points (*cf.* p. 31) must be taken into consideration and attempts should be made to avoid them as far as possible.

(1) Depending on the type of measurement, the source of light should be kept in a more or less stable state. In the case of accurate measurements, it is preferable to use a buffer-battery for supplying current to the source, since voltage stabilisers are often inadequate.

In addition, before commencing the measurement it is necessary to wait until the lamp has reached its operating temperature. The tendency to ageing is checked by determining the optical density of a known solution, and is appropriately corrected.

(2) Temperature of the photocell must be kept constant within limits which depend on the type of measurement envisaged. In certain unfavourable cases, the temperature should be kept constant to within 0.1°.

(3) In the case of devices incorporating filters, the temperature of the latter should be kept constant—to within 1° in dubious cases.

(4) The temperature of the solution should also be controlled.

It is best to place the apparatus in a constant-temperature chamber, (which simultaneously fixes the temperature of the photocell), to prevent heating of the filters (by thermal insulation), and to place a filter opaque to infrared radiation in front of them; this greatly reduces heating by radiation.

Whatever the precautions adopted however, the apparatus should still be checked regularly—for example, every hour—by means of a measurement using a known, perfectly stable solution possessing a constant colour and contained in a given cell. The measurement should always yield the same value for the optical density. If not, the causes of variations must be sought and eliminated.

With equipment which includes filters, when these precautions are observed, results with a precision of 0.2–0.5% can be obtained. The major disadvantage of such equipment is the frequent non-additivity of the optical densities.

(IV) CAUSES OF ERRORS OF CHEMICAL ORIGIN

Even if the instrument used is perfect, the accuracy of a colorimetric estimation may be limited by various chemical conditions in the solution.

Most of the existing colorimetric reactions are not reproducible with a very high degree of accuracy, so that their use is limited to the estimation of trace amounts. The accuracy may only be significantly increased by careful selection of the chemical reactions used. For example, Mn(II) may be estimated by oxidation to permanganate, fluoride by diluting down the colour of ferric sulphosalicylate, and nickel and cobalt by forming their respective complexes with EDTA, etc. In many cases, the absorption of the ions themselves is used without any intermediate chemical reaction. *e.g.* UO_2^{2+}, Cu^{2+}, Co^{2+}, Ni^{2+}. In these latter cases, the extinction coefficient is small and concentrated solutions must therefore be used.

(i) Influence of time. The optical densities of many coloured compounds change as a function of time. It is necessary when a new estimation is being developed, to follow the variations in optical density from the moment when the coloration is formed.

In many cases, the latter develops slowly from the instant when the reagents are mixed, and becomes stabilized after varying lengths of time, for example, 10 minutes or 1 hour. This is the case with the silicomolybdate, and with the ferrous *o*-phenanthroline complexes. In other cases, the coloration fades after a certain time; examples are the cadmium dithizonate and ferric thiocyanate complexes.

(ii) The case of lakes. Many colouring materials are adsorbed by certain colloidal substances, with a change of colour. The resulting mixtures are known as lakes. Their nature is complex since, it is possible that other bonds exist in addition to those responsible for adsorption.

When light passes through such a solution, it is partly diffracted, partly adsorbed, and partly reflected. The optical density is then greater than would be the case if only absorption were operative. However, the deviation is often proportional to the concentration of the colloid, so that the experimental optical density continues, within certain limits, to be proportional to the concentration.

Nevertheless, given states of colloidal agglomeration, and following this, the same adsorption of the colouring materials are not easily reproducible. Besides, the colour generally varies slowly in the course of time.

(iii) The case of a coloured reagent. When a reagent not completely transparent to the radiations used for the estimation is present in excess, it must be taken into account in the determination of the optical density.

For example, this may be done by carrying out all the estimations at the same total concentration of the reagent, C_R. The optical density then remains a function of the concentration c alone. The calibration curve should also be plotted under the same conditions. The maintenance of identical conditions during the measurement and during calibration also eliminates possible deviations due to the non-additivity of the optical densities.

If however the latter are in fact additive, it is also possible to treat the solution as a mixture of two materials to be estimated separately. The measurement is simplified, if necessary, by using a nomogram. Of particular interest is the case in which the reagent in excess is not transparent to the radiation used for the estimation, but the compound formed is almost so. The amount of reagent present after reaction may then be estimated. It is in such cases sufficient to use a known total concentration of the reagent and simply to estimate the excess. This method is often adopted for estimating dithizone.

(iv) Impurities in the reagents. Elements which appear in the reagents used, as impurities present in amounts which may not be neglected, may be estimated by a method similar to that used for the test solutions themselves. The appropriate correction is then made.

It is sometimes easier to purify the reagents—for example, by extraction.

(v) OPERATING PRECAUTIONS

The correct application of the various devices used for colorimetric estimations proceeds by trial and error, and requires operating precautions in the same way as do volumetric or gravimetric determinations.

The principal operating precautions will be summarized in this section.

(i) Cells. The cells used should be perfectly clean and in particular, free from dust, which is often difficult to detect. The surfaces traversed by the light must not be touched with fingers, and should be scrupulously clean and dry. The faces of a cell may be wiped with a chamois cloth.

Cleanliness of the cell can be checked by illuminating it laterally in front of a black background.

The compounds adsorbed by the cell may increase its optical density. The cell can be cleaned with various solvents, such as with nitric acid or aqua regia (cleaning agents containing chromic acid leave traces of chromium adsorbed on to the glass).

Finally, the cell should always be placed in the same position with respect to the light beam, as otherwise the thickness traversed may not remain constant, the emergent beam may not fall on the same area of the photocell, and the proportion of reflected light may vary.

It is often necessary to have two identical cells, whatever the method used. This is particularly the case with double-beam equipment or when the differential method is employed. Their identity must be checked very carefully with solutions of high optical density and under the conditions of maximum precision.

(ii) Solution. The solution must be perfectly clear. If it requires filtration,

sintered glass is often preferable to paper, which may leave fine fibres in suspension.

It is also necessary to ensure the absence of air bubbles, which may become attached to the walls of the cell.

IMPORTANT NOTE. During a series of several measurements, it is necessary to measure the known optical density of an indefinitely stable solution, as a regular check. The solution should be placed in the measurement cell itself, so that the cleanliness of the latter and the characteristics of the apparatus may be checked simultaneously.

Standard solutions are available for this purpose, which absorb in various regions of the spectrum; these consist of chromates, cobalt salt, copper salts, etc.

(iii) Influence of the temperature. Calibrations are only valid at a given temperature, and the optical density of a solution may often vary considerably as a function of temperature, particularly if a solvent with a high coefficient of expansion is used.

If the spectrophotometer cell is thermostatically controlled, the optical density is measured repeatedly until a constant figure is obtained, indicating that thermal equilibrium has been reached. In addition, once a temperature has been selected, care must be taken that it is not altered by any step in the operating procedure, such as, for example, dilution.

In most cases, the solution is allowed to attain thermal equilibrium with the ambient atmosphere, whose temperature has been determined, and the measurement is then carried out rapidly so that the temperature of the cell has no time to change.

The influence of temperature variations, due to which calibration curves for each temperature and several precautions during the measurement become necessary, may be neglected when two solutions of very similar compositions are compared.

It is advantageous to use a wavelength for which $d\varepsilon/d\lambda$ is small (as at the maxima and minima of the $\varepsilon = f(\lambda)$ curve), since temperature variations always involve displacements of such curves and the variations in ε are correspondingly large on the steeply inclined parts of the absorption curves.

(2) Elimination of the effect of interfering materials without separation

The value obtained for the optical density may be affected by the presence of impurities in the solution. This is particularly the case when the latter are coloured, when they react with the reagent used or with the material to be estimated itself, or simply when their concentration is too high (molecular interactions, variations in the ionic strength). Strictly speaking, the only way to avoid this would be to calibrate the instrument used under identical conditions. This would however present difficulties and furthermore, practically as many calibration curves would be required as there were solutions to be analysed. In general, a calibration curve is plotted using solutions of the material under study in the pure solvent, and an attempt is then made to eliminate the effect of the impurities with a greater or lesser degree of approximation, and without attempting to separate them. Certain techniques may be adopted for this purpose, but if these should fail, the separation of the interfering materials becomes obligatory, with all the inconveniences which it involves (loss of time, new sources of error, etc.).

(2a) *Selection of the wavelength*

If the $\varepsilon = f(\lambda)$ curves of the material to be estimated and the absorbing impurities are known, it is sometimes possible to select a wavelength for which the latter have an absorption coefficient sufficiently low for their influence to be negligible.

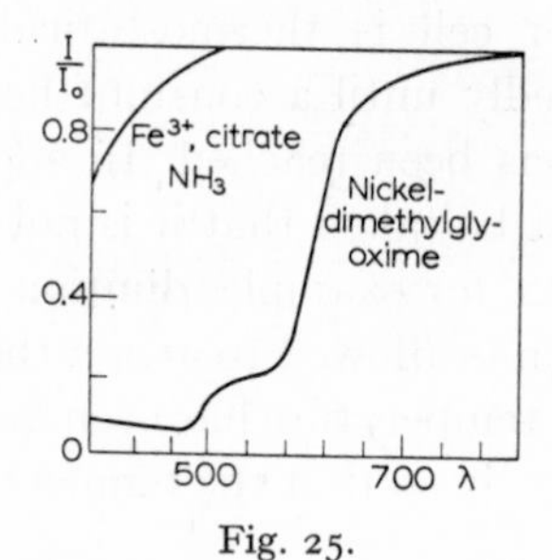

Fig. 25.

EXAMPLE. The estimation of nickel in a steel by means of nickel(III) dimethylglyoxime. In ammoniacal solution, in the presence of dimethylglyoxime and an oxidizing agent, Ni^{2+} gives a Ni(III) compound, the soluble red nickel(III) dimethylglyoxime; the Fe(III) citrate complex has a yellow colour.

The curves of $\varepsilon = f(\lambda)$ (Fig. 25) show that the ferric complex does not interfere above 500 mμ. In practice, the wavelength used is in the neighbourhood of 530 mμ.

(2b) *Blank tests*

The use of blank tests, skilfully carried out to eliminate the absorption of interfering compounds, is one of the most powerful techniques in colorimetric determination. We have seen that in the usual direct method, in order to eliminate the absorption due to the solvent and the cell, I_0 is taken as the intensity of the light emerging from the same cell filled with the pure solvent.

In an analogous manner, the absorption due to impurities can be eliminated by performing a blank test with a solution of the latter, of the same concentration as in the test specimen. Fig. 26 (compare Fig. 11, p. 26) shows how, using this procedure, the optical density of the material to be estimated may in fact be obtained. If necessary, it is possible to determine the degree of approximation with which the optical densities of the materials under consideration behave additively, by means of trials, using known solutions. For example, the optical densities of the material to be estimated and the impurities may be determined separately, and the optical density of the combination subsequently found by placing the two cells in series in the path of the light beam. In monochromatic light, this source of error is generally eliminated.

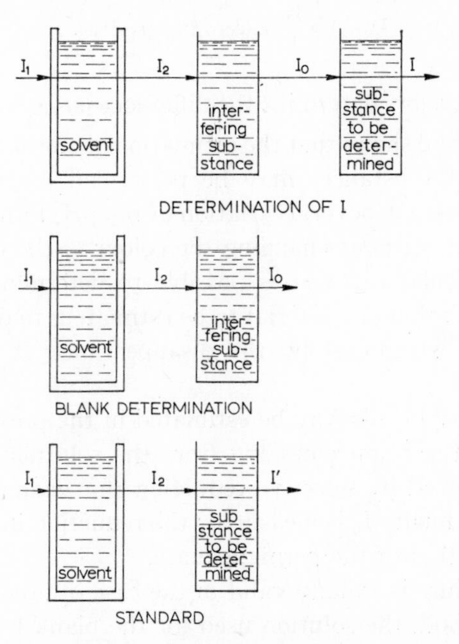

Fig. 26. Use of a blank test. Calibration is only valid if log I_0/I is equal to log I_2/I' for the same concentration of the material to be estimated. Thus, either I_0 and I_2 must have the same spectral composition, or the material to be determined should possess an absorption coefficient which is constant over the range of wavelengths considered.

Practical procedure. In practice, the optical density of the absorbing impurities (coloured materials or high concentrations of 'colourless' compounds) may qe determined using synthetic solutions, provided their concentrations are known.

Although in general this is not the case, it is often possible to adopt the following procedure:

(1) The intensity I_0 of the beam emerging from the solution of impurities is determined, the specific reagent which produces the material to be estimated is then added and the resulting intensity, I, is measured. From this, the increase in the optical density, and hence the required concentration, are calculated. Certain

precautions are of course necessary; thus the solutions must be brought to the same volume and account must be taken of the initial coloration, if any, of the material which is transformed, etc.

For example, in the estimation of titanium by the formation of its complex with hydrogen peroxide, I_0 is determined using the initial solution and I is then found for the same solution treated with hydrogen peroxide.

When the material to be estimated is itself coloured, it must play the same part during calibration as it does in the blank test. For example, in the estimation of chromium after oxidation to chromate, the value obtained for the optical density is given by:

$$D = \log \frac{I_0}{I} = (\varepsilon_{CrO_4^{2-}} - \varepsilon_{Cr^{3+}}) lc.$$

Calibration must therefore determine the difference $(\varepsilon_{CrO_4^{2-}} - \varepsilon_{Cr^{3+}})$ not $\varepsilon_{CrO_4^{2-}}$. This is very general and shows that the estimation of one of the components of a mixture of coloured substances may be performed merely by changing its colour by any means whatsoever, (variation of the pH, formation of a complex, oxidation–reduction), without changing the colour of the other materials.

(2) In certain particular cases, it is possible to determine I first, and then, having removed the coloured material to be estimated, to determine I_0, so that the optical density is reduced by the disappearance of the material under investigation.

Permanganate may in this way be estimated in the presence of dichromate. The intensity I of the beam emerging from the solution is first determined. MnO_4^- is then removed by selective reduction (for example, by oxalic or by hydrazoic acid) and finally I_0 is measured; the reduction in the optical density, $\log I_0/I$, is equal to that of the permanganate.

The above procedure is equally valid in the case of other methods. Thus, in the differential method, the solution used for the blank test must contain the same concentrations of each of the impurities, as that containing a known concentration c_0 of the material to be estimated. Either of the two outlined above may then be used, by adding to the blank solution a known concentration c_0 of the test substances.

In methods involving the equalization of two optical densities (visual or photoelectric), the reference solution should contain the same interfering ions as the test solution. In this case, it is generally simpler to place the cell containing a solution of the impurities (obtained by one of the methods described above) in series with another cell containing a solution of a known amount of the material to be estimated in the pure solvent. Under these conditions, the optical densities need no longer be additive.

(2c) *Elimination of interfering ions by controlling the pH, by oxidation-reduction, or by the formation of complexes*

The chemical behaviour of a solution may be fully exploited in the elimination of interference due to the presence of impurities. Thus, oxidation, reduction, the formation of complexes, or neutralization may be used to transform the absorption curves of the interfering materials in such a way that they become transparent in the band of wavelengths used. For example, interference due to the yellow ferric sulphate complex in the estimation of dichromate may be avoided by adding oxalic acid, which gives rise to a practically colourless ferric oxalate complex. Similar methods may in certain cases, be used to prevent foreign materials from reacting with the reagent or with the compound to be estimated. Thus, dissolved silica produces a yellow colouration with molybdates in an acid medium, owing to the formation of a silicomolybdate complex. This reaction is used for the colorimetric estimation of silica. In the presence of fluoride ions however, the estimation is prevented, because a more stable, colourless silicofluoride complex is formed. In consequence, a reagent must be found which gives with F^- a complex of greater stability than the silicofluoride. Boric acid, which gives a fluoborate, satisfies this condition. The silica is then liberated, and the silicomolybdate complex can form.

(2d) *Determinations at several wavelengths*

If the impurities cannot be prevented from reacting with the reagent to give coloured compounds, the procedure adopted is the same as that for the simultaneous estimation of several materials. The optical density is determined using as many different wavelengths as there are materials to be estimated. It is necessary to know the absorption coefficients of each material, for each wavelength. If this is impossible, the influence of the impurities can still be more or less eliminated by determining the optical density using three different wavelengths (and determining only the concentration of the main tests ubstance). This method is a particular case of the simultaneous estimation of several materials, and will be treated as such.

(3) Simultaneous estimation of several materials

In principle, two coloured compounds can be estimated simultaneously by determining the optical density of the mixture at two wavelengths.

If ε_1 and ε_2 are the respective extinction coefficients at a wavelength λ, and ε_1' and ε_2' those at a wavelength λ', and D and D′ are the corresponding optical densities of the mixture, three possibilities may occur:

(i) The optical densities are proportional to the concentrations and are additive.

If each of the materials obeys Beer's law and if the optical densities are additive (the two materials do not interact), then

$$D = \log \frac{I_0}{I} = \varepsilon_1 l c_1 + \varepsilon_2 l c_2, \tag{7a}$$

and

$$D' = \log \frac{I_0'}{I'} = \varepsilon'_1 l c_1 + \varepsilon'_2 l c_2. \tag{7b}$$

The concentrations c_1 and c_2 may be calculated by solving these two equations.

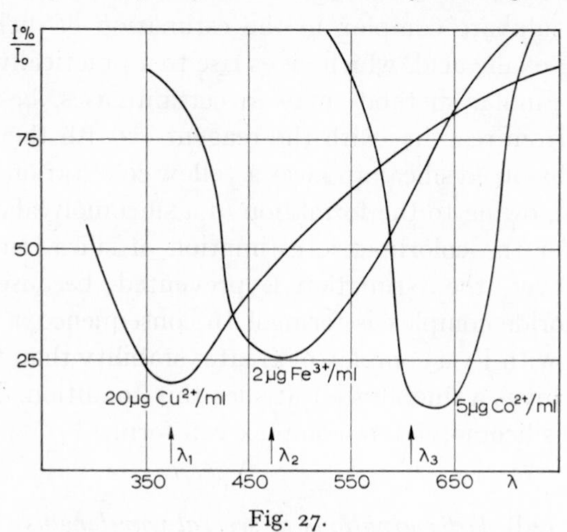

Fig. 27.

When the two optical densities have been determined with the same accuracy, the accuracy of the values obtained for c_1 and c_2 is greatest when the ratios $\varepsilon_1/\varepsilon_2$ and $\varepsilon_2'/\varepsilon_1'$ are as large as possible. Thus, if the curve $\varepsilon_1/\varepsilon_2 = f(\lambda)$ is plotted and two wavelengths are taken such that the ratio $\varepsilon_1/\varepsilon_2$ passes through a maximum for one and through a minimum for the other, the accuracy is the best obtainable.

The appropriate wavelengths are those for which the separation of the curves $\log \varepsilon = f(\lambda)$ are greatest and least. This factor is also important in connection with considerations other than the choice of wavelength alone.

EXAMPLE. The transmission curves of the thiocyanate complexes of iron, cobalt, and copper in a mixture of water and acetone are shown above, in Fig. 27 (after R. E. KITSON, *Anal. Chem.*, 22 (1950) 664). The latter author was able to estimate these three ions by determining the optical density of the solution at three wavelengths, namely 380, 480, and 625 mμ.

For precision colorimetry, it is necessary to check the additivity of the optical densities concerned, using mixtures of known composition covering the entire range of concentrations required.

In practice, and in common with all methods of indirect estimation, the above procedure should only be used with caution, since the errors involved in calculating c_1 and c_2 may be considerable.

The accuracy of the method or the simplicity of the calculations involved may be increased by certain procedures. Thus, if one of the coloured materials is removed by chemical reaction, the two determinations may be performed separately.

EXAMPLES. (1) Estimation of a mixture of $Cr_2O_7^{2-}$ and MnO_4^-. (S. LACROIX AND M. LABALADE, *Anal. Chim. Acta*, 3 (1949) 262). The absorption curves of the two materials are shown in Fig. 28. The optical density of the permanganate in the fully made up solution is determined at 520 mμ, and I_0 is then measured after reducing the permanganate by oxalic acid. This second solution is then used to estimate the dichromate ion in the normal manner, at 440 mμ.

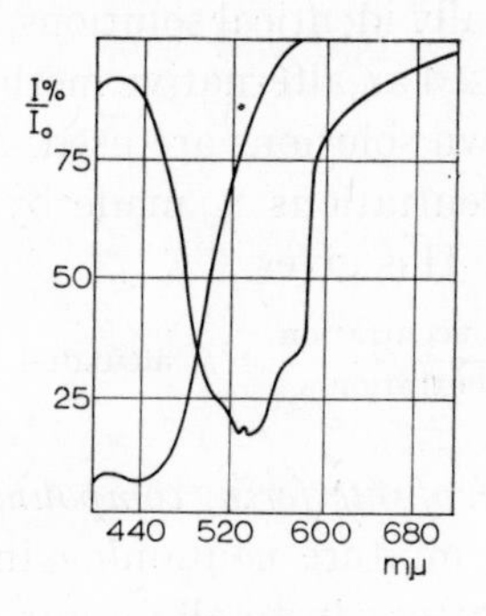

Fig. 28.

(2) Method of ratios of the extinction coefficients. (J. A. PERRY, R. G. SUTHERLAND AND N. HADDEN, *Anal. Chem.*, 22 (1950) 1122.)

(ii) The optical densities are not proportional to the concentrations, but are additive. If the materials do not obey Beer's law, but their optical densities are additive, equations of type (7) above may still be used, with the difference that ε_1, ε_2, ε_1', ε_2' are themselves concentration-dependent. If the four curves of type $\varepsilon = f(c)$ are known for ε_1, ε_2, ε_1', and ε_2', the problem may be solved by successive approximations. For example, the equations may first be solved with respect to average values of ε_1, ε_2, ε_1' and ε_2'. The approximate values of c_1 and c_2 obtained in this way correspond to new coefficients ε, which permit c_1 and c_2 to be calculated to a second approximation. This is repeated until two successive calculations give the same respective values of c_1 and c_2.

Other methods of approximation may be used (see for example GORDON AND POWELL, *J. Inst. Petrol.*, 31 (1945) 191; NUSBAUM AND RANDALL, *J. Appl. Phys.*, 17 (1946) 150).

(iii) The optical densities are not additive. Equations of type (7) may no longer be used.

In such a case, the only available expedient is to compare the optical densities obtained using two wavelengths, with corresponding pairs of optical densities measured under the same conditions, using mixtures of known composition. This can be done graphically. A double lattice of curves $D = f(D')$ is plotted by keeping first c_1 and then c_2 constant. If the experimental points fall between two curves, for example c and $c + \Delta c$, a linear interpolation is made to find the value c_1. The accuracy is increased by making Δc as small as possible.

Linear interpolation within a quadrilateral bounded by the curves given by c_1 and $c_1 + \Delta c_1$ on the one hand, and c_2 and $c_2 + \Delta c_2$ on the other hand, is in essence, the same procedure as the 'method of increments' of E. I. STEARNS (*Analytical Absorption Spectroscopy*, ed. G. MELLON, J. Wiley and Sons, New York, 1950). A single quadrilateral may suffice in the case of measurements made with a series of practically identical solutions, as shown by E. I. STEARNS.

The author has also proposed an alternative method, in which pairs of optical densities obtained with known solutions are used. An intermediate calculation of the sum of the two concentrations is made by successive approximations from a lattice of the curves. This gives

$$\frac{\text{calculated total concentration}}{\text{true total concentration}} = f(\text{calculated \% of material 1}).$$

(iv) Approximate elimination of interfering compounds. If it is desired to estimate the major constituent of a mixture containing impurities whose absorption curves are not known, the latter can to all intents and purposes be eliminated by determining the optical density at three neighbouring wavelengths λ_0, λ_1, and λ_2, such that $\lambda_1 < \lambda_0 < \lambda_2$; λ_0 corresponding to the absorption maximum of the estimated. In this case, we obtain:

$$D_{\lambda_0} = \varepsilon_0 c_0 + D_{\lambda_0}'$$
$$D_{\lambda_1} = \varepsilon_1 c_0 + D_{\lambda_1}'$$
$$D_{\lambda_2} = \varepsilon_2 c_0 + D_{\lambda_2}',$$

where D' represents the optical density of the impurities. If it is assumed that in the range between λ_1 to λ_2 the absorption curve of the impurities is linear, it also follows that

$$\frac{D_{\lambda_0}' - D_{\lambda_1}'}{D_{\lambda_2}' - D_{\lambda_1}'} = \frac{\lambda_0 - \lambda_1}{\lambda_2 - \lambda_1}.$$

We thus have four equations containing four unknowns D_{λ_1}', D_{λ_2}', D_{λ_0}', and c_0, from which the latter may be obtained. Several variations of this method may be used, (for example, the double impurity index method of E. I. STEARNS, in M. G. MELLON, *Analytical Absorption Spectroscopy*, J. Wiley and Sons, New York, 1950).

Bibliography

GENERAL

N. L. ALLPORT AND J. W. KEYSER, *Colorimetric Analysis*, 2nd ed., Chapman and Hall, 1957.
G. CHARLOT AND R. GAUGUIN, *Dosages colorimétriques*, Masson et Cie, 1952.
G. KORTÜM, *Kolorimetrie, Photometrie und Spektrometrie*, Springer, Berlin, 1962.
B. LANGE, *Kolorimetrische Analyse*, Berlin, 1956.
G. F. LOTHIAN, *Absorption Spectrophotometry*, Hilger and Watts, 2nd. ed., London, 1958.
F. X. MAYER AND A. LUSZCZAK, *Absorptions-Spektranalyse*, de Gruyter, Berlin, 1951.
M. G. MELLON, *Analytical Absorption Spectroscopy*, Wiley, New York, 1950.
E. B. SANDELL, *Colorimetric Determination of Traces of Metals*, Intersc. Publ., New York, 1959.
F. D. SNELL AND L. T. SNELL, *Colorimetric Methods of Analysis*, in 4 volumes from 1948 to 1954, Van Nostrand.
W. WEST, *Chemical Applications of Spectroscopy*, Interscience, New York, 1956.
H. H. WILLARD, L. L. MERRITT AND J. A. DEAN, *Instrumental Methods of Analysis*, 3rd. ed,. Van Nostrand, 1958.
D. F. BOLTZ, *Colorimetric Determination of Non-metals*, Intersc. Publ., New York, 1958.
R. P. BAUMAN, *Absorption Spectroscopy*, Wiley, 1962.

COLLECTIONS OF DATA

C. DUVAL, *Colorimétrie minérale*, 1954.
H. M. HERSHENSON, *Ultra-violet and visible Absorption Spectra*, Index for 1930–1954, Academic Press, 1956.

APPLICATIONS TO ALLOYS

E. ALLEN AND W. RIEMAN, III, *Anal. Chem.*, 23 (1953) 1325.
F. W. HAYWOOD AND A. A. R. WOOD, *Metallurgical Analysis*, A. Hilger, London, 1944.
M. JEAN, *Précis d'analyse chimique des aciers*, Dunod, Paris, 1950.
F. VAUGHAN, *Metallurgical Analysis*, A. Hilger, London, 1942.

Reviews.
M. G. MELLON, *Anal. Chem.*, 21 (1949) 3; 22 (1950) 2; 23 (1951) 2; 24 (1952) 2; 26 (1954) 2.
M. G. MELLON AND D. F. BOLTZ, *Anal. Chem.*, 28 (1956) 559; 30 (1958) 554.

ULTRAVIOLET. (See also the aforementioned works).

R. A. FRIEDEL AND M. ORCHIN, *Ultraviolet Spectra of Aromatic Compounds*, Wiley, 1951.

Review.
R. C. HIRT, *Anal. Chem.*, 28 (1956) 579; 30 (1958) 589.

DIFFERENTIAL COLORIMETRY

A. G. JONES, *Analytical Chemistry, Some New Techniques*, Butterworths, London, 1959.
H. H. WILLARD, L. L. MERRITT AND J. A. DEAN, *Instrumental Methods of Analysis*, 3rd. ed., Van Nostrand, 1958, pp. 20–64.

Theory.

G. Kortüm, *Angew. Chem.*, 50 (1937) 193.
R. Bastian, *Anal. Chem.*, 21 (1949) 972.
C. F. Hiskey, *Anal. Chem.*, 21 (1949) 1440.
A. Ringbom and K. Österholm, *Anal. Chem.*, 25 (1953) 1798.
C. N. Reilley and C. M. Crawford, *Anal. Chem.*, 27 (1955) 716.

With several components.

M. Beroza, *Anal. Chem.*, 25 (1953) 112.
R. Bastian, R. Weberling and F. Palilla, *Anal. Chem.*, 22 (1950) 160.
C. F. Hiskey and D. Firestone, *Anal. Chem.*, 24 (1952) 342.
S. D. Ross and D. W. Wilson, *Analyst*, 85 (1960) 276.

Nd, Pm, Sm, Er

C. V. Banks, J. L. Spooner and J. W. O. Laughlin, *Anal. Chem.*, 28 (1956) 1894; 30 (1958) 458.

Cr + Mn

C. F. Hiskey and D. Firestone, *Anal. Chem.*, 24, 1952. 342,

Influence of the temperature.

R. Bastian, *Anal. Chem.*, 25 (1953) 259.

Applications

Al D. K. Banerjee, *Anal. Chem.*, 29 (1957) 55.
J. O. Meditsch, *Rev. Quim. Ind.* (Rio de Janeiro), 26 (1957) 185, 193.
Be J. C. White, A. L. Meyers, Jr. and D. L. Manning, *Anal. Chem.*, 28 (1956) 956.
CN^- O. A. Ohlweiler and J. O. Meditsch, *Anal. Chem.*, 30 (1958) 450.
Cr R. Bastian, R. Weberling and F. Palilla, *Anal. Chem.*, 22 (1950) 160.
Cu R. Bastian, *Anal. Chem.*, 21 (1949) 972.
C. F. Hiskey, J. Rabinowitz and J. G. Young, *Anal. Chem.*, 22 (1950) 1464.
A. Ringbom and K. Österholm, *Anal. Chem.*, 25 (1953) 1798.
F^- J. J. Lothe, *Anal. Chem.*, 28 (1956) 949.
Mn R. Bastian, R. Weberling and F. Palilla, *Anal. Chem.*, 22 (1950) 160.
J. C. Young and C. F. Hiskey, *Anal. Chem.*, 23 (1951) 506.
C. F. Hiskey and J. C. Young, *Anal. Chem.*, 23 (1951) 1196.
Mo A. Bacon and G. W. C. Milner, *Anal. Chim. Acta*, 15 (1956) 573.
Nb, Ta
F. C. Palilla, N. Adler and C. F. Hiskey, *Anal. Chem.*, 25 (1953) 926.
B. M. Dobkina and T. M. Malioutina, *Zavodsk. Lab.*, 24 (1958) 1336.
Ni R. Bastian, *Anal. Chem.*, 23 (1951) 580.
P(V) A. Gee and V. R. Deitz, *Anal. Chem.*, 25 (1953) 1320.
Pt G. H. Ayres and A. S. Meyers, Jr., *Anal. Chem.*, 23 (1951) 299.
Pu G. Phillips, *Analyst*, 83 (1958) 75.
Ti R. A. G. de Carvalho, *Anal. Chem.*, 30 (1948) 1124.
G. W. C. Milner and P. J. Phennah, *Analyst*, 79 (1954) 414.
W. T. L. Neal, *Analyst*, 79 (1954) 403.
U C. E. Crouthamel and C. E. Johnson, *Anal. Chem.*, 24 (1952) 1780.
M. Jean, *Chim. Anal.*, 34 (1952) 226, 250.

C. D. SUSANO, O. MENIS AND C. K. TALBOTT, *Anal. Chem.*, 28 (1956) 1073.
A. BACON AND G. W. C. MILNER, *Analyst*, 81 (1956) 456.
T. W. STEEL, *Analyst*, 83 (1958) 414.
U,Nb C. V. BANKS, K. E. BURKE, J. W. O'LAUGHLIN AND J. A. THOMPSON, *Anal. Chem.*, 29 (1957) 995.
T. W. STEEL, *Analyst*, 83 (1958) 414.
V M. Q. FREELANT AND J. S. FRITZ, *Anal. Chem.*, 27 (1955) 1737.
Zr H. FREUD AND W. F. HOLBROOK, *Anal. Chem.*, 30 (1958) 462.
D. L. MANNING AND J. C. WHITE, *Anal. Chem.*, 27 (1955) 1389.
H_2O_2 R. BAILEY AND D. F. BOLTZ, *Anal. Chem.*, 31 (1959) 117.

Indirect differential colorimetry

J. J. LOTHE, *Anal. Chem.*, 27 (1955) 1546.
C. N. REILLEY AND G. P. HILDEBRAND, *Anal. Chem.*, 31 (1959) 1763.

MICROSPECTROPHOTOMETRY (10 μl)

D. F. H. WALLACH AND D. M. SURGENOR, *Anal. Chem.*, 30 (1958) 1879.

PERIODICAL

Journal of Molecular Spectroscopy.

Chapter 5

Fluorimetry

When a fluorescent material is placed in the path of a beam of light, it emits radiation, the intensity of which is a function of its concentration.

(1) Introduction

Although few inorganic compounds are fluorescent, many organic or organic-inorganic compounds exhibit this characteristic. In general, the light emitted by a fluorescent material has an intensity maximum at a wavelength greater than that for which the absorption of the material is greatest (the difference being in general between 20 and 30 mμ), the spectral composition of the emitted light being independent of the wavelength of the incident beam intensity.

According to the fundamental law of fluorimetry, the emission intensity, F, of radiation emitted by fluorescence is proportional, for a given volume of solution, to the luminous intensity absorbed, *i.e.* $F = kI_{abs}$.
If, in addition, Beer's law holds for the absorption of the incident intensity, the following derivation may be made:

$$\frac{I}{I_0} = 10^{-\varepsilon lc}$$

$$1 - 10^{-\varepsilon lc} = 1 - \frac{I}{I_0} = \frac{I_0 - I}{I} = \frac{I_{abs}}{I_0} = \frac{F}{kI_0};$$

$$F = kI_0(1 - 10^{-\varepsilon lc}).$$

It is thus seen that F tends asymptotically towards kI_0 when the concentration is increasing. In practice, F becomes constant above a certain concentration.

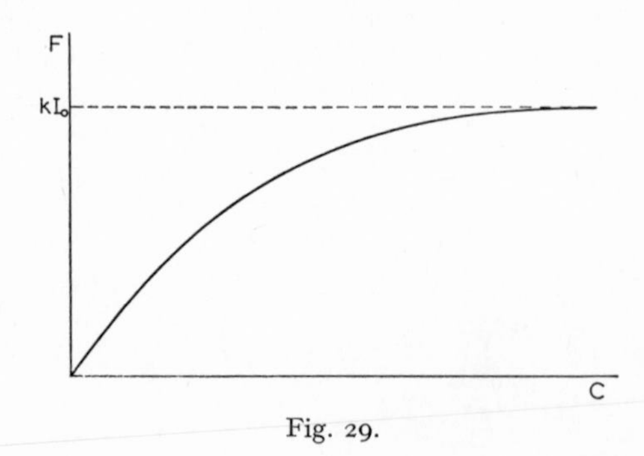

Fig. 29.

NOTE. The law describing the variation of F with c can be expressed in another way. If the maximum fluorescent intensity is denoted by F_0, so that $F_0 = kI_0$, then

$$\log \frac{F_0}{F_0 - F} = \varepsilon lc.$$

(i) Characteristics of the efficiency coefficient k. For substances in solution, the coefficient k is always less than 1 and is obviously greater the higher the degree in which the substance in question exhibits fluorescence.

The value of k may depend on the solvent. For eosin, for example, k has the values of 0.15 or 0.40, depending on whether the solvent used is respectively water or ethyl alcohol; no corresponding difference is found in the case of fluorescein.

The variation of k with temperature may be illustrated by the case of the UO_2^{2+} ion in sulphuric acid solution, where k changes from 0.26 to 0.21 when the temperature is increased from 18.5 to 27° C; this represents a reduction of 2.5% per degree. Changes with temperature usually occur in this direction.

k may also be affected by impurities present in the solution, but although this effect varies widely from case to case, three general possibilities can be distinguished.

(a) The impurities are responsible for some degree of absorption of the incident beam or the emitted radiation.

(b) The impurities react with the fluorescent compound to produce *e.g.* a complex, which exhibits a fluorescence different from that of the initial substance. For example, the very weak fluorescence of the Tl^+ ion is greatly increased by the addition of a large amount of free Cl^-, which forms a complex with Tl^+. In the same way, the pH of aqueous solutions may affect their fluorescence by promoting or opposing the dissociation of the emitting materials. This may be accompanied by a change in the spectral composition of the fluorescence.

(c) Traces of certain substances which do not appear to react with the fluorescent material, nevertheless diminish its fluorescence to some extent. For example, a concentration of I^- ions of 10^{-4} *M*, is sufficient to halve the fluorescence of the UO $^+$ ion.

The nature of the inhibitor is of great significance in determining the extent of this effect, but it is difficult to establish general rules. Thus, for fluorescein, quinine sulphate, and uranyl sulphate, the effectiveness of the inhibiting ions decreases in the order: I^-, SCN^-, Br^-, Cl^-, $C_2O_4^{2-}$, SO_4^{2-}, NO_3^-, F^-, while for α-naphthol and sulphanilate or naphthionate ions, the NO_3^- ion is much more active than I^-. The most efficient cations appear to be Cu^{2+}, Ni^{2+}, and Fe^{2+}. Finally, traces of oxygen dissolved in the solution also inhibit the fluorescence.

(ii) Validity of the law $F = f(c)$. The efficiency coefficient k of a substance may change in highly concentrated solutions owing to interactions between the fluorescent particles. Such variations of k and of F as a function of c are shown in Fig. 30, for the case of Tl^+ ions. In addition, the relation connecting F and c holds only when Beer's law is obeyed, which ceases to be the case at concentrations which are too high.

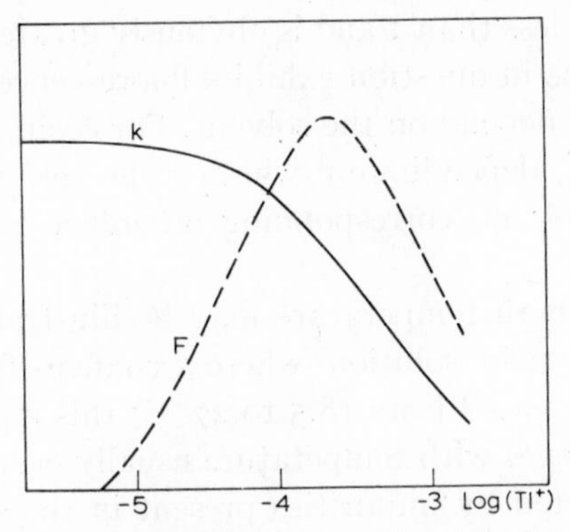

Fig. 30. The concentration dependence of k and F, for thallous ions.
(After P. PRINGSHEIM, *Fluorescence and Phosphorescence*, Interscience Publishers, New York, 1949)

The polymerization of certain organic compounds may be accompanied by the appearance of new fluorescence bands.

(2) Procedure adopted during measurements

A beam of light is passed through the solution to be estimated, and a definite fraction $1/n$ of the total fluorescence emitted in all directions (Fig. 32, *a* and *b*) is measured.

Several precautions are necessary.

(1) The receiver (eye or photoelectric cell) should not be affected by the incident light. The latter is frequently composed of ultraviolet radiation since these are generally absorbed by fluorescent substances to a sufficient degree. Since the eye is not sensitive to ultraviolet, no special operating precautions are necessary.

NOTE. It is nevertheless advisable to shield the eye from ultraviolet rays, *e.g.* by placing a filter opaque to ultraviolet between the solution and the eye when the latter observes in the direction of the incident beam.

With a photoelectric colorimeter, a system of two filters (Fig. 31, *a*) may be employed. The first of these, f_1, is placed between the light source and the solution, so that those radiations present in the incident beam which are necessary for excitation are transmitted, while those of the same wavelength as the radiation emitted are withheld. The second filter, f_2, which should have characteristics inverse with regard to f_1, is placed on the other side of the cell.

Provided the filters exactly fulfil the above requirements, this arrangement is ideal since the measured intensity $F_{meas.}$ is then a constant fraction of the total intensity emitted.

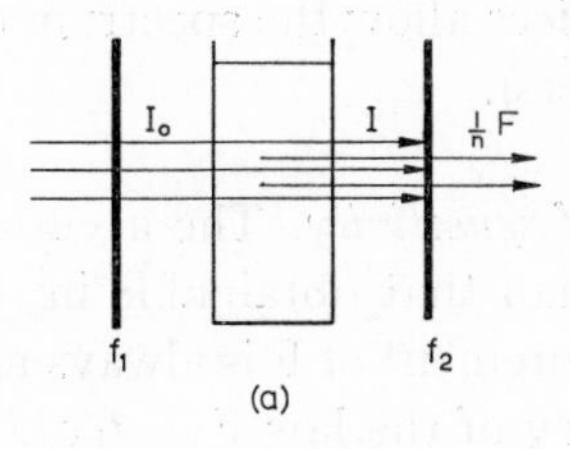

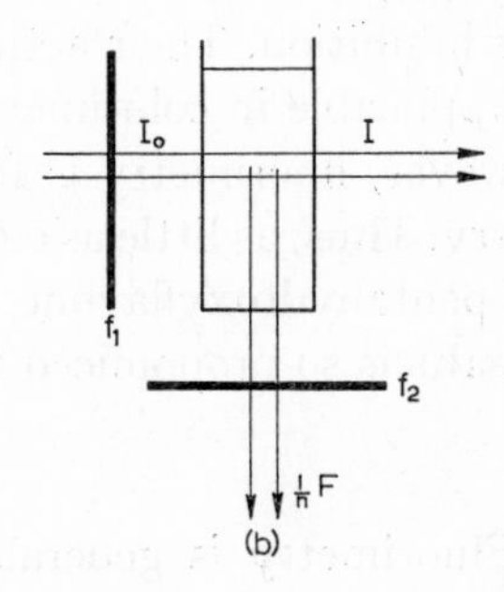

Fig. 31.

If the filters are not perfect, the difficulty may be partially circumvented by observing the fluorescence perpendicularly to the direction of the light beam (Fig. 31, *b*); in this case however, the spatial distribution of the fluorescence received by the cell depends on the concentration (the intensity absorbed and consequently that emitted by each successive cross-sectional element of solution decreases in step with the intensity of the primary beam during the passage of the latter through the solution). This may introduce additional errors, since the surface sensitivity of a photoelectric cell varies from one point to another.

(2) It is always necessary to carry out a blank test in order to determine the possible fluorescence of the cell, the solvent, or of any impurities. The result of the blank test may not however be used to correct for the other effects of impurities.

(3) An accumulation of theoretical and practical difficulties generally renders the use of a calibration curve $F = f(c)$ preferable to that of the theoretical equation considered above. The calibration curve should be plotted under identical conditions to those of the experiment, particularly with respect to the concentrations of inhibitors, if the results are to be sufficiently accurate.

(i) Apparatus. The types of equipment used are analogous to those used in colorimetric determinations. All that need be added is a supplementary filter and a device permitting the cell to be illuminated laterally, if this is required.

In order to increase the fraction of the fluorescent light received, certain instruments incorporate two photocells mounted in series and placed on either side of the cell.

Certain less common devices allow the spectrum of the fluorescent light to be obtained (spectrofluorimeters).

(ii) Accuracy and limits of sensitivity. The accuracy of fluorimetric measurements is generally lower than that obtainable in colorimetry. In practice, the error occurring in the measurement of F is always negligible in comparison with the possible lack of accuracy of the law $F = f(c)$.

As has already been mentioned, traces of impurities exert an enormous influence, owing to absorption or inhibition. Their action cannot be eliminated by methods as simple as those applicable in colorimetric determinations.

In spite of the above, however, fluorimetry is frequently more sensitive to trace amounts than colorimetry. Thus, as little as 0.002 ppm of beryllium can be perceived in the presence of pentahydroxyflavone, whilst in the molten state, the fluorescence of uranium salts is so pronounced that as little as 10^{-11} g may be detected.

(iii) Field of application. Fluorimetry is generally employed when no sufficiently sensitive or selective method of estimating an element colorimetrically has been found. In organic chemistry, the most frequent applications involve estimations of metallic ions (in the form of fluorescent organic complexes) either in solution, or after extraction.

A number of examples will be found in the second part of this work.

Bibliography

GENERAL (See general works, Chapter 4)

E. J. BOWEN AND F. WOKES, *Fluorescence of Solutions*, Longmans Green, London, 1953.

Fluorimetry in Physical Methods of Organic Chemistry, vol. I, 3rd edn., by A. WEISSBERGER, Interscience.

M. CURIE, *Fluorescence et Phosphorescence*, Hermann, Paris, 1946.

P. W. F. DANCKWORTH AND J. EISENBRAND, *Lumineszenz-Analyse*, 7th. edn., Geest et Portig, Leipzig, 1963.

P. PRINGSHEIM, *Fluorescence and Phosphorescence*, Interscience Publishers, New York, 1949.

J. A. RADLEY AND J. GRANT, *Fluorescence Analysis in Ultraviolet Light*, van Nostrand, New York, 1954.

REVIEWS

C. E. WHITE, *Anal. Chem.*, 26 (1954) 129; 28 (1956) 621; 30 (1958) 729.

Chapter 6

Turbidimetry and nephelometry

When the material to be estimated can be precipitated to form a sufficiently stable suspension, two methods related to colorimetry, both of very restricted application, can be used. In each case the solution containing the suspended precipitate is traversed by a beam of light of an initial intensity I_0. Intensity of the beam emerging from the solution, either in the direction of the incident beam (turbidimetry), or in a different direction (nephelometry) may then be determined. These intensities are respectively denoted by I and I_D.

By analogy with colorimetry and fluorimetry, the ratio I_0/I and the intensity I_D increase with the concentration of the precipitated material.

The phenomenon is governed by numerous factors. If the particles are very small (*e.g.* of diameter less than one tenth of the wavelength used), and if they are transparent, the incident beam is diffracted (Tyndall effect). This is revealed by a diminution in the intensity of the light in the direction of the incident beam due to its scattering in other directions.

(1) If it were possible to keep the mean volume of the particles constant, I_D would be proportional to *c* and would be the larger, the smaller the wavelength used. On the other hand, a law analogous to Beer's law could be established:

$$D = \log \frac{I_0}{I} = \tau lc.$$

This law is obeyed in the case of some 'colourless' colloidal precipitates.

(2) If in addition the particles absorb light, the intensity I will be lower than if this were not the case. For solutions obeying Beer's law, however, the optical density is always proportional to the concentration *c*, so that

$$D = \log \frac{I_0}{I} = (\tau + \varepsilon)lc.$$

This is the case with certain lakes. In fact, the method may be made more sensitive by covering the colloidal particles with an absorbed layer of a colouring agent. D may be increased considerably in this way. If D is to remain proportional to *c*, ε must remain constant, *i.e.* the amount of colouring material adsorbed on each particle must always be the same. This is approximately so as long as the excess of the colouring matter is sufficiently large, particularly at low concentrations of the material to be estimated.

(3) When the particles are sufficiently large, a reflection phenomenon occurs in addition to diffraction and absorption, causing a deviation from the above rules.

(4) When the particles are very close to one another, *i.e.* when the initial solution is too concentrated, molecular interactions become responsible for deviations from the first two laws.

It is clear that no sharp distinction can be made between turbidimetry and colorimetry. As the dimensions of the particles decrease, the phenomena of reflection and diffraction successively disappear, finally leaving an almost pure absorption effect.

Two conclusions follow from the above:

(a) It is not possible to establish a theoretical formula connecting D or I_D with c in every case. Consequently, it is in general essential to use calibration curves obtained under the same conditions of measurement.

(b) The volume of the precipitated particles affects to a very large extent the results obtained.

(i) Apparatus and procedure used for measurements. Colorimeters and spectrophotometers may be used for turbidimetry without any modification.

Nephelometry can be performed with a fluorimeter, in which the diffused light is observed laterally.

(ii) Accuracy and sensitivity. The sensitivity of these methods may be quite high, particularly if the particles absorb light (are coloured) or are opaque. If they are transparent, the best results are obtained by using short-wavelength radiation and taking the sensitivity of the receiver into account. If absorption takes place as well, it is necessary to work near the absorption maximum.

The accuracy of nephelometric and turbidimetric techniques is not very great. The errors occurring during the photometric measurement are generally negligible in comparison with those due to the lack of reproducibility of the properties of the suspension. In fact, it is almost impossible to ensure that the particle size will be the same during each measurement as during calibration. Numerous factors come into play: concentration of the material to be estimated, rate of addition of the reagent, stirring, temperature, presence of foreign materials, etc. In addition, the particle size varies in the course of time. Crystal growth can be prevented by the addition of surface-active materials, such as gelatin. The operating procedures used for the preparation of the various standards and the solution to be estimated must then be made as nearly identical as possible.

(iii) Field of application. Turbidimetry and nephelometry may be of use in certain particular cases, either where a sparingly soluble compound may be precipitated (with greater sensitivity than that obtainable with any known colorimetric method) as is the case with Al(III) and cupferron, or because no good colour reaction is available, as is the case for SO_4^{2-}, which is estimated in the form of barium sulphate.

Bibliography

GENERAL (See also p. 65)

J. H. Yoe and H. Kleinmann, *Photometric Chemical Analysis*, vol. II, Wiley, New York, 1929.

F. D. Snell and C. T. Snell, *Colorimetric Methods of Analysis*, van Nostrand, New York, 1949.

Chapter 7

Photometric and spectrophotometric titrations

Since the optical density of a solution is a linear function of the concentrations of the various absorbing materials present in it, it is possible to follow the variations in concentrations occurring during a reaction, or simply to detect the end point thereof. Volumetric (or possibly coulometric) titrations may thus be performed by plotting D against the amounts of reagent added.

(1) Characteristics of the titration curves

In common with other methods involving properties which vary linearly with concentration, (amperometric, conductometric, etc. titrations), the above procedure is distinguished by the following characteristics*.

(1) When the reaction is quantitative, the curves may generally be represented by straight-line segments (Figs. 32 to 35 for the reaction $A+B \rightarrow C+D$).

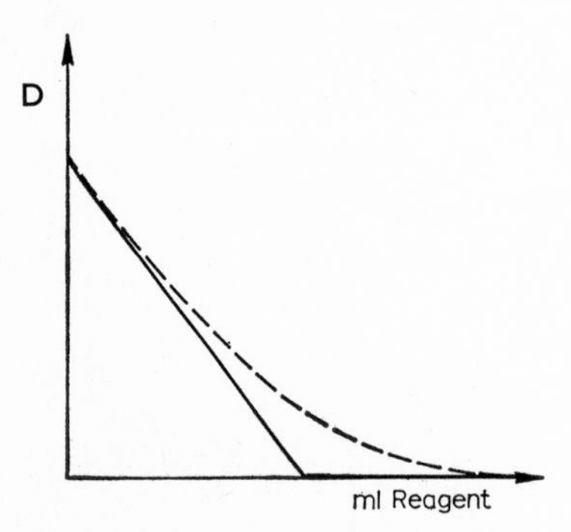

Fig. 32. Only A is absorbent.

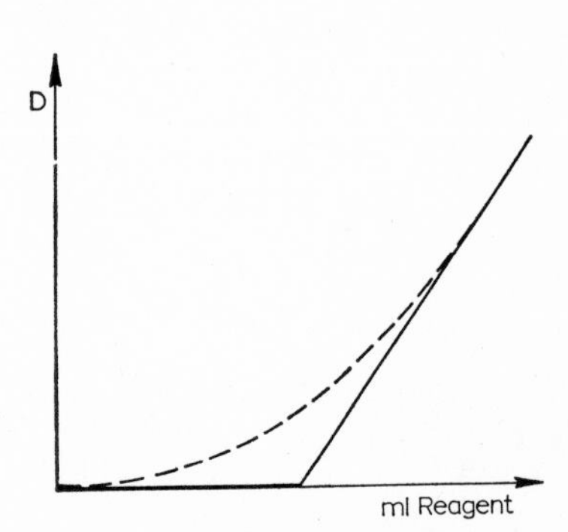

Fig. 33. Only B is absorbent.

Deformation of the curves owing to the progressive dilution caused by the continuous addition of the reagent during the titration may be avoided by adjusting the solution to a known volume before each measurement (discontinuous method). Alternatively, a concentrated solution of the reagent, contained in a microburette, may be used for the titration. In such a case, the results must be

* See G. CHARLOT. *Les méthodes de la chimie analytique,* 4th Edn., Masson et Cie, 1961.

corrected for the dilution factor, by being multiplied by $V + v/V$, where V is the initial volume and v the volume of the reagent added.

Such precautions are no longer necessary if the reagent is generated electrochemically within the solution itself.

(2) The reaction need not be highly quantitative at the equivalence point itself, since the result may be obtained by extrapolation from measurements made at any points which conform to a quantitative relationship. The latter may in fact be quite far removed from the equivalence point. The shapes of curves representing nonquantitative reactions are shown by the broken lines in Figs. 32 to 35.

In actual fact, the curves may be plotted accurately using only a few points. Thus, in the case of Fig. 32, in principle only two measurements are required: points may, for example, be determined only for the initial solution and at one other stage during the reaction.

Similarly, in the case of Fig. 34, one measurement before and one measurement past the equivalence point are sufficient.

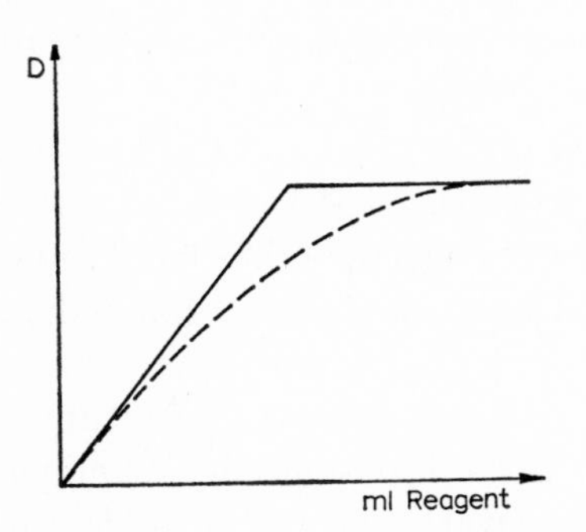

Fig. 34. Either C or D is absorbent.

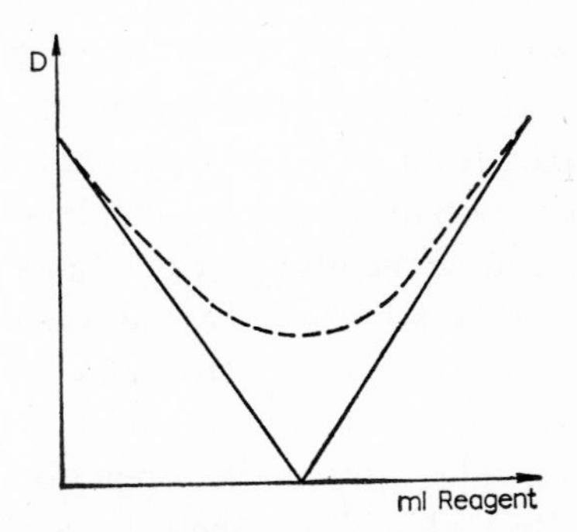

Fig. 35. Both A and B absorbent.

In practice it is however advisable to make several determinations and to fit the straight lines as closely as possible to the experimental points, in order to compensate partially for the scatter due to the measuring instrument.

If the method of measurement is very sensitive, extremely dilute solutions may in many cases be used: the reaction frequently ceases to be quantitative in such cases.

(2) Characteristics of the method

The various advantages of colorimetry over other methods of analysis are all reflected in the technique in question. They will be outlined below.

The selectivity of the method is high, since reagents which absorb or give absorbing compounds, and the wavelength used (including ultraviolet), may be selected to give the maximum advantage.

The effects of interfering compounds may be eliminated by a blank test.

The precision of measurements may be high under suitable conditions.

Two solutions may be compared with great accuracy, particularly if a double-beam instrument is used.

Owing to the high sensitivity, titrations may be carried out in extremely dilute solutions.

The main advantage of the method over direct colorimetry is that high accuracy does not necessarily entail the use of an expensive (monochromatic) spectrophotometer and, in addition, calibration becomes unnecessary. A standardized solution must however be available, except in the case of coulometric measurements. Furthermore, in many cases the equivalence point can be located graphically with great accuracy.

Moreover, the titration can be made automatic, since certain instruments record the titration curve directly.

Whereas visual titrations employing indicators may only be performed with solutions of concentrations *l*, 10^{-1} and on rare occasions 10^{-2}, the above method offers the possibility of working with much lower concentrations, of the order of 10^{-2}, 10^{-3} and sometimes 10^{-4}.

Apart from its sensitivity, (which is sometimes higher than those of amperometric, potentiometric or conductometric methods), spectrophotometry deals with a different group of reactions and solvents. On the other hand, however, the method requires an apparatus incorporating a readily accessible cell, and in particular a microburette or a coulometric device.

Particular attention should be paid to variations in the temperature of the system caused by heats of reactions, even when dilute solutions are employed.

If Beer's law is not obeyed, the theoretical titration curve can be determined with the aid of a calibration curve $c = f(D)$.

(3) Simplification of the method

Where the reaction under consideration is sufficiently quantitative, and if a sensitive instrument is used, the equivalence point may be located simply by noting the sharp variation in the optical density, which coincides with it. It is then no longer necessary to take precautions in relation to dilution effects and the temperature, and Beer's law need not necessarily be obeyed.

(4) Examples of applications

The titration curves are known and the accuracy of determinations can be calculated from the equilibrium constants*.

* See, for example, G. CHARLOT, *Les méthodes de la Chimie analytique*, 4th edn., Masson et Cie, 1961.

(1) Titrations may be performed by the usual method, in which the concentration of one of the substances involved is followed: thus, As(III) can be estimated volumetrically by means of its reaction with Ce(IV):

$$As(III) + 2\,Ce(IV) \rightarrow As(V) + 2\,Ce(III)$$

The optical density is measured at 320 mμ, since only Ce(IV) absorbs at this wavelength (Fig. 36).

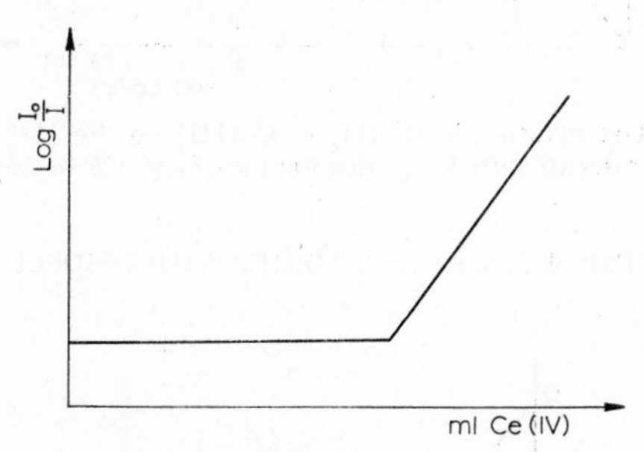

Fig. 36. The reaction is: As(III) + 2 Ce(IV) → As(V) + 2 Ce(III).

(2) If several materials absorb, a suitable wavelength is selected. Thus, Fe(II) can be titrated against Co(III), according to the reaction Fe(II) + Co(III) → Fe(III) + Co(II).

A suitable wavelength may be selected with the aid of the D = $f(\lambda)$ curves of the various materials present (Fig. 37). At 360 mμ, the absorption of Fe(III) is high (titration curve, Fig. 38), while at 610 mμ, the light is mostly absorbed by Co(III).

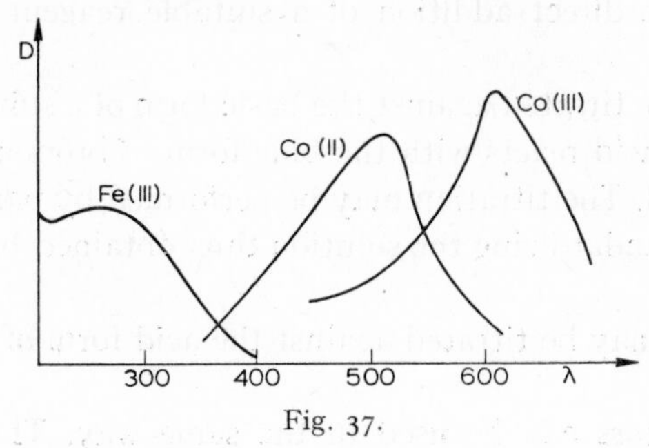

Fig. 37.

(3) A material may be made absorbent by adding a reagent to it. Thus, Bi(III) can be titrated against EDTA according to

$$Bi^{3+} + Y^{4-} \rightarrow BiY^{-}.$$

The concentration of Bi(III) may be followed by adding an excess of thiourea,

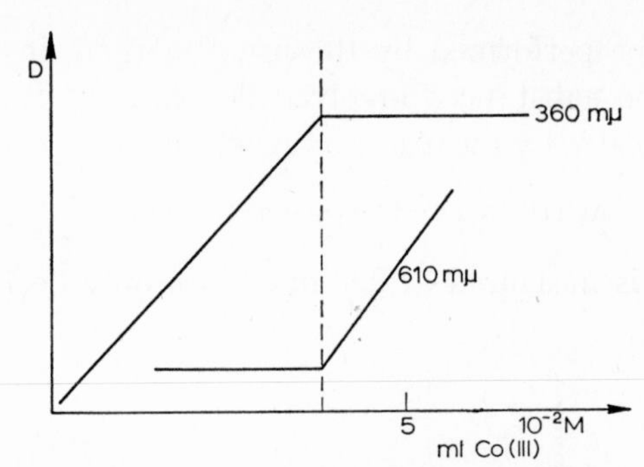

Fig. 38. The reaction is: Fe(II) + Co(III) → Fe(III) + Co(II).
(After C. E. BRICKER AND L. J. LOEFFLER, *Anal. Chem.*, 27 (1955) 1419.)

which gives a yellow complex whose stability with respect to BiY^- is sufficiently low (Fig. 39).

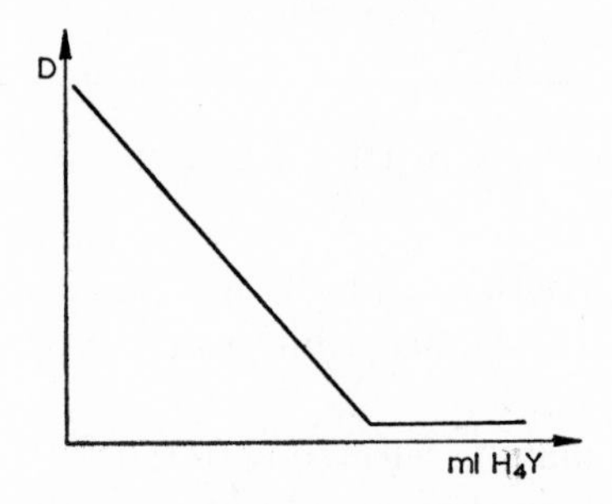

Fig. 39. The reaction is: $Bi^{3+} + H_4Y \rightarrow BiY^- + 4H^+$.
An excess of thiourea is present.

(4) Titration by the direct addition of a suitable reagent is possible in the following cases:

(a) an acid may be titrated against the basic form of a suitable pH indicator. For example, boric acid reacts with the blue form of bromothymol blue in the presence of mannitol. The titration may be performed by partially neutralizing the indicator and standardizing the solution thus obtained before the measurement is begun.

Similarly, a base may be titrated against the acid form of the corresponding indicator.

(b) Redox indicators can be used in the same way. Thus, Ce(IV) can be titrated against ferrous *o*-phenanthroline.

(c) Various reagents may be used specifically with certain ions: thus, Mg^{2+} can be titrated against Eriochrome Black T, etc.

(5) Successive reactions

It is often possible to titrate several compounds successively.

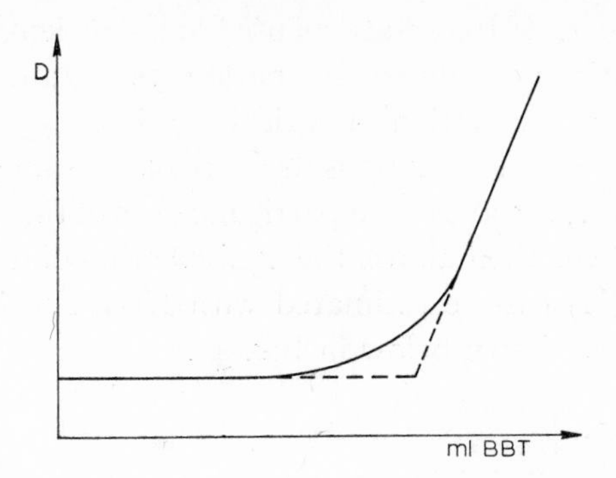

Fig. 40. Estimation of boric acid in mannitol, using the basic form of bromothymol blue.

EXAMPLE. Titration of Fe(III) + Cu(II) against EDTA. At pH = 2, the complexes FeY^- and CuY^{2-} are formed in succession. At 745 mμ, only CuY^{2-} absorbs. This leads to the titration curve of Fig. 41.

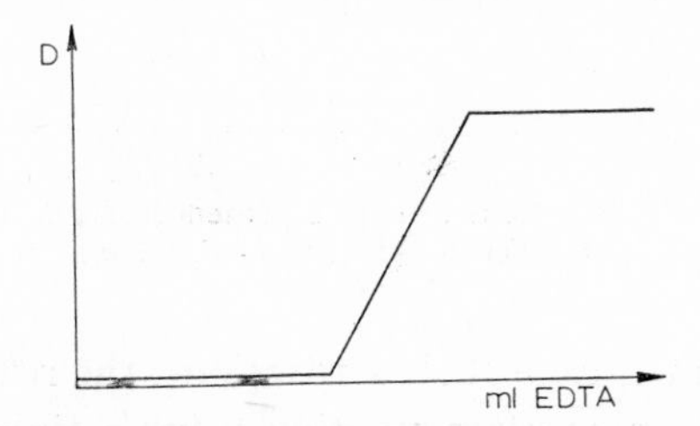

Fig. 41. Estimation of Fe(III) and Cu(II) with EDTA.
(After A. L. UNDERWOOD).

When it is not possible to follow a reaction A + B $\rightarrow$ C directly, the solution of A may be treated with a substance A′ which will react with excess B at the conclusion of the first reaction. A′ should be selected so as to form a detectable reaction product with B. The inception of the second reaction then indicates the equivalent point of the first. In many cases, A′ can be suitably selected indicator of B.

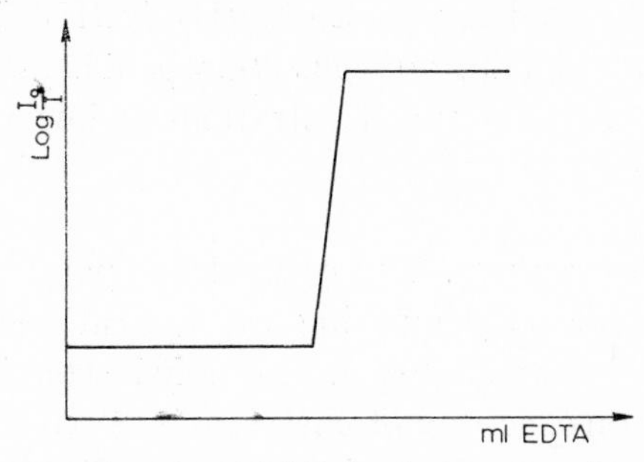

Fig. 42. Titration of Ca(II) against EDTA in the presence of ammonium purpurate.

(i) Use of an indicator. If the solutions used are sufficiently concentrated, the end point of the reaction may quite easily be detected by automatically recording the corresponding sharp variation in optical density.

If the concentration of the solutions used are of the same order of magnitude as that of the indicator, the above is a particular case of the method of successive reactions. The indicator then allows the equivalence point to be detected.

For example, Ca(II) can be estimated with EDTA in the presence of ammonium purpurate, as shown below in Fig. 42.

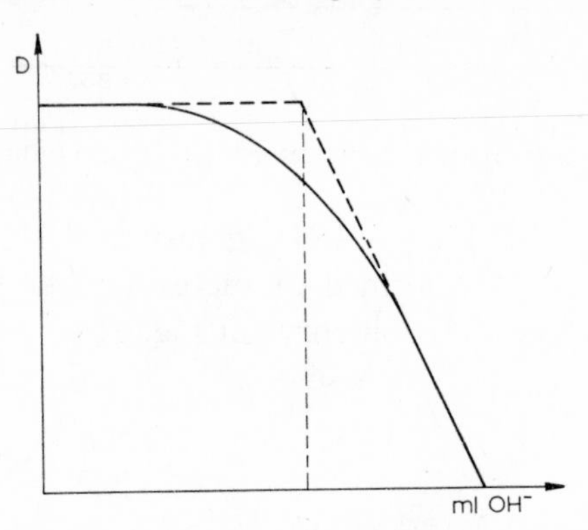

Fig. 43. Titration of 10^{-3} *M* boric acid in the presence of 4×10^{-5} *M* *p*-nitrophenol. (After R. F. GODDU AND D. N. HUME, *Anal. Chem.*, 26 (1954) 1740).

(ii) Back-titrations. In the case of slow reactions, the reagent may be added in excess, the amount remaining after reaction being estimated by spectrophotometric titration.

(iii) Solvents other than water. A number of materials can be estimated with the aid of organic reagents dissolved in aqueous alcohol or acetone, as the reagents themselves and their respective reaction products are often soluble in such media. This is the case for example, with dithizone and the dithizonates:

$$M^{2+} + 2\,HDz \rightarrow MDz_2 + 2\,H^+.$$

Similarly, oxine may be used to estimate Cu(II), Zn(II), Cd(II), Ni(II), Pb(II), and U(VI) in aqueous dioxan, aqueous *n*-propanol, and aqueous dimethylformamide solvents. In this way concentrations as low as 3×10^{-5} *M* may be determined, to within 0.1–0.2%[1].

(iv) Null methods (differential). The equivalence point of a reaction may be determined by continuously comparing the optical density of the reacting solution with that of a known standard. This may be conviently performed with the aid of a double-beam spectrophotometer, and obviates the need for plotting a titration curve. It was already pointed out in Chapter 3 that two nearly

identical solutions could be compared with a high degree of precision. Furthermore, Beer's law need not necessarily hold true.

EXAMPLE. Estimation of traces of boric acid.

Partially neutralized bromothymol blue is added to the neutralized solution to be analyzed. The former consists of a buffered mixture of the yellow (acid) HI form and the blue (basic) I form. The solution is distributed between the two cells of a double-beam spectrophotometer. A little solid mannitol is added to one of the cells; the boric acid ($pK_A = 9.2$), which was at first unable to react with the indicator ($pK_A = 7$), is now complexed and behaves as a much stronger acid ($pK_A \approx 5$). It then reacts with the indicator:

$$I + HB \rightarrow HI + B.$$

A difference in the optical density appears between the two solutions. Two small electrodes, (made respectively of Pt and Ag), are then immersed in the reaction cell and coulometry is carried out at constant intensity with a simplified apparatus. The reaction at the silver electrode, in which the chloride present in the solution takes part, is $Ag\downarrow + Cl^- - e \rightarrow AgCl\downarrow$, whilst at the platinum electrode, the corresponding reaction

$$2\,HI + 2e \rightarrow H_2 + 2I$$

takes place.

Electrolysis is continued until the difference in optical densities again becomes zero.

The method is exceedingly accurate; it is *e.g.* possible to estimate 10^{-7} H^+ to within 1% (Platzer, C.E.A.).

Of the impurities present, those not capable of being electrochemically oxidized or reduced will be in identical state in the two solutions when the pH values of the latter become equal.

(v) Derived curves. Certain automatic instruments directly record the $\Delta D/dV$ curve during the titration, thus showing up variations in the slope of the normal titration curve.

(6) Turbidimetric ('heterometric') titrations

The possibility of continuously following the variation in optical density occurring during a precipitation reaction is at present being investigated.

If precautions are taken to ensure that at any stage, the appropriate precipitation occurs instantaneously rather than by gradual development, (*e.g.* by adding certain substances to the solution), it is possible to obtain approximately linear curves which intersect at the equivalence point (see Nephelometry and Turbidimetry, p. 73).

It is also possible to carry out the titration without waiting for each stage to develop a constant optical density, provided that constant amounts of reagent are added at regular intervals. This method yields curves which are often linear, particularly past the equivalence point. In all cases, the latter is often indicated by a discontinuity in the curve.

Numerous estimations of this type have been proposed[1]. The method is sometimes accurate and often very sensitive. It is thus possible to titrate solutions ranging from 10^{-1} to 10^{-5} and sometimes even 10^{-6} M, though the majority of such estimations are performed in the range of $10^{-3} - 10^{-4}$ M.

The procedures used and the advantages of the method are analogous to those of colorimetric titrimetry.

Bibliography

(1) W. G. Boyle, Jr. and R. J. Robinson, *Anal. Chem.*, 30 (1958) 958.

REVIEWS

R. H. Osburn, J. H. Elliott and A. F. Martin. *Ind. Eng. Chem., Anal., Ed.*, 15 (1943) 642.
R. F. Goddu and D. N. Hume, *Anal. Chem.*, 22 (1950) 1314.
R. F. Goddu and D. N. Hume, *Anal. Chem.*, 26 (1954) 1740 (review up to 1954 with 105 references).
J. B. Headridge, *Talanta*, 1 (1958) 293 (review up to 1957 with 119 references).

AUTOMATIC TITRIMETERS

T. L. Marple and D. N. Hume, *Anal. Chem.*, 28 (1956) 1116.
H. V. Malmstadt and C. B. Roberts, *Anal. Chem.*, 28 (1956) 1408.
C. A. Weilley and R. A. Chalmers, *Analyst*, 82 (1957) 329.
H. V. Malmstadt and D. A. Vassallo, *Anal. Chem.*, 31 (1959) 862.
P. W. Mullen and A. Anthon, *Anal. Chem.*, 32 (1960) 103.
J. M. Thoburn, C. M. Jankowski and M. S. Reynolds, *Anal. Chem.*, 31 (1959) 124.
J. Phillips, *Automatic Titrators*, Academic Press, 1959.

COULOMETRIC TITRATIONS

E. N. Wise, P. W. Gilles and C. A. Reynolds, Jr., *Anal. Chem.*, 26 (1954) 779.
G. W. Everett and C. N. Reilley, *Anal. Chem.*, 26 (1954) 1750.
N. H. Furman and A. J. Fenton, Jr., *Anal. Chem.*, 28 (1956) 515.
H. V. Malmstadt and C. B. Roberts, *Anal. Chem.*, 27 (1955) 741.
H. V. Malmstadt and C. B. Roberts, *Anal. Chem.*, 28 (1956) 1412.
H. V. Malmstadt and C. B. Roberts, *Anal. Chem.*, 28 (1956) 1884.
C. B. Roberts, *Dissert. Abstr.*, 16 (1956) 1798.
J. W. Miller and D. D. de Ford, *Anal. Chem.*, 29 (1957) 475.
R. Gauguin, *Chim. Anal.*, 36 (1954) 92.

dD/dV

H. V. Malmstadt and C. B. Roberts, *Anal. Chem.*, 28 (1956) 1408, 1412.
R. C. Chalmers and C. A. Walley, *Analyst*, 82 (1957) 329.

USE OF INDICATORS

Theory

A. Ringbom, in I. M. Kolthoff and P. J. Elving, *Treatise on Analytical Chemistry*, Interscience, 1959.

T. Higuchi, C. Rehm and C. Barnstein, *Anal. Chem.*, 28 (1956) 1506.

A. Ringbom and F. Sundman, *Z. Anal. Chem.*, 116 (1939) 104.

A. Ringbom and E. Vanninen, *Anal. Chim. Acta*, 11 (1954) 153.

HETEROMETRY

M. Bobtelsky, *Heterometry*, Elsevier, 1960.

Review

A. Ringbom, *Z. Anal. Chem.*, 122 (1941) 263.

Chapter 8

Separation of trace amounts

No mention has been made in this book of the common techniques of separating out appreciable amounts of various substances, since they have all been described repeatedly in works on analytical chemistry, and no account has been given of the familiar methods of separation by precipitation*. The techniques used for separating out trace amounts, some of which have been developed quite recently, will however be considered in this and subsequent chapters.

Separation by adsorption

The following quantitative relationship may be applied in this case:

$$C_{A\downarrow} = \frac{kC_A}{1 + k_A C_A + k_B C_B + \ldots},$$

where $C_{A\downarrow}$ is the concentration of the ions of A adsorbed on the solid, C_A their concentration in the solution, while $C_B, \ldots,$ are the concentrations of all the other materials in the solution.

The above relationship may be manipulated to yield a definition of the distribution coefficient, D_A, of the substance A between the solid and the solution:

$$D_A = \frac{C_{A\downarrow}}{C_A} = \frac{k}{1 + k_A C_A + k_B C_B + \ldots}$$

Furthermore, in dilute solutions, $D_A = k$ (Fig. 44).

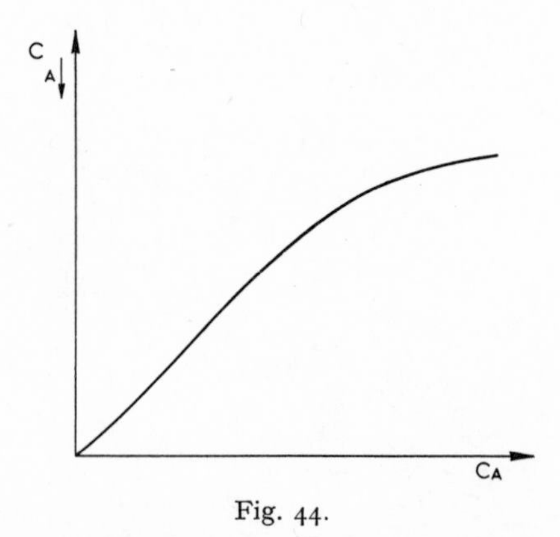

Fig. 44.

* See, for example, G. Charlot, *Méthodes de la Chimie analytique*, 4th. edn., Masson et Cie, 1961.

NOTES. (1) k is generally small and varies with the physical state of the precipitate.
(2) The adsorbed matter may consist of ions or polar (or polarizable) molecules.
(3) The value of C_A may be affected if A participates in any equilibria occurring in the solution. For example, if A is strongly complexed, its concentration will be low and it will cease to be adsorbed; the complex formed may or may not be adsorbed as well.

(i) Separation by precipitation. It is not generally practicable to isolate trace amounts of a substance in solution (for future estimation) by precipitating out any other interfering substances present, since the material in question will in most cases be totally or partially adsorbed on to the precipitate. It is, however, sometimes possible to 'protect' the trace element by adding large amounts of an element as similar to it as possible; the latter is then adsorbed under the same conditions as the trace element, and because of its much higher concentration, it almost entirely prevents the adsorption of the other. For example, solutions of radium may be 'protected' by large additions of barium salts.

More often, the trace elements are themselves precipitated in the form of a sparingly soluble compound. In many cases, however, there is a tendency for the latter to appear in colloidal form or to remain in supersaturated solution; it should therefore be coprecipitated by adding a large amount of a substance which precipitates under the same conditions. This is known as separation by entrainment.

(ii) Separation by entrainment. The mechanism of entrainment may consist either of adsorption, or the formation of mixed crystals.

The entrained salt need not itself be sparingly soluble. This technique is particularly useful in connection with the chemistry of the radio-elements.

Entrainment is favoured when the two compounds formed have analogous compositions; they should possess similar formulae, and preferably contain a common ion—for example, $PbSO_4$ and $SrSO_4$.

EXAMPLES. Titanium hydroxide may be entrained from very dilute solutions by coprecipitation with aluminium hydroxide; solutions of titanium of as little as 1 μg of the metal per litre, which would not otherwise precipitate at all, may be treated in this way.

CuS entrains HgS down to 0.02 μg/l, as well as MoS_3, ZnS, PbS, etc.

Titanium cupferrate entrains traces of zirconium.

(iii) Adsorbents. Certain solids particularly such as powdered cellulose, 'chromatographic' alumina, active charcoal, etc. exhibit a marked tendency to adsorb other substances.

(iv) Operating procedure. (1) The adsorbing precipitate may be formed in situ and should then be separated from the solution.
(2) The solution to be treated may be filtered through the adsorbing precipitate.
(3) The most effective and most general method consists of passing the solution through a column packed with an adsorbing material, *e.g.* cellulose powder wetted with water.

Chapter 9

Separations by extraction

(1) Principle

Compounds dissolved in two immiscible solvents in contact with each other, become distributed between them according to the partition law. An equilibrium is eventually set up as a result of solute exchange at the common interface of the two liquids.

The general equilibrium is of the form

$$aA + bB + \ldots \rightleftarrows mM + nN + \ldots$$

where, of the materials concerned, some are in one solvent and the others in the other solvent.

EXAMPLES. Dithizone HDz and zinc dithizonate are soluble in carbon tetrachloride but not in water. Thus:

$$\underset{\text{water}}{Zn^{2+}} + \underset{CCl_4}{2HDz} \rightleftarrows \underset{CCl_4}{ZnDz_2} + \underset{\text{water}}{2H^+}$$

on the basis of which we may write

$$\frac{[ZnDz_2]_{CCl_4}[H^+]_{\text{water}}}{[Zn^{2+}]_{\text{water}}[HDz]^2_{CCl_4}} = K.$$

Acetic acid distributes itself between water and benzene, dimerizing in the latter solvent:

$$\underset{\text{water}}{2\ CH_3CO_2H} \rightleftarrows \underset{\text{benzene}}{(CH_3CO_2H)_2} \quad \text{whence,} \quad \frac{[(CH_3CO_2H)_2]}{[CH_3CO_2H]^2} = K.$$

(i) Distribution coefficient. If a given material (whether in the form of ions or molecules) exists in two solvents in contact with each other, then at equilibrium

$$A_{\text{solv. 1}} \rightleftarrows A_{\text{solv. 2}} \quad \text{and} \quad \frac{[A_{\text{solv. 2}}]}{[A_{\text{solv. 1}}]} = D,$$

where $A_{\text{solv. 1}}$ and $A_{\text{solv. 2}}$ represent the activities of the material A in the two solvents. The ratio D is known as the distribution coefficient of A between solvent 2 and solvent 1, and is constant at a given temperature, whatever the concentrations in the two solvents may be. When the solute A is in the molecular state, if one of the solvents is saturated, (*i.e.* is in contact with excess A in the solid, liquid, or gaseous state), the other solvent must necessarily be saturated as well. Thus, D is in general equal to the ratio of the solubilities of A in the two solvents.

NOTES. (1) In the majority of cases, water is one of the solvents. If the other solvent is an inert organic substance, or one with a low dielectric constant, which is frequently the case, any solute present in this second solvent will be in molecular form.

(2) The addition of a high concentration of ionized salts to the aqueous solution if often used to facilitate the passage of compounds into the organic phase ('salting-out').

A number of factors contribute to, and cause this effect. Among these are, for example, the changes in the activity coefficients of the substances present, occurring as a result of the increased ionic strength of the aqueous solution. The degrees of hydration of the ions are also affected, since very high concentrations of ions reduce the dielectric constant of the aqueous solution, thus stabilizing the hydration complexes. Predictions of the result of an extraction may, in such cases, be invalid (see Fig. 45).

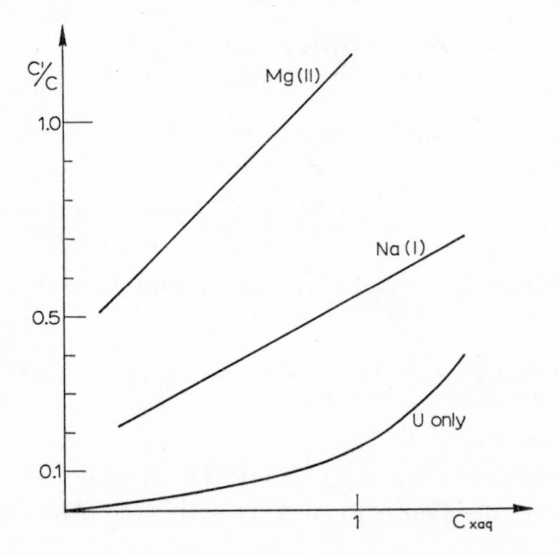

Fig. 45. Distribution ratic c'/c of uranyl nitrate in the presence of high concentrations of different nitrates. Solvents: water, hexone.

(3) The presence of colloids such as $Zr(OH)_4$, $Si(OH)_4$, etc. may prevent the extraction of certain materials, owing to the occurrence of adsorption phenomena.

(4) It is difficult to predict the course of extractions of some inorganic compounds into certain solvents, such as the ethers, esters, alcohols, and ketones. This is because such solvents interact with many of the solutes, so that the mutual solubilities of these solvents with water are often considerably modified in an unpredictable fashion.

(ii) Distribution ratio and extraction efficiency. If a substance A is distributed between two solvents S′ and S, then

$$A_S \rightleftarrows A_{S'} \text{ and } \frac{[A]_{S'}}{[A]_S} = D.$$

A may assume several forms in aqueous solution, for example AB, AC, etc., and from the practical point of view it is of interest to define an apparent partition coefficient D′, equal to the ratio between the total concentrations of A in all its forms in two solvents S′ and S.

In addition, the extraction efficiency, R, may be defined as the ratio between the total amounts of A taken up by the solvent, and present in the initial solution.

If the volume of the aqueous solution is denoted by V and that of the solvent by V′, R (percentage extraction) is given by the expression:

$$R\% = \frac{[A]_{solv.}V'}{[A]_{solv.}V' + \Sigma[A]_{water}V} \times 100.$$

(1a) *Extraction and acidity*

The effect of pH on the extraction efficiency will be described in terms of several examples, which are presented below.

(i) The extraction of dithizone as a function of pH. The first example considered is that of dithizone or diphenylthiocarbazone, which may be denoted by HDz. In the molecular state, this substance is insoluble in water and soluble in carbon tetrachloride. When the two solvents are in contact, the following equilibrium is set up

$$\underset{CCl_4}{HDz} \rightleftarrows \underset{water}{Dz^- + H^+} \text{ whence, } \frac{[Dz^-][H^+]}{[HDz]} = K_A = 10^{-8.7}$$

$$D' = \frac{[HDz]_{CCl_4}}{[Dz^-]_{water}} = \frac{[H^+]}{K_A}.$$

It is evident that the addition of an acid will displace the above equilibrium to the left, and the dithizone will pass into the organic solvent. Conversely, the solute is displaced into water when the latter is in the form of Dz^-. The following scale may thus be drawn:

$$\underset{\hspace{2cm}}{HDz_{(CCl_4)}} \qquad [HDz] = [Dz^-]_{(H_2O)} \qquad Dz^-_{(H_2O)}$$

$$\xrightarrow[8.7 \hspace{6cm} pH]{}$$

PERCENTAGE EXTRACTION

This is given by:

$$R\% = \frac{[HDz]V'}{[HDz]V' + [Dz^-]V} \times 100,$$

where V′ and V are the volumes of the organic and aqueous phases respectively.

From this, using the equation for K_A,

$$R\% = \frac{[HDz]V' \times 100}{[HDz]V' + [HDz]\dfrac{K_A}{[H^+]}V} = \frac{V'/V \times 100}{V'/V + \dfrac{K_A}{[H^+]}}$$

$$\text{if } V' = V,\ R\% = \frac{1}{1 + \dfrac{K_A}{[H^+]}} \times 100.$$

Thus, for $[H^+] = K_A$, of $pH = pK_A$, the extraction efficiency is 50%, *i.e.* $[HDz]_{CCl_4} = [Dz^-]_{water}$. The percentage extraction is plotted as a function of the pH in Fig. 46, from which it is evident that for $pH \leqslant pK_A - 3$, dithizone is entirely retained in the organic solvent (to within 0.1%), while for $pH \geqslant pK_A + 3$, it passes entirely into the aqueous phase.

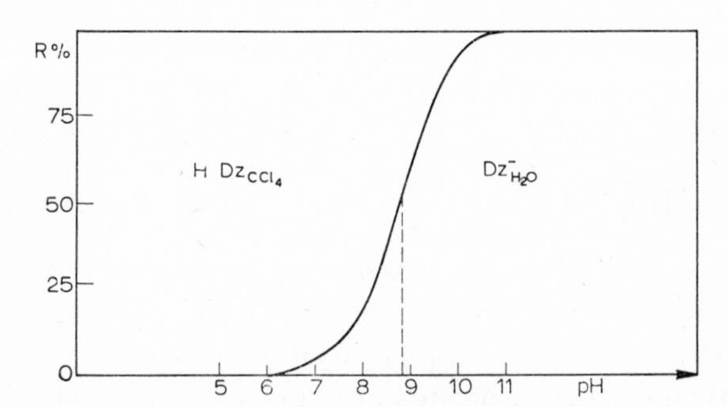

Fig. 46. Extraction of dithizone by carbon tetrachloride plotted as a function of pH (V = V′).

(ii) The extraction of oxine as a function of pH. Denoting a molecule of oxine by HOx, the equilibria are set up and may be described by

$$HOx + H^+ \rightleftarrows \underset{\text{oxinium ion}}{H_2Ox^+}, \quad pk_2 = 5.0$$

$$HOx \rightleftarrows \underset{\text{oxinate ion}}{Ox^-} + H^+, \quad pk_1 = 9.7.$$

If chloroform is used as the extraction solvent, then

$$\underset{\text{water}}{\text{HOx}} \rightleftarrows \underset{\text{CHCl}_3}{\text{HOx}} \text{ whence, } \frac{[\text{HOx}]_{\text{CHCl}_3}}{[\text{HOx}]_{\text{water}}} = \text{D} = 720.$$

It is clear that provided the aqueous phase is sufficiently acidic, *i.e.* of pH < 5, the oxine will be in the form of H_2Ox^+, contained in this phase; very little oxine can persist in the molecular state, and its concentration in chloroform is therefore small. Similarly, in an alkaline medium of pH > 10, the oxine exists mainly in the form of Ox^- anions. On the other hand, in a neutral aqueous medium, most of the oxine present transforms to the molecular state and may consequently be extracted by chloroform.

(iii) Separation by extraction. The distribution ratio D′ is given by

$$\text{D}' = \frac{[\text{HOx}]_{\text{CHCl}_3}}{[\text{HOx}] + [\text{H}_2\text{Ox}^+] + [\text{Ox}^-]} = \frac{\text{D}[\text{HOx}]}{[\text{HOx}]\left[1 + \frac{\text{H}^+}{10^{-5.0}} + \frac{10^{-9.7}}{[\text{H}^+]}\right]} = \frac{720}{1 + \frac{[\text{H}^+]}{10^{-5.0}} + \frac{10^{-9.7}}{[\text{H}^+]}}$$

Fig. 47 is a plot of log D′ = f(pH).

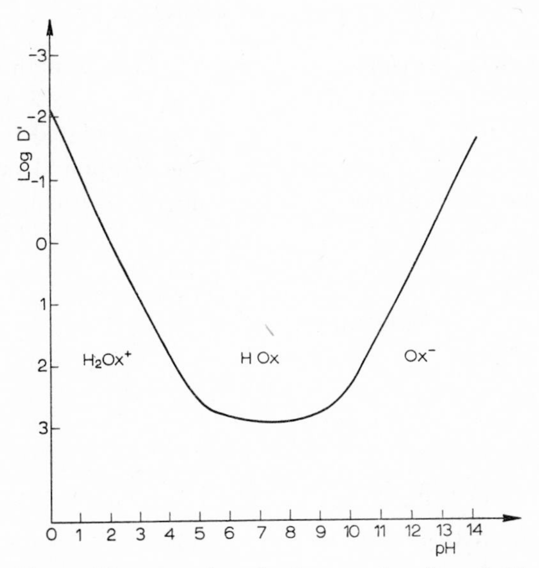

Fig. 47. The extraction of oxine, as a function of pH.

The extraction efficiency is given by

$$\text{R}\% = \frac{[\text{HOx}]_{\text{CHCl}_3}\text{V}'}{[\text{HOx}]_{\text{CHCl}_3}\text{V}' + [\text{HOx}]\left[1 + \frac{[\text{H}^+]}{10^{-5}} + \frac{10^{-9.7}}{[\text{H}^+]}\right]\text{V}} = \frac{\text{DV}'/\text{V}}{\text{DV}'/\text{V} + 1 + \frac{[\text{H}^+]}{10^{-5}} + \frac{10^{-9.7}}{[\text{H}^+]}}$$

It is evident from the above curves that chloroform will extract all but 0.2% of the oxine from an aqueous solution whose pH lies between 5.5 and 9.4

($V = V'$). The oxine may however be returned to the aqueous phase if the latter is acidic or highly alkaline.

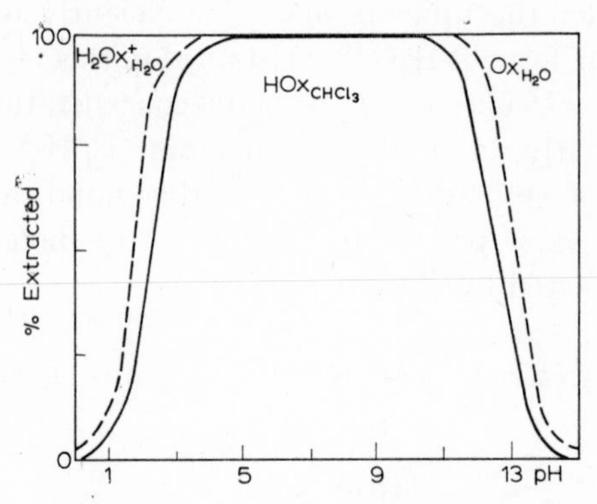

Fig. 48. The extraction of oxine by chloroform. The broken curve corresponds to the yield obtained after two successive extractions (after S. LACROIX).

(1b) *Extraction and complexes*

(i) Extraction of tetraphenylarsonium chloride. This compound, $(C_6H_5)_4AsCl$, which will, for the sake of simplicity be denoted by R_4AsCl, is soluble in both water and chloroform. In water, it is completely dissociated to R_4As^+ and Cl^-, at all values of pH. In chloroform, however the compound probably exists in the molecular state. The following equilibrium is thus set up:

$$\underset{\text{(water)}}{R_4As^+ + Cl^-} \rightleftarrows \underset{(CHCl_3)}{R_4AsCl}$$

This gives the equilibrium constant:

$$\frac{[R_4As^+] \cdot [Cl^-]}{[R_4AsCl_{(CHCl_3)}]} = K = 10^{2.3} \tag{8}$$

As in the preceding examples, it is impossible to measure the partition coefficient $D = \frac{[R_4AsCl_{(CHCl_3)}]}{[R_4AsCl_{(H_2O)}]}$, since $[R_4AsCl_{(H_2O)}]$ cannot be determined. In practice, it is therefore once more necessary to define an apparent partition coefficient, D', which is given by

$$D' = \frac{[R_4AsCl_{(CHCl_3)}]}{[R_4As^+]} = \frac{[Cl^-]}{K}$$

This is entirely analogous to the system

$$\underset{\text{water}}{Dz^- + H^+} \rightleftarrows \underset{CCl_4}{HDz}$$

which was considered above. Here however, the $[Cl^-]$ ion takes part in the equilibrium instead of $[H^+]$. The reasoning involved is however the same.

The influence of anions. The equilibrium can be displaced adding an excess of chloride ions to the aqueous phase; this promotes the extraction of the reagent into chloroform. Conversely, if Cl^- is removed in the form of a sparingly soluble compound, the reagent is retained in the aqueous phase.

(ii) Extraction of tetraphenylarsonium perrhenate. The ion ReO^-_4 gives a tetraphenylarsonium perrhenate, which is sparingly soluble in water and soluble in chloroform according to the equilibrium

$$\underset{\text{water}}{ReO_4^- + R_4As^+} \rightleftarrows \underset{(CHCl_3)}{R_4AsReO_4}$$

for which,

$$\frac{[ReO_4^-] \cdot [R_4As^+]}{[R_4AsReO_4]_{(CHCl_3)}} = K' = 10^{-6.3}.$$

This equilibrium in unaffected by pH since $HReO_4$ is a strong acid.

The distribution ratio of rhenium between the two phases may be defined as

$$D' = \frac{[R_4AsReO_4]_{(CHCl_3)}}{[ReO_4^-]} = \frac{[R_4As^+]}{[K']} = \frac{K}{K'} \frac{[R_4AsCl]_{(CHCl_3)}}{[Cl^-]}$$

On the basis of the above, the conditions which favour quantitative extraction of rhenium from the aqueous phase, and also those which enable the recovery of the ReO_4^- reagent, may be calculated. Qualitatively, it may be seen that while an excess of reagent favours the extraction of rhenium, an excess of Cl^- ions produces the opposite effect.

The most frequent application of this method consists in the recovery of traces of rhenium from solutions of various molybdenites.

COMPARE

S. Tribalat, *Anal. Chim. Acta*, 3 (1949) 113.

(1c) *Complexes and pH*

(i) Extraction of dithizonates. Many dithizonates do not dissolve in water but are soluble in carbon tetrachloride. For example, below pH 7, the cupric ion forms a dithizonate according to the following equilibrium:

$$\underset{CCl_4}{2\,HDz} + \underset{\text{water}}{Cu^{2+}} \rightleftarrows \underset{CCl_4}{CuDz_2} + \underset{\text{water}}{2\,H^+}$$

The equilibrium constant is given by

$$\frac{[CuDz_2]\,[H^+]^2}{[Cu^{2+}]\,[HDz]^2} = K'.$$

Thus, the distribution ratio of the copper will be:

$$D' = \frac{[CuDz_2]_{CCl_4}}{[Cu^{2+}]_{water}} = K' \frac{[HDz]^2}{[H^+]^2}.$$

The extraction of copper is governed by the excess of the reagent HDz and the pH. Thus, Cu(II) is retained in the acidic aqueous phase, while if the pH of the latter is high, the copper passes into the carbon tetrachloride.

The percentage extraction is given by

$$R\% = \frac{[CuDz_2]V' \times 100}{[CuDz_2]V' + [Cu^{2+}]V}.$$

Thus, if the values of [HDz] and V'/V are fixed, we obtain:

$$R\% = \frac{k\ 100}{k + [H^+]^2}.$$

Fig. 49 is a plot of the experimental results obtained with conditions fixed in this way.

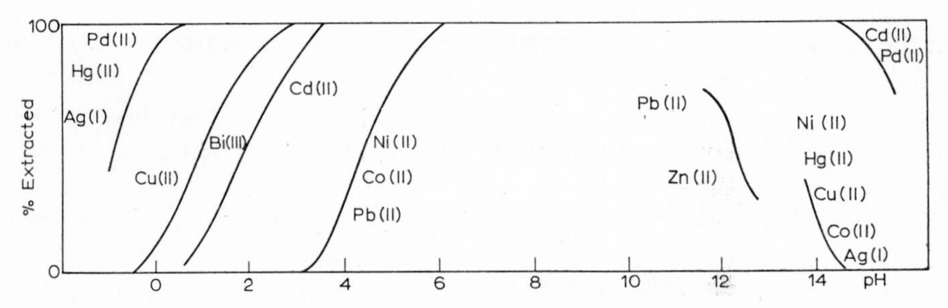

Fig. 49. Extraction of various dithizonates as a function of pH.

(ii) Separations by extraction at a controlled pH. From Fig. 49 it may be seen that Hg(II) may be extracted by maintaining the pH of the aqueous phase at 0.5–2.5. At pH 6.5–10.0, Zn(II) would in turn pass into the carbon tetrachloride. The tetrachloride solution of zinc dithizonate could then be shaken with an aqueous solution of pH 2.5. This would return the Zn^{2+} entirely to the aqueous phase.

(iii) Extraction of aluminium oxinate with chloroform. When oxinate ions are added to a solution containing aluminium, a precipitate of aluminium oxinate is formed, according to the reaction

$$Al^{3+} + 3\ Ox^- \rightleftarrows Al(Ox)_3\downarrow.$$

The oxinate $Al(Ox)_3$ is very sparingly soluble in water, as shown by its solubility product:

$$[Al^{3+}]\cdot[Ox^-]^3 = 10^{-32.3},$$

but relatively soluble in chloroform. Thus, $S_{CHCl_3} = 4.5 \times 10^{-2}$ M.

The concentration of oxinate in the aqueous phase is immeasurably small; consequently, the succession of equilibria:

$$\underset{(H_2O)}{3\,Ox^-} + \underset{(H_2O)}{Al^{3+}} \rightleftarrows \underset{(H_2O)}{Al(Ox)_3\downarrow} \rightleftarrows \underset{(CHCl_3)}{Al(Ox)_3}$$

yield an equilibrium constant K directly, where

$$K = \frac{[Al^{3+}]\cdot[Ox^-]^3}{[Al(Ox)_{3\ CHCl_3}]}.$$

The two phases are saturated when in equilibrium with an excess of solid oxinate; in this case, the above equation is valid, and according to S. LACROIX *(loc. cit.)*,

$$K = \frac{s}{S_{CHCl_3}} = 10^{-31.1}.$$

The ratio $\frac{[Al(Ox)_{3\ CHCl_3}]}{[Al^{3+}]}$ represents the apparent partition coefficient; the latter is also given by $\frac{[Ox^-]^3}{K}$.

In an acid medium, the Ox^- ions are removed from the aqueous phase by the formation of oxine and H_2Ox^+ ions; the above equilibrium is then displaced towards the left, and aluminium remains in the aqueous phase. On the other hand, if more chloroform or more oxine is added, the equilibrium is displaced towards the right. If the acid-base equilibria of oxine and of the Al^{3+} ions are taken into account, the form of the extraction curve of aluminium (Fig. 50) may be predicted by a calculation analogous to that given above. This is

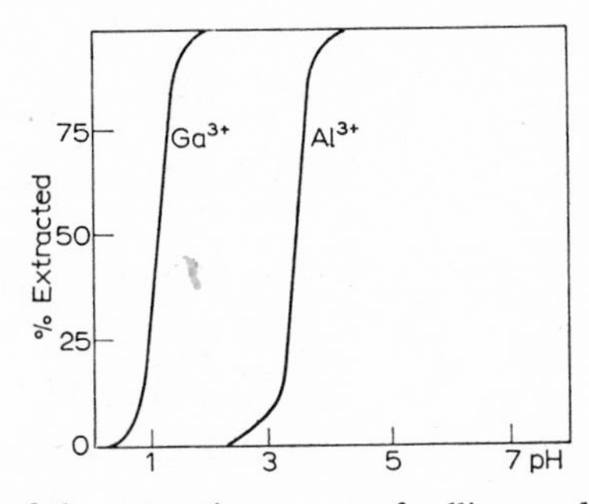

Fig. 50. Comparison of the extraction curves of gallium and aluminium oxinates.

however the case only if the values of each of the ratios V'/V and $\frac{\text{total amount of oxine}}{\text{total amount of aluminium}}$ are fixed.

(iv) The general case. In many cases, the organic reagent is of the form HR (dithizone, oxine, cupferron, etc.), and equilibria of the following type are set up:

$$\underset{\text{org. solvent}}{n\text{HR}} + \underset{\text{water}}{\text{M}^{n+}} \rightleftarrows \underset{\text{org. solvent}}{\text{MR}_n} + \underset{\text{water}}{n\text{H}^+} \qquad \text{K.}$$

In addition,

$$\underset{\text{org. solvent}}{\text{HR}} \rightleftarrows \underset{\text{water}}{\text{R}^- + \text{H}^+}.$$

The distribution ratio is then of the form:

$$\text{D}' = \frac{[\text{MR}_n]_s}{[\text{M}^{n+}]} = \text{K}\,\frac{[\text{HR}]_s^n}{[\text{H}^+]^n}.$$

THE APPARENT EQUILIBRIUM CONSTANT. If a substance M exists in aqueous solution in several forms, the distribution ratio for M may be defined by the expression $\text{D}' = \frac{[\text{MR}_n]_s}{\text{C}_\text{M}}$, where C_M represents the concentration of M in all its forms in the aqueous phase. It is also possible to show that $[\text{M}^{n+}] = \text{C}_\text{M}\cdot f_\text{M}(\text{pH})$.

In addition, HR can exist in the forms HR and R^-, and if its (fictitious) total free concentration with respect to the solvent is denoted by C_R, we obtain:

$$\text{C}_\text{R}\text{V}' = [\text{HR}]\text{V}' + [\text{R}^-]\text{V} = [\text{HR}]\left[\text{V}' + \text{V}\,\frac{\text{K}_\text{A}}{\text{H}^+}\right],$$

whence, $[\text{HR}] = \text{C}_\text{R}\cdot f_\text{R}(\text{pH})$.

The apparent equilibrium constant K' can now be defined by the expression

$$\text{K}' = \frac{\text{C}_{\text{M(solv.)}}[\text{H}^+]^n}{\text{C}_{\text{M(water)}}\text{C}_\text{R}^{\,n}}$$

or, from the above,

$$\text{K}' = \text{K}\cdot f_{\text{R}^n}(\text{pH})\cdot f_\text{M}(\text{pH}).$$

The corresponding apparent partition coefficient D'' is given by:

$$\text{D}'' = \frac{\text{C}_{\text{M(solv.)}}}{\text{C}_{\text{M(water)}}} = \text{K}\cdot f_{\text{R}^n}(\text{pH})\cdot f_\text{M}(\text{pH})\,\frac{\text{C}_\text{R}^{\,n}}{[\text{H}^+]^n}$$

Curves of the extraction yields of gallium and aluminium oxinates as functions of the pH have been calculated and plotted in Fig. 50. Similar relationships have been determined experimentally for many other cases.

REFERENCES

Numerous calculations of general interest have also been made. In addition to the books mentioned on page 124, the reader is referred to:

H. M. Irving and Williams, *J. Chem. Soc.*, (1949) 1841.
D. Dyrssen and L. G. Sillén, *Acta Chem. Scand.*, 7 (1953) 663.
J. Rydberg, *Rec. Trav. Chim.*, 75 (1956) 737.
R. M. Diamond, *J. Phys. Chem.*, 61 (1957) 69, 75.

OXINE, OXINATES OF Ga(III) AND Al(III)

S. Lacroix, *Anal. Chim. Acta*, 1 (1947) 260.
M. Oosting, *Anal. Chim. Acta*, 21 (1959) 397.

(v) Extraction of uranium by long-chain amines. The reaction is as follows:

$$(1)\quad UO_2^{2+} + SO_4^{2-} + 2\underset{\text{solvent}}{(R_3NH)_2SO_4} \rightleftarrows \underset{\text{solvent}}{UO_2(SO_4)_3(R_3NH)_4}$$

In addition

$$(2)\quad \underset{\text{solvent}}{(R_3NH)_2SO_4} \rightleftarrows \underset{\text{solvent}}{2\ R_3N} + 2H^+ + SO_4^{2-}$$

Several ions influence the extraction; among these are: H^+, SO_4^{2-}, UO_2^+.

The overall influence of the acidity has many component mechanisms. Thus:

$$(a)\quad UO_2^{2+} + H_2O \rightleftarrows UO_2OH^+ + H^+$$

followed by

$$UO_2OH^+ + H_2O \rightleftarrows UO_2(OH)_2\downarrow + H^+,$$

$$(b)\quad SO_4^{2-} + H^+ \rightleftarrows HSO_4^-$$

followed by

$$\underset{\text{solvent}}{(R_3NH)_2SO_4} \rightleftarrows \underset{\text{solvent}}{R_3N} + 2H^+ + SO_4^{2-}$$

$$(c)\quad R_3N + H^+ \rightleftarrows R_3NH^+$$

Because of the first three of the above equilibria, both too strongly acidic and too weakly acidic media should be avoided. This involves the selection of an amine which will comply with these conditions.

Uranium can be extracted quantitatively from its aqueous solution without any loss of reagent. After separation, it may be returned into solution in a neutral aqueous phase, and the amine left behind in the solvent in its R_3N form.

COMPARE

C. Boirie, *Bull. Soc. Chim.*, (1958) 1088.

(1d) *The effect of complex formation occurring in the aqueous solution*

Reverting to the example of zinc dithizonate, let it be assumed that the aqueous solution has also been treated with cyanide. The latter complexes with Zn^{2+} ions,

$$Zn^{2+} + 4CN^- \rightleftarrows Zn(CN)_4^{2-}$$

diminishes their concentration in the aqueous phase, and thus tends to prevent the migration of zinc into the organic solvent.

The formation of complexes may therefore be used to develop new methods of separation. For example, whereas the dithizonates of Pd(II), Hg(II), Cu(II), Au(III), and Pt(II), may all be extracted from dilute acidic media, the presence of cyanide prevents the extraction of those of Au(I) and Pt(II).

(1e) *The effect of precipitation reactions on extraction*

If a substance forms compounds which are sparingly soluble in one of the phases, its extraction into the other phase is impeded. For example, silver dithizonate may be extracted by carbon tetrachloride over a large range of pH. If however, Cl^- or SCN^- ions are added to the aqueous phase, the extraction process is impeded by the precipitation of AgCl or AgSCN respectively, even if the pH remains the same.

EXAMPLE. Consider the partition equilibria of zinc, cadmium, and lead dithizonates dissolved in an aqueous phase on contact with carbon tetrachloride. As we have just seen, the general equilibrium

$$\underset{(CCl_4)}{2\,HDz} + \underset{(water)}{Zn^{2+}} \rightleftarrows \underset{(CCl_4)}{ZnDz_2} + \underset{(water)}{2\,H^+}$$

leads to the relation

$$\frac{[ZnDz_2]}{[Zn^{2+}]} = K_{Zn}\frac{[HDz]^2}{[H^+]^2}\text{, from which } K_{Zn} = \frac{K_A^2}{K_C'} = 10^{2.7}.$$

For the other elements, the corresponding constants are $K_{Cd} = 10^{2.1}$ and $K_{Pb} = 10^{1.3}$. These differences are too low to permit the separation of, for example, zinc from cadmium, by altering the acidity of the aqueous phase.

The equilibria can however be displaced by precipitating the sulphides of the metals concerned, since the differences between their respective K values are thereby accentuated (A. K. BABKO AND A. T. PILIPENKO). If hydrogen sulphide is dissolved in the aqueous solution, the following equilibrium is set up in addition to the preceding ones:

$$H_2S + ZnDz_2 \rightleftarrows ZnS\downarrow + 2HDz.$$

In this case,

$$\frac{[HDz]^2}{[H_2S]\cdot[ZnDz_2]} = K'_{Zn}.$$

An analysis of the constant K'_{Zn} reveals it to be the product of a number of factors:

$$K'_{Zn} = \frac{[H^+]^2[S^{2-}]}{[H_2S]} \times \frac{[HDz]^2[Zn^{2+}]}{[ZnDz_2]\,[H^+]^2} \times \frac{1}{[S^{2-}]\,[Zn^{2+}]} = \frac{k_{H_2S}}{K_{Zn}S_{ZnS}}$$

where k_{H_2S}, the overall acidity constant of H_2S, has the value 10^{-22}, and S_{ZnS} is the solubility product of zinc sulphide. From this it follows that:

$$[ZnDz_2] = \frac{[HDz]^2}{[H_2S]} \times \frac{1}{K'_{Zn}} = \frac{[HDz]^2}{[H_2S]} \times \frac{K_{Zn}S_{ZnS}}{k_{H_2S}}$$

It is now evident that the concentration of the dithizonate no longer depends on the pH of the aqueous phase; furthermore, the variation of K is now supplemented by that of S, which increases the selectivity of the extraction. If the various constants are known and if it is assumed that $[H_2S] = 10^{-2}$ and $[HDz] = 10^{-4}$, the concentration of the dithizonate extracted can be calculated.

	K_M	S_{MS}	$1/K'_M$	MDz_2
Zn^{2+}	$10^{1.2}$	$10^{-22.9}$ to $10^{-25.2}$	$10^{1.8}$	$10^{-4.2}$
Cd^{2+}	$10^{1.1}$	$10^{-27.2}$ to $10^{-28.4}$	$10^{-3.1}$	$10^{-9.1}$
Pb^{2+}	$10^{2.3}$	$10^{-27.1}$ to 10^{-29}	$10^{-4.1}$	$10^{-10.1}$

This considerably facilitates the separation of zinc from the two other metals.

COMPARE

A. K. Babko and A. T. Pilipenko, *Zh. Analit. Khim.*, 2 (1947) 33.

(1f) *Importance of the method*

The method described above offers considerable possibilities, which will be discussed below.

(1) While a few inorganic substances are soluble in organic solvents, an ever-increasing number of organic reagents are becoming available, which combine with inorganic ions to give compounds which dissolve either in a single solvent or in mixtures of solvents.

(2) By modifying the equilibria occurring in the aqueous phase by conventional means, (pH, formation of complexes, oxidation and reduction), it is possible to adjust the distribution of the material between the two solvents, and thus to obtain highly selective separations.

The method has consequently attained considerable importance as a technique of separation. In addition, it possesses certain advantages in comparison with separation by precipitation, being very rapid, since the equilibria are often established in a few minutes and the separation of the two phases by decantation may be performed very quickly. Furthermore, side effects such as adsorption etc., do not occur. Finally, the method offers ideal conditions for the separation and concentration of trace amounts.

(2) Technique of extraction

The procedure involved is very simple. Extraction is generally carried out in a separating funnel fitted with accurately ground taps and stoppers in order to give air-tight seals without lubrication, since greases frequently dissolve in the solvents employed. Plastic stoppers may often be used.

In general, the two liquid phases only reach equilibrium after vigorous agitation; the latter is often performed by mechanical means. It is advisable to determine the time of agitation required by means of preliminary trials with known solutions, since occasionally equilibrium is reached only after prolonged periods. Thus for example, although aluminium oxinate cannot be extracted from aqueous solution by chloroform at pH = 2, it passes back from the chloroform into the aqueous phase at the same pH only with difficulty.

It may be occasionally be necessary to centrifuge in order to effect the separation of the layers, although conditions which favour the formation of stable emulsions are generally avoided.

If the determination is to be precise, the aqueous solution should, after separation, be washed with a little solvent to collect any droplets scattered over the walls of the container or on the surface of the aqueous phase. In addition, it is often advisable to wash the organic phase with a suitable aqueous solution.

If the organic solution obtained is itself to be estimated colorimetrically, the last droplets of water should be separated by filtration through a plug of hydrophilic cotton or glass wool; it may be necessary to dry the solvent in order to clarify the solution (*e.g.* with 1 g of anhydrous sodium sulphate per 10 ml of solution). In any case, the solution should always be brought to a known volume and an aliquot then taken.

SUCCESSIVE EXTRACTION. Let c_0 be the initial concentration of the material to be extracted, V the volume of the aqueous solution, and V′ the volume of the solvent. Let c_1 and c'_1 be the equilibrium concentrations of the solute in the aqueous phase and in the organic solvent respectively. The following equations describe the partition of the solute:

$$c_0\mathrm{V} = c_1\mathrm{V} + c'_1\mathrm{V}' \text{ with } \frac{c'_1}{c_1} = \mathrm{D},$$

It follows from these that

$$c_1 = c_0 \frac{1}{1 + \dfrac{\mathrm{DV}'}{\mathrm{V}}}$$

If the value of the extraction coefficient is inadequate, the solvent should be separated and the operation repeated with pure solvent.

After n operations of this type, the residual concentration of the solute in the aqueous phase will be given by:

$$c_n = c_o \left(\frac{V}{V + DV'} \right)^n = c_o \left(\frac{1}{1 + \frac{DV'}{V}} \right)^n \quad (9)$$

A very small value of c_n indicates that the extraction of A is almost quantitative.

Equations of type (9) above show the influence of D and V'/V, which have a similar effect.

The variation of c_n with $D\frac{V'}{V}$ is shown in Table 3.

TABLE 3

$D\frac{V'}{V}$	1	10	100	1000
c_0	100	100	100	100
c_1	50	9	1	0.1
c_2	25	0.8	0.01	
c_3	12.5	0.07		

When V'/V = 1 and D = 1000, the extraction is quantitative to within 0.1% after a single operation. On the other hand, if D = 100, two operations are needed unless V'/V is made equal to 10. Thus, if D is not sufficiently high, this may be compensated for by increasing the V'/V ratio. It is not, however, always advisable to do this, since the extract obtained is often intended for direct colorimetric estimation, and it may be necessary to avoid diluting the substance to be estimated as far as possible. In such cases, the method of successive extraction is useful, since it can be shown that the sum of the volumes of solvent used for different successive extractions is less than the volume necessary to achieve the same result in one step. There is thus no longer any advantage in increasing V'/V. For example, with D = 10, a separation which is quantitative to within 0.1% may be achieved either by three extractions with V'/V = 1, or one extraction with V'/V = 100. In the first case, the total volume of the solvent, S', is 3 V, *i.e.* 33 times smaller than in the second case.

A variation of this method consists in dividing the solution into several fractions, and extracting these successively with a given volume of solvent. This is equivalent to the passage of a given volume of the solvent through a column within which the aqueous phase is contained and prevented from circulating by a suitable obstacle—for example, powdered cellulose.

COUNTERCURRENT EXTRACTION. In this method, the solution is divided into several fractions contained in separate funnels, and these are treated successively with several portions of the solvent. An equivalent operation can be carried out in a column.

Finally, and in the case of industrial rather than analytical extraction, a continuous operation can be carried out by circulating the two solvents in a column, in opposite directions. This operation can be considerably improved (reflux). The literature should be consulted for further details.

The following example is presented as a summary of the above.

For the case $V = V'$, $D = 2$, we obtain:

	1 Extraction	*3 Successive extractions*	*Same total volume in 3 fractions*	*3 Fractions counter-current*	*Continuous counter-current (3 plates)**
% extracted	67	78	74	88	94
% remaining	33	22	26	12	6

* See *Chromatography*.

(3) The principal general reagents used in separation by extraction

A considerable number of organic reagents are soluble in various solvents. Furthermore, it is often observed that compounds of these substances with inorganic ions are more soluble in organic solvents than in water.

In this section, an account will be given only of some of the most important reagents.

(I) CHELATES

The reagents are generally of the form HR, and the extraction equilibria are often of the type

$$\underset{\text{solvent}}{n\text{HR}} + \text{M}^{n+} \rightleftarrows \underset{\text{solvent}}{\text{MR}_n} + n\text{H}^+$$

The equilibrium position depends on the pH of the aqueous phase. The coordinate bonds formed between the organic reagent and the inorganic ion are generally one of the following:

$$=\text{S}\rightarrow,\ =\text{O}\rightarrow,\ \text{or} \equiv\text{N}\rightarrow$$

(3a) *Dithizone*

(i) Properties. 'Dithizone' (diphenylthiocarbazone) possesses the structural formula: $\text{S}=\text{C}\begin{matrix}\diagup\text{NH—NH—C}_6\text{H}_5\\ \diagdown\text{N=N—C}_6\text{H}_5\end{matrix}$

and is soluble in carbon tetrachloride and chloroform (green solutions), nitrobenzene, toluene, and in other organic solvents. When in the molecular state, it is practically insoluble in acidic or neutral aqueous media; if the latter are alkaline, the apparent solubility increases owing to the formation of anions:

$$\underset{\text{solvent}}{HDz} \rightleftarrows \underset{\text{water}}{Dz^- + H^+}$$

Many metallic ions form coloured dithizonates which are sparingly soluble in water but soluble in organic solvents such as carbon tetrachloride or chloroform.

Depending on the acidity of the medium, such compounds have the following formulae:

ketonic form: $S=C\langle^{NH-N\searrow-C_6H_5}_{N=N\nearrow-C_6H_5}\ Me^I$ (in acidic or neutral media)

enolic form: $Me^I-S-C\langle^{N-N\searrow-C_6H_5}_{N=N\nearrow-C_6H_5}\ Me^I$ (occasionally present in alkaline media: often sparingly soluble in CCl_4 and $CHCl_3$)

The fundamental equilibrium is of the form

$$\underset{\text{(water)}}{M^{n+}} + \underset{\text{(solvent)}}{nHDz} \rightleftarrows \underset{\text{(solvent)}}{MDz_n} + \underset{\text{(water)}}{nH^+}$$

A subsidiary equilibrium is

$$\underset{\text{(water)}}{H^+} + \underset{\text{(water)}}{Dz^-} \rightleftarrows \underset{\text{(solvent)}}{HDz} \qquad pK_A = 8.7$$

The extraction of dithizonates currently constitutes the most general method of separating and estimating trace amounts. It is suitable for quantities of elements lying between 0.1 and 200 μg.

Properties of the several important dithizonates are summarised above. The extraction pH's indicated are only approximate, since the exact values depend on the operating conditions such as the ratio of the volumes of the two phases, the excess of reagent, the anions present, and the ionic strength of the solution.

Separations are only possible with the given pH values if the concentrations of the various elements are of the same order of magnitude. A different procedure must be adopted if it is desired to separate trace amounts in the presence of an appreciable quantity of any other element.

The metals of the Fe(III) group do not give rise to stable compounds with dithizone. Hence, when extractions are to be carried out in ammoniacal or alkaline media, these metals are complexed by additions of tartrate, citrate or other ions, in order to prevent interference by the precipitation of their hydroxides.

TABLE 4

		pH for total extraction	
Ag(I)	acid: yellow	−1 to 7	CCl_4
	basic: violet-red		
Au(III)	acid: yellow	2	CCl_4
	basic	7–8	CCl_4
Bi(III)	acid: orange-yellow	2–10	CCl_4
	basic: orange-red		
Cd(II)	red	6–14	CCl_4
Co(II)	acid: violet	6–8	CCl_4
	basic: brownish	12–14	
Cu(II)	acid: violet-red	2–5	CCl_4
	basic: yellow-brown	7–14	
Fe(II)	violet-red, unstable	7–9	CCl_4
Hg(II)	acid: yellow-orange	−1 to 4	CCl_4
	basic: purple	7 to 14	
In(III)	red	5–6	CCl_4
		8–9.5	$CHCl_3$
Mn(II)	unstable	5.2–6.3	CCl_4
Ni(II)	brownish	6–8	
Pb(II)	red	7–10	CCl_4
Pd(II)		−1 to 4	CCl_4
Pt(II)		−1 to 4	CCl_4
Po(IV)		0 to 3	CCl_4
Sn(II)	red	5–9	CCl_4
Tl(I)	red	9–12	CCl_4
Zn(II)	red	6–9	CCl_4

(ii) Examples of separations. 1. By the action of pH.
(a) in a medium consisting of a 0.1–0.5 *N* acid, only the dithizonates of Pd(II), Ag(I), Hg(II), Cu(II), Au(III), and Pt(II) may be extracted;
(b) at pH 3, Cu(II) and Bi(III) may be extracted, and thus separated from Pb(II) and Zn(II).

In *N* NaOH, Cd(II) is converted to a dithizonate and may then be separated from Pb(II) and Zn(II).

2. By the formation of complexes. At pH 4.7, in the presence of EDTA, Ag(I) may be extracted and so separated from Cu(II), Bi(III), and Pb(II).

3. By the formation of sparingly soluble compounds. Silver dithizonate can be decomposed by the precipitation of silver, as AgCl or AgSCN.

(iii) Procedure used for the estimations. Solutions of dithizone in chloroform or carbon tetrachloride are green. The dithizonates exhibit a range of colours, and are frequently red or orange.

An estimation may be performed in one of the following ways:

pH	*Complexing ion*	*Stable dithizonates*
Basic	CN^-	Pb(II), Sn(II), Tl(I), Bi(III)
Weakly acidic..	CN^-	Pd(II), Hg(II), Ag(I), Cu(II)
– – ..	SCN^-	Hg(II), Au(III), Cu(II)
– – ..	$SCN^- + CN^-$	Hg(II), Cu(II)
– – ..	$Br^- + I^-$	Pd(II), Au(III), Cu(II)
pH 5..........	$S_2O_3^{2-}$	Pd(II), Sn(II), Zn(II), Cd(II)
pH 4.5	$S_2O_3^{2-} + CN^-$	Sn(II), Zn(II)

(after E. B. SANDELL)

(a) Colorimetry of the dithizonate. The wavelengths corresponding to maximum adsorption are given below:

Dithizone		620 mμ (second maximum at 450 mμ)
Dithizonates of	Bi(III)	500 - - - -
-	Cd(II)	520 - - - -
-	Cu(II)	510 - - - -
-	Pb(II)	525 - - - -
-	Hg(II)	530 - - - -
-	Ag(I)	460 - - - -
-	Zn(II)	535 - - - -

(after E. B. SANDELL)

1. Colorimetry in the presence of an excess of dithizone[1]. Since dithizone is itself coloured, the estimation should be carried out at a wavelength such that the absorption of dithizone is as low as possible, whilst that of the dithizonate is high. The selection of such a wavelength frequently presents problems.

The solution may however be treated as a mixture of two coloured materials, the optical density being determined at two wavelengths. This is the most satisfactory method[2].

In most cases, a calibration curve is plotted using an accurately known total quantity of dithizone.

2. Colorimetry after separation of the excess dithizone. The dithizonate of the element to be estimated is extracted once, using an excess of dithizone and the best conditions available. The excess of dithizone is removed by shaking the extract with a 0.2 N (1/70) ammonia solution or with dilute caustic soda. Under these conditions, the free dithizone passes into the aqueous phase; the method is only suitable if the dithizonate to be estimated is stable under these conditions. The latter is then estimated, either by comparison with a series of standards prepared under the same conditions, or by determining its optical density at a suitable wavelength.

NOTE. When this method is used, it must be noted that the alkaline aqueous solution may sometimes convert the ketonic into the enolic form, thereby changing its colour.

(b) Colorimetry of the dithizone. 1. In the presence of the dithizonate, a wavelength is selected at which the latter does not absorb. This procedure is often simpler than that used for estimating the dithizonate in the presence of dithizone. For example, zinc dithizonate does not absorb above 600 mμ, while dithizone itself absorbs strongly at 620 mμ. If the total amount of the latter is known, the concentration of dithizonate may thus be deduced from it. The disadvantage of this method lies in the necessity of knowing the precise concentration of the reagent.

2. After the extraction has been carried out and the excess of dithizone has been removed by the method described above, the dithizonate is the only substance present in the solution. It is then decomposed, by shaking either with a solution of suitable pH, or with one of a complexing ion. The metallic ion returns to the aqueous phase and the dithizone is then estimated.

(c) The 'sensitive shade' method. This method, for which a visiual technique is adequate, is both relatively precise, and very rapid. In addition, it avoids errors caused by the separation of the excess dithizone. Use is made of the fact that a mixture of dithizone and dithizonate in suitable proportions possesses a grey coloration, which is particularly sensitive to variations in composition. The method is thus, in effect, a titration.

It has been shown that between 5 and 50 μg of copper(II) can be estimated in this way, to within 1.5%, and the same amount of zinc(II) to within 2.5%.

(d) Titration by extraction. The metallic ion to be estimated is extracted with small fractions of a standardized solution of dithizone, separation being carried out after each extraction. As long as extraction is not complete, the extracts exhibit the characteristic (generally red) colour of the corresponding dithizonate. The extraction is complete when a green colour is observed, indicating the presence of excess dithizone. The dithizone solution used is standardized by a similar operation in which a known solution of the metallic ion is used.

Precision of this method may be improved by first performing a rough operation, and following this up with a second, more precise one.

The method is only practicable if the extraction of the metal is quantitative even with no excess of dithizone, and provided the latter is insoluble in the aqueous phase under the conditions of extraction.

REFERENCES

(1) H. A. Liebhafsky and E. H. Winslow, *J. Amer. Chem. Soc.*, 59, (1937) 1966.
(2) E. A. Brown, *Anal. Chem.*, 18, (1946) 493.
(3) H. Grubitsch and J. Sinigoj, *Z. Anal. Chem.*, 114, (1938) 30.

(iv) Preparation of the reagents. Solution of dithizone. A stock solution of 250 mg per litre of dithizone in carbon tetrachloride or in chloroform is prepared. The solution can be stored over a considerable period, preferably at 0 °C, but should be protected from light. It is covered with a layer of 0.2 *M* sulphurous acid to protect it from oxidation, and is diluted as required immediately before use.

Purification of the dithizone. Commercial dithizone always contains its oxidation products, which are of an intense brown colour, and consequently interfere greatly in the course of estimations. Purification is therefore necessary.

Under alkaline conditions, dithizone passes into aqueous solution in the form of anions, while the oxidation products remain in the organic phase.

Dithizone (0.25 g) is dissolved in 50 to 75 ml of chloroform and the solution is filtered. It is then shaken 4 times in succession with 100 ml of pure dilute ammonia (1/100). The ammoniacal solution is then filtered through cotton wool and acidified with pure hydrochloric acid (1/2), and the dithizone is extracted 3 times in succession with 20 ml of chloroform. It is then washed three times with water and diluted to 1 litre with chloroform containing 0.5% of alcohol to ensure its preservation.

Solutions of dithizone in carbon tetrachloride may be prepared in a similar manner.

Recovery of the chloroform. The impure chloroform is washed 5–10 times with an equal volume of concentrated sulphuric acid. The solvent should then be colourless. It is then neutralized with lime water in a flask, and distilled. Finally, 0.5% of alcohol are added to stabilise the distillate.

Precautions. The solution to be analyzed occasionally contains oxidizing agents, such as Fe^{3+}, capable of reacting with dithizone; these should be reduced in advance, with for example, hydroxylamine hydrochloride.

The other reagents used in the course of the separation and estimations should also be specially tested for impurities. It is therefore normal practice to carry out a blank test with identical amounts of all the reagents. If however, the quantity of interfering impurities present is small in proportion to the amount of the material to be estimated, and if the precision need not be excessively high, adequate results may be obtained by making an appropriate correction for the impurities present in the reagent.

The reagents can sometimes be purified by shaking with dithizone. For example, distilled water can be purified in this way, and a solution of citrate can be freed from most of its impurities in a similar manner; the small amount of dithizone remaining in the aqueous phase may then be extracted by means of the pure solvent. If high precision is required, many precautions are necessary. Thus, in the estimation of Zn(II), pyrex funnels are suitable but not Jena glass.

Generally speaking, the solvent should never be allowed to come into contact with grease (the tap and the stopper of the separating funnels should not be greased), or with rubber; porcelain containing lead and lead glass equipment (pipettes, burettes) should be avoided; in addition, filtration should be carried out by centrifuging, and not through paper etc.

NOTE. When the equilibrium between the aqueous and organic phases is attained only very slowly, the process may be accelerated by emulsifying the solvent with a surface-active product (100 ml of xylene + 10 ml of 1% lauryl sulphonate).

COMPARE

Works on colorimetry and G. IWANTSCHEFF, *Das Dithizon*, Verlag Chemie, Weinheim, 1958.

(3b) *Oxine*

Oxine and the majority of metallic oxinates, which are sparingly soluble in water, are relatively soluble in chloroform, carbon tetrachloride, ethyl acetate, hexone, etc.

In water the following equilibria are set up:

$$HOx + H^+ \rightleftarrows H_2Ox^+$$

$$HOx \rightleftarrows Ox^- + H^+$$

$$n\,HOx + M^{n+} \rightleftarrows MOx_n\downarrow + n\,H^+$$

In addition, in the presence of an organic solvent

$$HOx_{(H_2O)} \rightleftarrows HOx_{(solvent)}$$

$$M^{n+}_{(H_2O)} + n\,HOx_{(solvent)} \rightleftarrows MOx_{n(solvent)} + n\,H^+$$

Optimum pH for the extraction of the oxinates. Table 5 given below represents a summary of currently available data. It should however be borne in mind that the pH's indicated depend on the operating conditions, such as the ratio of the volumes of the two solvents, the amount of excess oxine present, and the number of extractions.

TABLE 5

	pH for total extraction	
Al(III)	> 4.6	$CHCl_3$
Bi(III)	4.0–5.2	$CHCl_3$
Ca(II)	> 13	butyl cellosolve + $CHCl_3$
Ce(III)	9.9–10.5	$CHCl_3$
Co(II)	> 6.8	$CHCl_3$
Cu(II)	2.8–12.0	$CHCl_3$
Fe(II)	1.9–3.0	$CHCl_3$
Fe(III)	> 2.5	
Ga(III)	> 2.0	$CHCl_3$
In(III)	3.0–4.5	$CHCl_3$
Mg(II)		butyl cellosolve + $CHCl_3$
Mo(VI)	1.6–5.6	$CHCl_3$
Ni(II)	4.5–9.5	$CHCl_3$
Pu(IV)	~ 8	amyl acetate
Pu(VI)	4–8	amyl acetate
Sc(III)	9.7–10.5	C_6H_6
Sn(IV)	2.5–5.5	$CHCl_3$
U(IV)	5.8–8.0	$CHCl_3$
V(V)	3.5–4.5	$CHCl_3$
Zn(II)		$CHCl_3$
Oxine	5.6–10.3	V′ = V; $CHCl_3$

The oxinates of Cd(II), Pb(II), Mn(II), Ti(IV), Pd(II), and Zr(IV) are also soluble in chloroform, and can therefore be extracted under certain conditions of pH.

At pH 2.0, even trace amounts of Ga(III) can be separated from Al(III); similarly, Fe(III) may be separated from Al(III) and U(VI).

At pH 4.5, Fe(III) and Al(III) can be extracted and separated from Be(II), Mg(II), and Ca(II). Similarly, V(V) can be extracted and separated from Cr-(VI).

An example of major practical interest is the extraction of aluminium oxinate at pH = 9 in the presence of tartrate, cyanide, and hydrogen peroxide; under these conditions, it may be separated from: Cu(II), Co(II), Ni(II), Zn(II), Cd(II), Fe(III), Ti(IV), V(V), V(VI), U(VI), Mn(II), Cr(III), Mo(IV), Sn(IV), and Ag(I), and small amounts of Zr(IV) and Nb(V).

TABLE 6

$pH_{\frac{1}{2}}$, *or the pH at which 50% is extracted* ($CHCl_3$)	
10^{-1} *M* total oxine; V = V':	
Al(III)	3.4
Ga(III)	1.0
In(III)	2.1
Hf(IV)	1.3
La(III)	6.5
Sm(III)	5.7
Th(IV)	3.1
U(VI)	2.6
0.07 *M* total oxine; V = 5 V'; $CHCl_3$	
Al(III)	3.8
Cu(II)	2.1
Fe(III)	2.0
Mn(II)	6.4
Mo(VI)	1.0
Ni(II)	3.7
Sn(IV)	0.0
10^{-2} *M* oxine; 4 successive extractions:	
Al(III)	4.2
Bi(III)	3.0
Co(II)	6.5
Cu(II)	2.0
Fe(III)	1.6
In(III)	2.2
Ni(II)	6.1

cf. D. Dyrssen and V. Dahlberg, *Acta Chem. Scand.*, 7 (1953) 1186.

Colorimetry of the oxinates. The chloroform solutions are suitable for direct colorimetric estimation. Their absorption maxima are located at the following wavelengths:

TABLE 7

Oxine	320 mμ	Ga(III)	392 mμ
Al(III)	395 -	In(III)	400 -
Bi(III)	395 -	Ni(II)	395 -
Ce(III)	505 -	Tl(III)	400 -
Co(II)	420 -	U(VI)	400 to 425 mμ
Cu(II)	410 -	V(V)	550 -
Fe(III)	470 and 570 mμ		

COMPARE

General works on extraction referred to on p. 124.
R. G. W. HOLLINGSHEAD, *Oxine and its Derivatives*, Butterworths, 1954, 3 vols.
S. LACROIX, *Anal. Chim. Acta*, 1, (1947) 260.
T. MOELLER, *Ind. Eng. Chem., Anal. Ed.*, 15, (1943) 346.
C. H. R. GENTRY AND L. G. SHERRINGTON, *Analyst*, 75, (1950) 17.

IN THE PRESENCE OF EDTA

R. P. TAYLOR AND N. H. FURMAN. *Anal. Chem.*, 27 (1955) 309.

Various elements are discussed individually in the second part of this work.

(3c) *Cupferron*

Cupferron itself and most of its inorganic derivatives are extremely soluble in many water-immiscible organic solvents, such as ether, ethyl acetate, chloroform, hexone, etc., while being very sparingly soluble in water.

The overall equilibrium for the formation of the complex is of the form:

$$\underset{\text{(solvent)}}{n\text{HCup}} + \underset{\text{(water)}}{M^{n+}} \rightleftarrows \underset{\text{(solvent)}}{\text{M Cup}_n} + \underset{\text{(water)}}{n\text{H}^+}$$

As in the examples already quoted, the value of the apparent partition coefficient of the solute in the two phases is governed by the pH of the aqueous phase.

Optimum pH for extraction. Few figures are in fact, available. In addition, the pH values at which total precipitation occurs are often close to those suitable for extraction.

TABLE 8

Values of pH for the extraction of cupferrates

	Solvents	*pH*
Ag(I)	insoluble	neutral
Al(III)	chloroform, ether	2 to 5
As(III) and As(V) ...	no reaction	
Bi(III)	toluene, hexone	1
Cd(II)	ether	neutral
Ce(III)	insoluble	neutral
Ce(IV)	numerous solvents	2
Co(II)	esters, ethers, alcohols	dilute acetic acid
Cu(II)	chloroform	1
Fe(III)	ether, chloroform, esters	1
Ga(III)	chloroform	2
Hg(I)	insoluble	0.5 *N* HCl
Hg(II)	alcohol, benzene, chloroform	neutral
In(III)	alcohol, benzene, chloroform	dilute acid
Mn(II)	ether	neutral

[continued]

TABLE 8 *(Continued)*

	Solvents	pH
Mo(VI)	alcohols (MoO_3 Cup)	acidic
	esters, benzene, chloroform (MoO_3 Cup_2)	> 1
Nb(V), Ta(V)	insoluble	acidic
Nd(III)	insoluble	
Pb(II)	insoluble	slightly acidic
Sb(III)	chloroform, numerous solvents	3.6 *N* H_2SO_4
Sb(V)	no reaction	
Sn(IV)	esters, chloroform	*N* HCl
Sn(II)	benzene, chloroform	1.5 *N* HCl
Th(IV)	ether, esters	*N* HCl
Ti(IV)	chloroform, ether, esters	*N* HCl
U(IV)	chloroform	3–4 *N* H_2SO_4
U(VI)	insoluble	neutral
V(V) and V(IV)	ether, esters	*N* HCl
W(VI)	ethyl acetate	*N* HCl (partially)
Zn(II)	ether	neutral
Zr(IV)	ethyl acetate	3–4 *N* H_2SO_4

COMPARE

N. H. Furman, W. B. Mason and J. S. Pekola, *Anal. Chem.*, 21 (1949) 1325.
E. B. Sandell and P. F. Cummings, *Anal. Chem.*, 21 (1949) 1356.
D. Dyrssen and V. Dahlberg, *Acta. Chem. Scand.*, 7 (1953) 1186.

(3d) *Diethyldithiocarbamate*

Derivatives formed by this compound with the following materials can be extracted by means of esters, alcohols, carbon tetrachloride, etc., at pH 3: Ag(I), Hg(II), Pb(II), Bi(III), Cu(II), Cd(II), Mo(VI), Se(IV), Te(IV), Fe(III), Mn(II), Ni(II), V(V), Co(II), Zn(II), In(III), Ga(III), Tl(I), Pd(II), Sb(III), and Tl(III); at pH 1–1.5: W(VI); in conc. HCl: Re(VI).

In an ammoniacal medium and in the presence of citrate and EDTA, only Cu(II), Hg(II), Ag(I), and Bi(III) may be extracted. The colorimetric estimation of Cu(II) can then be carried out.

In the presence of CN^-, only Bi(III), Cd(II), Pb(II), and Tl(III) may be extracted.

With diethylammonium dithiocarbamate in an acid medium, As(III), Sb(III), Sn(II), Cu(II), Bi(III), and Hg(II), may be extracted, but not As(V).

COMPARE

H. Bode, *Z. Anal. Chem.*, 143 (1954) 182; 144 (1955) 165.
H. Bode and K. J. Tusche, *Z. Anal. Chem.*, 157 (1957) 415
G. Eckert, *Z. Anal. Chem.*, 155 (1957) 23.

Review

M. Delépine, *Bull. Soc. Chim.*, (1958) 5.

Other reagents. The properties of numerous other reagents will be mentioned in the second part of this work, and are also discussed in the general books mentioned on page 124.

(II) Extraction by 'association of ions'

When one of the components of an equilibrium occurring in an aqueous solution forms a stable complex in a suitable solvent, the equilibrium is displaced.

If the reagent used is a colouring material, the compound formed is generally coloured as well, and can therefore be estimated by spectrophotometry.

(3e) *Organic cations*

(i) Tetraphenylarsonium. Tetraphenylarsonium $(C_6H_5)_4As^+$ salts are completely dissociated in aqueous solution. Their compounds with MnO_4^-, IO_4^-, ClO_4^-, ReO_4^-, TcO_4^-, and BF_4^-, are sparingly soluble in water but exhibit a particularly high solubility in chloroform. The same applies to $HgCl_4^{2-}$, $SnCl_3^{2-}$, $CdCl_6^{2-}$, and $ZnCl_4^{2-}$ and to the anions of certain dyes such as bromothymol blue. Reagents of the types R_4Sb^+, R_4P^+, R_3S^+, and R_3Se^+ have similar properties.

(ii) Quaternary ammonium compounds. Quaternary ammonium cations corresponding to the general formula R_4N^+ may also combine with many anions to give extractable compounds. This is particularly so when the number of carbon atoms in the R group is large, as for example, in the cases of cetyltrimethylammonium, and tetra-*n*-propyl ammonium.

If the anion is itself a compound of an inorganic cation and an organic reagent, the efficiency is considerably greater.

COMPARE

Extraction of ReO_4^- by R_4As^+ and R_4Sb^+.

S. Tribalat, *Anal. Chim. Acta*, 3 (1949) 113; 4 (1950) 228; 5 (1951) 115.
S. Tribalat, I. Pamin and M. L. Jungfleisch, *Anal. Chim. Acta*, 6 (1952) 142.

TcO_4^- by R_4As^+.

S. Tribalat and J. Beydon, *Anal. Chim. Acta*, 8 (1953) 22.
S. Tribalat, *Rhénium et Technécium*, Gauthier-Villars, 1957.

50 ions by R_4As^+.

K. Ueno and C. Chang, *J. Atom. Energy Soc. Japan*, 4 (1962) 457.
R. Bock and G. M. Beilstein, *Z. Anal. Chem.*, 192 (1963) 44.

P(V), As(V), V(V,) and POLYVANADATES; Fe(III) + FERRON; Co(II) and Fe(II) + NITROSALT R, by R_4As^+ and R_4P^+.

M. Ziegler and O. Glemser, *Angew. Chem.*, 68 (1956) 522.

BF_4^- by R_4As^+.

J. Coursier, J. Huré and R. Platzer, *Anal. Chim. Acta*, 13 (1955) 379.
L. Ducret and P. Seguin, *Actas do Congresso*, Lisbon, 1957, p. 88.

Co^{2+} + SCN^- by R_4As^+ and R_4P^+.

L. P. Pepkowitz and J. L. Marley, *Anal. Chem.*, 27 (1955) 1330.

Co^{2+}, Zn^{2+} + SCN^-; Bi^{3+}, Sb^{3+}, Hg^{2+} + I^- by R_3Se^+ and dodecyltrimethylammonium.

M. Shinagawa, H. Matsuo and M. Yoshida, *Japan Analyst*, 3 (1955) 139; 3 (1955) 213.

Various exemples with R_4N^+.

57 ions W. J. Maeck, G. L. Booman, M. F. Kussy and J. E. Rein, *Anal. Chem.*, 33 (1961) 1774.
A. M. Wilson, L. Churchill, K. Kiluk and P. Hovsépian, *Anal. Chem.*, 34 (1962) 203.
M. Ziegler and O. Glemser, *Angew. Chem.*, 68 (1956) 411, 620.

(iii) Cations of amines and various bases. Many basic organic materials, among which are certain amines and basic dyes, are capable of forming cations of the form HB^+, which combine with inorganic anions to give compounds soluble in organic solvents.

LONG CHAIN AMINES. For example, amines of the type R_3N may be used for the extraction of many inorganic anions under suitable pH conditions:

$$\underset{\text{solvent}}{2R_3N} + H^+ + HSO_4^- \rightleftarrows \underset{\text{solvent}}{(R_3NH)_2SO_4}$$

$$\underset{\text{solvent}}{4R_3N} + 3SO_4^{2-} + 4H^+ + UO_2^{2+} \rightleftarrows \underset{\text{solvent}}{UO_2(SO_4)_3\,(R_3NH)_4}$$

Amines such as tri-*n*-octylamine, methyl dodecylamine, etc. are used for the extraction of uranium from its solutions in sulphuric acid.

COMPARE

K. B. Brown, C. F. Coleman, D. J. Crouse, J. O. Denis and J. G. Moore, *Atomic Energy Commission*, ORNL (1954) 1734.
C. Boirie, *Bull. Soc. Chim.*, (1958) 980, 1088; *Thesis*, Paris, 1959.
C. F. Coleman, C. A. Black Jr. and K. B. Brown, *Talanta*, 9 (1962) 297.

(iv) Oxine compounds. Certain oxinates can be extracted in the presence of amines; these are: $RNH_3[UO_2Ox_3]$, $RNH_3[BeOx_3]$, and the oxines of the alkaline-earth metals.

COMPARE

F. UMLAND, W. HOFFMANN AND K. U. MECKENSTOCK, *Z. Anal. Chem.*, 173 (1960) 211.

BASIC DYES. Extractions may similarly be performed with many dyes capable of adding H^+ to give coloured compounds; the latter may then be estimated colorimetrically.

This applies to methylene blue:

$$B + H^+ + BF_4^- \rightleftarrows \underset{\text{solvent}}{(BH)\,(BF_4)}$$

COMPARE

BF_4^- + methylene blue

L. DUCRET, *Anal. Chim. Acta*, 17 (1957) 213.

Various dyes

L. DUCRET AND H. MAUREL, *Anal. Chim. Acta*, 21 (1959) 74; 21 (1959) 79.
L. DUCRET AND M. DROUILLAS, *Anal. Chim. Acta*, 21 (1959) 86.
L. DUCRET AND M. RATOUIS, *Anal. Chim. Acta*, 21 (1959) 91.

BF_4^-, ReO_4^-, Ta + Fluoride by methyl violet

N. S. POPUEKTOV, L. I. KONONENKO AND R. S. LANER, *Zhur. Analit. Khim.*, 13 (1958) 396.
A. K. BABKO AND P. V. MARCHENKO, *Chem. Abstr.*, 53 (1959) 3843.

Zn^{2+} + SCN^- + Rhodamine B

G. MARTIN, *Bull. Soc. Chim. Biol.*, 34 (1952) 1174.

(3f) *Organic anions*

Certain organic anions combine with inorganic cations to give salts which are stable in organic solvents. This is particularly so in the case of long-chain acids such as perfluorooctanoic acid, alkyl sulphates, or sulphonates.

COMPARE

Perfluorocarboxylic acids

G. F. MILLS AND H. B. WHETSEL, *J. Am. Chem. Soc.*, 77 (1955) 4690.

Salicylic, cinnamic, etc acids

B. HÖK-BERNSTRÖM, *Acta Chem. Scand.*, 10 (1956) 163.

Anilic acids

S. K. DATTA, *Z. Anal. Chem.*, 148 (1955) 259, 267.

(3g) *Organic anions and cations*

A number of solvents are suitable for the treatment of combinations of this type. A dye is generally chosen as one of the two constituents of the complex, so that a coloured compound, which can be estimated spectrophotometrically, is obtained. Many detergents are estimated in this way.

COMPARE

E. L. COLICHMAN, *Anal. Chem.*, 19 (1947) 430.
J. LONGWELL AND W. D. MANIECE, *Analyst*, 80 (1955) 167.

Methyl violet

W. A. MOORE AND R. A. KOLBESON, *Anal. Chem.*, 28 (1956) 161.

Fuchsin

G. R. WALLIN, *Anal. Chem.*, 22 (1950) 616; etc.

Review

C. W. BALLARD, J. ISAACS AND P. G. W. SCOTT, *J. Pharm. and Pharmacol.*, 6 (1954) 971.

(4) Extraction of inorganic compounds

Some inorganic compounds may be dissolved in organic solvents. This is generally the case with the compounds of certain particular elements, such as iodine and bromine, and may even occur with numerous complexes which are unstable in aqueous solution. Examples of substances which behave in this way are the complexes of chloride, bromide, iodide, thiocyanate and other ions, certain heteropolyacids, phosphovanadic and phosphomolybdic acids, etc. Their solubilities in the organic solvents are sometimes very high, and separations of considerable amounts of material are thus possible. Direct colorimetric estimation of the organic solution obtained is sometimes practicable.

A distinction may be made between two possible types of extraction of inorganic compounds.

(1) Some substances, such as $HgCl_2$, Br_2, SO_2, etc. dissolve in organic solvents (including the inert solvents) in molecular, *i.e.* undissociated form.

(2) Certain other materials pass into solution by forming coordinate bonds with the solvent. This is the case with ferric chloride, uranyl nitrate, etc., when dissolved in ethers, esters, ketones and alcohols, which form compounds such as *e.g.* $(C_2H_5)_2O \rightarrow H^+FeCl_4^- \cdot mH_2O$; here the solvent behaves as a base.

(i) Extraction of chlorides. Certain chlorides may be extracted with ethers, esters, ketones, alcohols, etc. Thus, $AsCl_3$, and $GeCl_4$ are soluble in solvents such as carbon tetrachloride; $HMoO_2Cl_3$, $FeCl_3$, and $HFeCl_4$ are soluble in ethers, etc.

The concentration of the chloride or of hydrochloric acid in the aqueous phase plays a large part in determining the extraction equilibria:

$$Fe^{3+} + 4Cl^- + H^+ \rightleftarrows \underset{\text{solvent}}{HFeCl_4}$$

The extraction is generally the easier, the greater the stability of the complex chlorides in aqueous solution.

The extraction of large amounts of Fe(III) in concentrated HCl solution by successive operations with ethyl ether and then *iso*-propyl ether, is well established. Fe(III) may, in this way, be separated from Al(III), Ni(II), Fe(II), Co(II), Mn(II), etc. Ga(III) can similarly be separated from Al(III) and Zn(II). In this case, however, a phosphoric acid solution is used, and interference sometimes occurs owing to the precipitation of alkaline salts. In addition, some H_3PO_4 passes into the ether.

Extraction by ethyl ether in 8 *N* HCl (water 20 ml, alcohol-free ether 50 ml): The percentages of various elements extracted by a single operation are given in Table 9.

TABLE 9

	Extracted, %		*Extracted, %*		*Extracted, %*
Fe(III)	99	Cu(II)	0.05	Pb(II)	0
Ga(III)	97	In(III)	traces	Mn(II)	0
Au(III)	95	Hg(II)	0.2	Ni(II)	traces
Tl(III)	90–95	Pt(IV)	traces	Os(VIII)	0
Mo(VI)	80–90	Se(IV)	0	Pd(II)	0
Sb(V)	81	V(V)	0	Ln(III)	0
As(III)	68	Zn(II)	0.2	Rh	0
Ge(IV)	40–60	Al(III)	0	Ag(I)	0
Te(IV)	34	Bi(III)	0	Th(IV)	0
Sn(II)	15–30	Cd(II)	0	Ti(IV)	0
Sn(IV)	17	Cr(III)	0	W(VI)	0
Sb(III)	6	Co(II)	0	U(VI)	0
Ir(IV)	5	Be(II)	0	Zr(IV)	0
As(V)	2–4	Fe(II)	0		

COMPARE

E. H. SWIFT, *J. Am. Chem. Soc.*, 46 (1924) 2375.

With hexone

G. GOTO AND Y. KAKITA, *Sci. Rept. Res. Inst.*, Tohoku Univ., Ser. A., 11 (1958) 1.

Extraction of iron with iso-propyl ether. A solution of ferric chloride containing about 2 g of iron is reduced to 10 ml by evaporation, and is transferred to a 250 ml separating funnel together with 75 ml of concentrated hydrochloric acid (12 *N*) and 25 ml of water; 50 ml of *iso*-propyl ether are then added. The mixture is shaken and allowed to settle, and the aqueous layer, containing not more than 5 mg of iron, is separated.

NOTE. The operation should not be performed in bright light, which reduces the efficiency of the extraction.

COMPARE

With ether

R. W. DODSON, G. J. FORNEY AND E. H. SWIFT, *J. Am. Chem. Soc.*, 58 (1936) 2573.

With dichloroethyl ether

J. AXELROD AND E. H. SWIFT, *J. Am. Chem. Soc.*, 62 (1940) 33.

With iso-propyl ether

N. H. NACHTRIEB AND J. C. CONWAY, *J. Am. Chem. Soc.*, 70 (1948) 3547.
R. J. MYERS, D. E. METZLER AND E. H. SWIFT, *J. Am. Chem. Soc.*, 72 (1950) 3767.
D. E. METZLER AND R. J. MYERS, *J. Am. Chem. Soc.*, 72 (1950) 3776.

With amyl acetate

J. E. WELLS AND D. P. HUNTER, *Analyst*, 73 (1948) 671.

With other solvents

E. BANKMANN AND H. SPECKER, *Z. anal. Chem.*, 162 (1958) 18.

OTHER APPLICATIONS. In 6 *N* HCl, Fe(III), Tl(III), Ga(III), Au(III), Ge(IV), As(III), As(V), Sb(III), Sb(V), Te(IV), Mo(VI), Sn(II), and Sn(IV) may also be extracted by *iso*-propyl ether.

In 8–9 *N* HCl, $GeCl_4$ and $AsCl_3$ may be extracted by carbon tetrachloride and can thus be separated from numerous compounds including As(V). Se(IV) is extracted under the same conditions. If the HCl is diluted, As(III) passes back into the aqueous solution.

The chlorides of Ni(II) and Co(II) are soluble in 8-octanol.

Ga(III) chloride is soluble in tributyl phosphate. This may be made use of in separating it from Al(III), in particular.

Sb(V) can be extracted from 1–2 *N* HCl by *iso*-propyl ether.

Methyl *iso*-butyl ketone has been proposed for extracting Fe(III) from 5.5–7 *N* HCl. Al(III), Co(II), Ce(III) and Ce(IV), Cd(II), Cu(II), Ni(II), Cr(III), Be(II), U(VI), Zr(IV), the rare-earths(III), the alkaline-earths(II) and the alkalines(I) are not extracted.

LiCl is extracted by the higher alcohols and may thus be separated from the other alkali metals.

COMPARE

G. H. MORRISON AND H. FREISER, *Solvent Extraction in Analytical Chemistry*, Wiley, 1957.

(ii) Extraction of bromides. The conditions are somewhat similar to those mentioned above. In general, bromides may be extracted somewhat more readily than the chlorides.

In 4 *N* HBr, In(III), Tl(III), Fe(III), Ga(III), Au(III), Sn(II), Sn(IV), Sb(V), Mo(VI) and Pt(II) may be extracted with *iso*-propyl ether.

Au(III) may thus be separated from the metals of the platinum group, except for Os(VIII).

$HgBr_2$ is soluble in benzene.

COMPARE

With ether

R. BOCK, H. KUSCHE AND E. BOCK, *Z. Anal. Chem.*, 138 (1953) 167.

With iso-propyl ether

F. A. POHL AND K. KOHES, *Mikrochim. Acta*, (1957) 318.

With methyl iso-butyl ketone

A. R. DENARO AND V. J. OCCLESHAW, *Anal. Chim. Acta*, 13 (1955) 239.

(iii) Extraction of iodides. Many iodides *e.g.* those of Sb(III), Cd(II), Au(III), In(III), Pb(II), Hg(II), Sn(II), Tl(III), Bi(III), Sn(IV) etc., may be extracted by various solvents.

The following are not extracted: alkali and alkaline-earth metals, Fe(II), Ni(II), Cr(III), Co(II), Mn(II), Ti(IV), Zr(IV), Th(IV), Al(III), Be(II), U(VI), V(IV), etc. Thus, In(III) can be separated from Ga(III)[1]. Pb(II), Bi(III), Cu(II), Pd(II), and Cd(II) may be extracted with methyl *iso*-butyl ketone[2].

COMPARE

(1) H. M. IRVING AND F. J. C. ROSSOTI, *Analyst*, 77 (1952) 801.
(2) P. W. WEST AND J. K. CARLTON, *Anal. Chim. Acta*, 6 (1952) 406.
F. A. POHL AND K. KOHES, *Mikrochim. Acta*, (1957) 318.

(iv) Extraction of thiocyanates. Many thiocyanates can be extracted with ether[1], hexone, butyl phosphate[2], or 1-butanol[3].

Ether extracts Be(II), Zn(II), Co(II), Se(IV), Ga(III), Nb(V), In(III), Ti(III), Fe(III), Sn(IV), Mo(V), Cu(II), Bi(III), Sc(III). This may be used to separate Th(IV), Zr(IV), Mn(II), and Pb(II), etc. from the rare earth metals.

Using methyl *iso*-butyl ketone, Zn(II) can be separated from small amounts of Cd(II), Co(II), Cu(II), and Fe(III).

COMPARE

(1) R. BOCK, *Z. Anal. Chem.*, 133 (1951) 110.
(2) M. AVEN AND H. FREISER, *Anal. Chim. Acta*, 6 (1952) 412; *Anal Chem.*, 24 (1952) 597.
L. M. MELNICK, H. FREISER AND H. F. BEEGHLY, *Anal. Chem.*, 25 (1953) 856.
L. M. MELNICK, *Dissert. Abstr.*, 14 (1954) 760.

L. M. MELNICK AND H. FREISER, *Anal. Chem.*, 27 (1955) 462.
(3) G. E. MARKLE AND D. F. BOLTZ, *Anal. Chem.*, 26 (1954) 447.

Ga(III)/Al(III) ether-tetrahydrofuran
H. SPECKER AND E. BANKMANN, *Z. Anal. Chem.*, 149 (1956) 97.

(v) Extraction of fluorides. Many fluoride complexes are suitable for extraction. For example, Nb(V), Ta(V) and Re(VII) can be extracted into ethers and ketones, and under suitable conditions, they can, in consequence, be separated from other elements and from each other.

COMPARE
R. BOCK AND M. HERMANN, *Z. Anorg. Allgem. Chem.*, 284 (1956) 288.

Nb-Ta
G. R. WATERBURY AND C. E. BRICKER, *Anal. Chem.*, 30 (1958) 1007.
M. L. THEODORE, *Anal. Chem.*, 30 (1958) 465.
G. W. C. MILLNER, G. A. BARNETT AND A. A. SMALES, *Analyst*, 80 (1955) 380.

(vi) Extraction of nitrates. A certain number of nitrates can be extracted with ethers, ketones, butyl phosphate, etc.

The method is question is highly selective, and may be used to separate U(VI), Pu(VI), Np(VI), and Am(VI) from most other elements. The only other elements which are extracted are Th(IV), Ce(IV), and Zr(IV), and small quantities of Fe(III), Au(III), Sc(III), Pa(IV), Tl(III), As(V), Bi(III), Cr(VI), and V(V).

Thorium(IV) is commonly extracted by an extremely selective technique employing mesityl oxide.

COMPARE
A. NÖRSTROM AND L. G. SILLÉN, *Svensk Kem. Tidsk.*, 60 (1948) 227.
R. BOCK AND E. BOCK, *Z. Anorg. Allg. Chem.*, 263 (1950) 146.
H. A. C. MCKAY AND A. R. MATHIESON, *Trans. Faraday Soc.*, 47 (1951) 428.
E. GLUECKHAUF, H. A. C. MCKAY AND A. R. MATHIESON, *Trans. Faraday Soc.*, 47 (1951) 437.
K. LINDH, R. RYNNINGER AND E. SKÖRAEUS, *Svensk. Kem. Tidsk.* 61 (1949) 180.

Ce(IV)
A. W. WYLIE, *J. Chem. Soc.*, (1951) 1474.
J. C. WARF, *J. Am. Chem. Soc.*, 71 (1949) 3257.
H. A. C. MCKAY AND T. V. HEALY, *J. Inorg. Nuclear Chem.*, 4 (1957) 100; 4 (1957) 304, 315, 321. *Rec. Trav. Chim.*, 75 (1956) 730; *Trans. Faraday Soc.*, 52 (1956) 633.
J. RYDBERG AND B. BERNSTRÖM, *Acta Chem. Scand.*, 11 (1957) 86.

(vii) Extraction of perchlorates by 2-octanol

COMPARE

T. E. MOORE, R. J. LARAN AND P. C. YATES. *J. Phys. Chem.*, 59 (1955) 90.

General References

See the books given on page 124 and H. Irving, F. J. C. Rossoti and R. J. P. Williams, *J. Chem Soc.*, (1955) 1906.

(5) Principal solvents

(i) Ethers. In the past, the only ether used for extractions was ethyl ether· This however, suffers from the disadvantages of being too volatile, and is moreover slightly soluble in water. At present therefore, of the inexpensive water-insoluble ethers, *iso*-propyl ether is the most commonly used.

The use of mesityl oxide is also sometimes advocated (*cf.* thorium).

(ii) Ketones. These are generally equivalent to the ethers. The ketone most frequently used is methyl *iso*-butyl ketone or 'hexone'.

(iii) Esters. The properties of esters are analogous to those of the ethers and ketones. Ethyl acetate is often used, despite its volatility and susceptibility to hydrolysis. Extractions are also frequently performed with amyl acetate, or with esters derived from glycol; examples of the latter are: diethylcellosolve

$$C_2H_5—O—CH_2—CH_2—CH_2—O—C_2H_5$$

and dibutylcarbitol

$$C_4H_9—O—CH_2—CH_2—O—CH_2—CH_2—O—C_4H_9$$

Butyl phosphate and analogous compounds have certain special properties and are suitable for the extraction of very many compounds, some of which may even be inorganic. In principle, they give compounds soluble in an excess of solvent, or in numerous other solvents.

(iv) Alcohols. The higher alcohols sometimes permit the extraction of compounds which are not soluble in other solvents; the reactions involved are however complicated, since the alcohols have some slight mutual solubility with water; this is the case with *iso*-butanol and *iso*-amyl alcohol.

(v) Inert solvents. Where these are suitable for use, carbon tetrachloride, chloroform, benzene, etc., have the advantage that they separate easily from aqueous solutions. Chloroform is a better solvent than carbon tetrachloride, but has the disadvantage of being relatively unstable. In addition, although it can be stabilized by adding 1% alcohol, this is not always permissible.

Cyclohexanol and many other solvents are in common use at present.

(vi) Water-miscible solvents. Miscible solvents, which reduce the dielectric

constant and favour the formation of complexes, are occasionally added to the aqueous solution. Acetone, and to a greater extent, 1,4- dioxan, may be used for this purpose. Under these conditions, a large number of complexes such as dithizonates, oxinates, etc. are formed. The many advantages of this procedure have not yet been fully exploited.

In an analogous manner, solvents may be added to the organic phase in order to modify its properties. For example, certain compounds may be caused to return to the aqueous phase after extraction.

(vii) Solubility of certain solvents in water. Apart from the inert solvents, solubility of the majority of solvents used—ethers, alcohols, esters, ketones etc.—vary quite markedly with the composition of the aqueous phase. Conversely, the solubility of water itself in the organic phase varies according to the nature and the concentration of the substances dissolved in the latter. This is in consequence of the bonds which are established between the molecules of the solvent itself and the materials in solution. It is therefore difficult to define partition constants between water and these solvents which, owing to the above effect, frequently have a variable composition.

Inert solvents. The so-called inert solvents, among which are benzene, mesitylene, cyclohexane, hexane and carbon tetrachloride, are extremely sparingly soluble in water. The solubility of chloroform is 10 g per litre, that of 1,2-dichloroethane 9 g, and that of trichloroethylene 1 g.

Alcohols. *Iso*-amyl alcohol 30 g per litre, cyclohexanol 6 g.
Ethers. Ethyl ether 4 to 75 g; *iso*-propyl ether 4 to 6.5.
Ketones. Hexone 8 to 20 g.
Mesityl oxide 16 to 30.

General bibliography on extraction

Books

T. K. SHERWOOD AND R. L. PIGFORD, *Absorption and Extraction*, McGraw Hill, New York 1952.
R. E. TRAYBAL, *Liquid Extraction*, McGraw Hill.
G. KORTÜM AND H. BUCHHOLZ, Meisenheimer, *Die Theorie der Distillation und Extraktion von Flüssigkeiten*, Springer, Berlin, 1952.
M. GIBERT, *Echangeurs à contre-courant. Extraction.* Presses Documentaires, Paris, 1952.
DR. RAUEN, *Gegenstromsverleitung*, 1953.
L. ALDERS, *Liquid-liquid Extraction*, Elsevier, 1953.
M. VIGNERON, *Fractionnements par solvants*, Vigot, Paris, 1954.
G. H. MORRISON AND H. FREISER, *Solvent Extraction in Analytical Chemistry*, Wiley, 1957.
H. IRVING AND R. J. WILLIAMS, *Treatise on Analytical Chemistry*, ed. I. M. KOLTHOFF AND P. J. ELVING, Part. I, Vol. III, Interscience, 1961.

EXTRACTION TABLES OF INORGANIC COMPOUNDS

G. H. Morrison and H. Freiser, in C. L. Wilson and D. W. Wilson, *Comprehensive Analytical Chemistry*, vol. 1A, Elsevier, 1959.

Reviews

L. C. Craig, *Anal. Chem.*, 23 (1951) 41; 24 (1962) 66.
G. H. Morrison, *Anal. Chem.*, 22 (1950) 1388.
L. C. Craig, *Anal. Chem.*, 26 (1954) 110.
L. C. Craig, *Anal. Chem.*, 28 (1956) 723.
G. H. Morrison and H. Freiser, *Anal. Chem.*, 30 (1958) 632.
H. Irving, *Quart. Revs., London*, 5 (1951) 200.
V. I. Kuznetsov, *Uspekhi Khim.*, 23 (1954) 654.
F. S. Martin and R. J. W. Holt, *Quart. Revs., London*, 13 (1959) 327.

SUCCESSIVE EXTRACTIONS

T. W. Evans, *Ind. Eng. Chem.*, 26 (1934) 439.
E. L. Smith, *J. Soc. Chem. Ind.*, 47 (1928) 159T.

LIMITED COUNTER-CURRENT

T. G. Hunter and A. W. Nash, *Ind. Eng. Chem.*, 27 (1935) 836.

COUNTER-CURRENT EXTRACTIONS

L. C. Craig and O. Post, *Anal. Chem.*, 21 (1949) 500; 22 (1950) 1346.
H. L. Lochte and H. W. H. Meyer, *Anal. Chem.*, 22 (1950) 1064.
L. C. Craig, W. Hausmann, E. H. Ahrens, Jr. and E. J. Harfenist, *Anal. Chem.*, 23 (1951) 1236.
J. D. A. Johnson, *J. Chem. Soc.*, (1950) 1743.
J. D. A. Johnson and A. Talbot, *J. Chem. Soc.*, (1950) 1068.
G. Golumbic, *Anal. Chem.*, 23 (1951) 1210.
G. Kalopissis, *Chim. Ind. (Paris)* 64, (1950) 563.

Chapter 10

Separations by ion-exchange

(1) Principle

Ion-exchangers consist of insoluble solid materials, generally synthetic resins possessing a macromolecular lattice upon which ionizable radicals have been grafted, capable of exchanging their ions with those of a solution with which they are in contact.

The grafted group may *e.g.* be

$$-SO_3H, \ -CO_2H, \ \equiv NOH, \text{ etc.}$$

Ion-exchangers may be classified into those which exchange cations and those which exchange anions. As an example, a cation-exchanger possessing available ions A^+ will be considered. If this is placed in contact with a solution containing different cations B^+, the following equilibrium between the resin and the solution will be set up:

$$\underset{\text{resin}}{A_r^+} + \underset{\text{solution}}{B_s^+} \rightleftarrows \underset{\text{solution}}{A_s^+} + \underset{\text{resin}}{B_r^+}, \qquad (10)$$

As in the case of isomorphic crystals, it is possible to define the activities of A^+ and B^+ in the solid; if these are respectively denoted as $[A^+]_r$ and $[B^+]_r$, then, applying the law of mass action:

$$\frac{[A^+]_s \cdot [B^+]_r}{[A^+]_r \cdot [B^+]_s} = K \text{ or } \frac{[A^+]_s}{[A^+]_r} \Big/ \frac{[B^+]_s}{[B^+]_r} = K.$$

Resins have been successfully prepared, with which the exchange of ions proceeds extremely rapidly, since equilibrium is reached in a very short time.

Analogous equilibria exist in the case of an anion-exchanger:

$$C_r^- + D_s^- \rightleftarrows C_s^- + D_r^-$$

NOTE. No exchange can take place when an exchange resin comes into contact with pure water.

(i) Importance of ion-exchangers. The resin and the solution form two separate phases, between which ions are exchanged. The importance of ion-exchangers lies in the possibility of applying this to separation techniques.

For example, if equilibrium (10) lies on the right, it follows that the dissolved cation B^+ may be quantitatively extracted into the resin, while A^+ replaces it in solution.

In addition, if one of these ions is involved in some equilibrium occurring in the aqueous solution, the latter can be displaced and the relative concentrations of A^+ and B^+ altered. At the same time however, equilibrium (10), and with it the degree of separation of the dissolved ion, will be affected. The separation may therefore be made more or less selective by controlling the pH, by complex formation or by oxidation or reduction reactions, as in the case of separations by precipitation or extraction.

(ii) Capacity of the resin. The quantity of exchangeable ions present in a fixed amount of any resin is known as the capacity of that resin, and is frequently defined in milliequivalents per gram of resin.

(iii). General properties. 1. Resins which bear sulphonic groups or their derivatives ($—SO_3H$ or $—SO_3M$) are generally readily ionizable. The affinity of a cation for the resin is the greater the higher the charge which it possesses, since a high charge will enable it to form several bonds with the radicals of the resin. Thus for example, the order of affinities of the following ions is:

$$Th^{4+} > Al^{3+} > Ca^{2+} > K^+.$$

As a result of this, quantitative reactions of the following type are often observed:

$$2K_r^+ + Ca_s^{2+} \rightarrow 2K_s^+ + Ca_r^{2+}$$

For cations bearing a single charge, the order is as follows:

$$Ag^+ > Tl^+ > Cs^+ > Rb^+ > K^+, NH_4^+ > Na^+ > H^+ > Li^+ > (CH_3)_4H^+$$

Thus, almost any cation, whether univalent or not, will replace H^+ in a cationic resin.

For divalent cations, the relative affinities are in the order:

$$Ba^{2+} > Pb^{2+} > Sr^{2+} > Ca^{2+} > Co^{2+} > Ni^{2+} = Cu^{2+} > Zn^{2+} > Mg^{2+} > Mn^{2+} > Be^{2+} > Cd^{2+}$$

2. The properties of certain resins which bear $—CO_2H$ radicals differ from the above since many ions, such as H^+, have a high affinity for $—CO_2^-$. If therefore, the solution is acid, there is little chance of displacing H^+ from $—CO_2H$.

The respective affinities of alkali metal ions for such resins are in the order

$$Li^+ > Na^+ > K^+.$$

3. Certain anionic resins contain radicals which give completely ionizable hydroxide or quaternary ammonium salt groups. The order of affinities of a number of anions for each resin is presented below:

$$\text{citrate} > SO_4^{2-} > C_2O_4^{2-} > I^- > NO_3^- > CrO_4^{2-} > Br^- > CN^- > HSO_3^- > Cl^- > HCO_2^- >$$
$$> HO^- > F^- > CH_3CO_2^-.$$

Once again polyvalent anions combine more strongly than monovalent anions *e.g.*

$$Fe(CN)_6^{4-} > Fe(CN)_6^{3-} > SO_4^{2-} > Cl^-$$

In the case of the quaternary ammonium resins, the order of affinities is:

$$HO^- > SO_4^{2-} > CrO_4^{2-} > \text{citrate} > \text{tartrate} > NO_3^- > \text{arsenate} > \text{phosphate} > \text{molybdate} > > CH_3CO_2^-, I^-, Br^-, > Cl^- > F^-.$$

4. Numerous new resins contain complex radicals and this results in the group being combined as a non-ionic compound; for example, complex resins containing imino acetate, anthranilate, etc.
5. The weak complexes formed by certain combinations of ions in aqueous solution are reinforced by fixation in the resin:

$$Cl_r^- + Fe^{3+} + 3Cl^- \rightarrow FeCl_{4r}^-$$

It appears therefore, that the affinity of $FeCl^-_4$ for the resin is stronger than that of Cl^-.
6. Influence of the concentration. Regeneration of the resin. This will be illustrated by considering the following equilibrium as an example.

$$Ca_s^{2+} + 2H_r^+ \rightleftarrows 2H_s^+ + Ca_r^{2+} \quad (11)$$

Although the position of equilibrium lies far over to the right, if sufficiently concentrated acid is added to a resin containing Ca^{2+}, the increase in the concentration of $[H^+]_s$ displaces the equilibrium towards the left and Ca^{2+} is expelled from the resin.

(iv) Operational procedure. The resin and the solution are mixed in a flask agitated till equilibrium is established. The solid material former is then filtered and washed. Alternatively, the resin may be placed in a column and the solution allowed to filter through it gradually. When all the solution has passed down the column, the resin is washed with pure water. In the liquid recovered, the cations originally present have for example, been replaced by H^+ according to:

$$Ca_s^{2+} + 2H_r^+ \rightleftarrows Ca_r^{2+} + 2H_s^+$$

This is equivalent to the performance of several successive operations in a flask, being much more efficient than the latter method (see Chromatography). The resin may then be regenerated by passing a strongly acidic solution continuously down the column. Equilibrium (11) is established at many points within the latter, and eventually all of the Ca^{2+} in the resin will have been expelled.

(v) Simple separation. 1. Cations can be separated either from anions or from

substances in molecular form. If a cationic resin containing H^+ is used, H^+ can be replaced by cations from the solution. The separation is completed by washing the resin. In this way, Fe^{3+}, Cu^{2+}, etc. may be separated from boric acid, phosphoric acid, many organic molecules, or from Cl^-, Br^-, and the like.

2. In a similar way, it is possible to separate anions from cations or molecular substances by means of an anionic resin containing HO^- radicals.

3. A solution may in principle be purified from all of its ions by passing it successively through a cationic, and an anionic resin.

For example, in the case of a solution containing $Mg^{2+} + SO_4{}^{2-}$, the following exchanges may be used:

$$2H^+_r + Mg^{2+}_s \rightarrow Mg^{2+}_r + 2H^+_s,$$

followed by:

$$2HO^-_r + SO_4{}^{2-}{}_s \rightarrow SO_4^{2-}{}_r + 2HO^-_s$$

Molecules of water are then formed according to:

$$2H^+ + 2HO^- \rightarrow 2H_2O$$

The operation can also be performed in a single step by mixing the two exchangers (mixed resin).

(vi) Ion-exchangers and acidity. The equilibrium between the resin and the ion in solution is supplemented and affected by a second equilibrium, occurring within the aqueous phase.

1. With a solution containing Li^+ in contact with a H^+ resin, the exchange equilibrium is of the form

$$Li^+_s + H^+_r \rightleftarrows Li^+_r + H^+_s;\ K_r$$

In the case of a sulphonic acid resin, the equilibrium position lies over to the left, since Li^+ has a lesser affinity for the resin than H^+.

If however the solution contains a base—for example, lithium in the form of its acetate, a second acid-base equilibrium is set up:

$$CH_3CO_2 + H^+ \rightleftarrows CH_3CO_2H;\ K_A.$$

The overall equilibrium will be of the form

$$Li^+_s + CH_3CO^-_{2\,s} + H^+_r \rightleftarrows Li^+_r + CH_3CO_2H_s;\ K_r'$$

with

$$K'_r = \frac{K_r}{K_A}$$

In this case, the value of $K_A(10^{-4.7})$ is such that Li^+ is quantitatively extracted into the resin.

2. In the case of a resin in the NH_4^+ form to which a solution of Na^+ is added, we obtain:

$$NH_4{}^+{}_r + Na_s{}^+ \rightleftarrows Na_r{}^+ + NH_4{}^+{}_s;\ K_r$$

Na^+ exhibits a lower affinity for the resin than NH_4^+, except when the sodium ion comes from a solution of sodium hydroxide, in which case ammonia is produced in the aqueous phase:

$$NH_4{}^+{}_s + HO^- \rightleftarrows NH_3 + H_2O$$

The overall equilibrium is:

$$NH_4{}^+{}_r + Na_s{}^+ + HO_s{}^- \rightleftarrows NH_{3\,s} + Na_r{}^+ + H_2O;\ K'_r = \frac{K_r}{10^{-4.8}}$$

In this system the relative affinities of Na^+ and NH_4^+ for the resin are reversed.

3. If a solution of Zn^{2+} is placed in contact with an H^+ resin, we obtain:

$$Zn_s{}^{2+} + 2H_r{}^+ \rightleftarrows Zn_r{}^{2+} + 2H_s{}^+;\ K_r$$

Zn(II) may exist in various forms, depending on the pH. If the total concentration of Zn(II) in the aqueous phase is C_{Zn} then $[Zn^{2+}] = C_{Zn} \times f(pH)$.

From this, an apparent constant K'_r may be obtained as

$$K'_r = \frac{[Zn^{2+}]_r\,[H^+]_s{}^2}{C_{Zn}\,[H^+]_r{}^2} = \frac{K_r}{f(pH)}$$

(vii) The effect of complexing on ion-exchange. 1. If a complexing agent is added to a solution, the concentration of the free ion present is effectively diminished, and their apparent affinity for the resin is reduced.

This may be exemplified by considering the equilibrium:

$$Fe_r{}^{3+} + 3\,K_r{}^+ \rightleftarrows Fe_r{}^{3+} + 3\,K_s{}^+;\quad K_r$$

If ions which complex Fe^{3+}, for example F^-, are added, the concentration of free Fe^{3+} falls, and the position of equilibrium is displaced towards the left. Thus, whereas $Fe^{3+}{}_{free}$ will normally replace K^+ in the resin, in the presence of fluoride it is possible to reverse this procedure. This effect may be utilized for the purpose of highly selective separations, if suitable complexing ions are chosen.

If the complex is sufficiently stable—for example FeY^- (Fe^{3+}/EDTA),—there is effectively no cationic affinity for the resin, since the element to be extracted is now in an anionic form; it may however be extracted by an anion-exchanger.

2. When complex ions formed in the solution exhibit a strong affinity towards an appropriate resin, any equilibria in the solution involving complex-generating

ions are affected. For example, in aqueous solutions, Co^{2+} gives the following unstable complexes: $CoSCN^+$, $Co(SCN)_2$, $Co(SCN)_3^-$, and $Co(SCN)_4^{2-}$. The last compound is blue, but the blue colour is not visible in dilute solution. Since however the affinity of $Co(SCN)_4^{2-}$ for anionic resins is extremely large, all equilibria existing in the solution will be displaced in the direction favouring the formation of the complex and the resin will assume a blue coloration. The extraction equilibrium is then of the form:

$$2\,SCN_r^- + Co^{2+} + 2\,SCN^- \rightleftarrows Co(SCN)_{4r}^{2-}$$

Similar processes occur with many chloride complexes which are unstable in aqueous solution; these are bound into a complex according to *(e.g.)*:

$$Fe^{3+} + 3Cl^- + Cl_r^- \rightleftarrows FeCl_{4r}^-$$

The extraction of uranium in the form of its sulphate, acetate, nitrate, chloride, and other complexes may be performed in a similar manner.

$$2\,SO_4^{2-} + UO_2^{2+} + SO_{4r}^{2-} \rightleftarrows UO_2(SO_4)_{3r}^{4-}$$

The degree of separation is varied by modifying such equilibria according to the concentration of complexing ions introduced into the solution. Many important applications of this procedure have been developed.

(viii) Ion-exchangers and precipitation. Although barium sulphate is very sparingly soluble, there is nevertheless some degree of dissociation according to the equilibrium:

$$BaSO_4\downarrow \rightleftarrows Ba^{2+} + SO_4^{2-} \qquad (13)$$

A precipitate of barium sulphate in an aqueous phase may be treated with an ion-exchanger, when the following reaction will be established.

$$\underset{\text{water}}{Ba^{2+}} + \underset{\text{resin}}{2\,H^+} \rightleftarrows \underset{\text{resin}}{Ba^{2+}} + \underset{\text{water}}{2\,H^+}$$

Equilibrium (13) is thereby displaced towards the right, until the precipitated barium sulphate finally disappears. The overall equilibrium is of the form:

$$BaSO_4\downarrow + \underset{\text{resin}}{2\,H^+} \rightleftarrows \underset{\text{resin}}{Ba^{2+}} + \underset{\text{water}}{2\,H^+ + SO_4^{2-}}$$

(ix) Separation techniques. The resin should be in the form of fine granules, in order to provide a large contact surface while permitting a ready flow of the liquid; the size fraction may be isolated by sieve analysis. For efficient operation, the resin should be completely wetted by the solution, and no air-bubbles should be entrapped. To ensure this, the resin is agitated with water for some

time before use. The water-resin mixture is then charged into the column, which consists of a glass tube (possibly a burette) fitted with a glass-wool plug or a perforated plate at its lower end. The resin should always be kept completely immersed in liquid; this condition is easily achieved in a column ending in a syphon, such as that shown in Fig. 51.

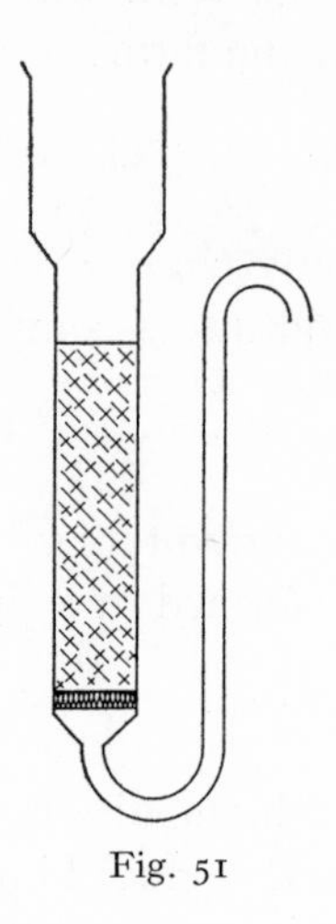

Fig. 51

Before the first operation, the resin is washed several times with dilute acids and alkalis alternately; details of solutions for this purpose are generally supplied by the manufacturer. The final solution used in the pretreatment is selected appropriately with respect to the application envisaged—for example, an acid solution is employed if the resin is to be used as an H^+ exchanger. After finally washing with water, the resin is ready for use.

When the resin is saturated, it may, if desired, be regenerated by reaction with a suitable solution. Cation exchangers for example, should be treated with acids. A single charge of resin may thus be used repeatedly. A further account of various practical details (volumes, rate of flow, etc.) will be presented in the second part of the work.

(2) Applications

A number of applications of a general nature will be discussed below.

(i) Anion/cation/molecule separations. A number of examples of separations of this type, which are in principle quite simple, will be found in the second part of this work; thus, H_3AsO_4 may be separated from Fe^{3+} and Cu^{2+} in an acid medium, by extracting the latter with a cationic resin.

Where a cation is bound by an anionic complex, the case is entirely analogous.

(ii) Separations in the presence of EDTA. In the presence of EDTA, and in solutions of suitable pH, a large number of ions give rise to anionic complexes which may be extracted from solutions of other substances in molecular form, or as simple cations, which may subsequently be collected on a cation exchanger. The isolation of H_3AsO_4, H_3PO_4, H_3BO_3 etc. may thus be carried out in acid media.

Ti^{4+}, Mn^{2+}, Ca^{2+}, Mg^{2+}, UO_2^{2+}, and Be^{2+}, which are not complexed in acid media, may be separated from AlY^-, FeY^-, ZnY^{2-}, etc. by this method.

Be^{2+} and UO_2^{2+} may be separated from most other cations at pH 3–4.

(iii) Separations in chloride media. (a) Anion-exchangers. Cations with a greater or lesser tendency to give anionic complexes in aqueous solutions may be collected by anionic exchange resins. Even if the complexes formed are unstable in water, their dissociation equilibria are often displaced in the direction favouring their formation in the presence of the resin.

$$\underset{\text{water}}{Fe^{3+} + 4\,Cl^-} \rightleftarrows \underset{\text{resin}}{FeCl_4^-}$$

If the resin is already in the Cl^- form, we obtain:

$$Fe^{3+} + 3\,Cl^- + Cl_r^- \rightleftarrows FeCl_{4r}^-$$

It is evident that these equilibria will be governed by the concentration of Cl^- in the aqueous solution. A large number of separations can be carried out by suitably adjusting this concentration.

NOTE. The dependence of ion-exchange extraction on a number of properties is in many cases analogous to that of solvent extraction.

The following substances are removed from very dilute HCl by appropriate resins: Zn(II), Cd(II), Hg(II), Tl(III), Pb(II), Sb(III), Bi(III), Ru(IV), Os(IV), Rh(III), Ir(IV), Ir(III), Pd(II), Pt(IV), Ag(I), and Au(III).

In 2 *N* HCl, the above list is augmented by: Mo(VI), W(VI), U(VI), Fe(III), In (III), Rh(III), and Ir(III).

In 4 *N* HCl, Ga(III), and Sb(V) may also be added.

In 9 *N* HCl, substances retained by resins are: Fe(III), Sn(IV), Sb(V), U(VI), Zr(IV), Mo(VI), Zn(II), Sn(II), Fe(II), Cu(II), Cd(II), Co(II), W(VI), and Ga(III); these may therefore be separated from Al(III), Cr(III), Ni(II), Ti(IV), Th(IV), Ln(III) [rare-earth metals], Be(II), V(IV), Mg(II), In(III), and Pb(II).

If elution is performed with 4 *N* HCl, the less stable complexes are decomposed and the corresponding elements pass into the aqueous solution, as is the case with Co^{2+}, which may thereby be separated from Cu(II), Fe(III), and Zn(II). In the latter case, a 12 *N* HCl solution of Co^{2+} is passed through the column; Ni(II) is not retained by the resin. The column is subsequently washed with 12 *N* HCl and the solution is recovered. It is then diluted to 6 *N* HCl and recirculated; under these conditions, Mn(II) goes back into solution. Then, in 4 *N* HCl, Co(II) stays in the solution, followed by Cu(II) in 2.5 *N* HCl, Fe(III) in 0.5 *N* HCl, and Zn(II) in 0.005 *N* HCl.

COMPARE

K. A. KRAUS, *J. Am. Chem. Soc.*, 77 (1955) 4508.
F. NELSON AND K. A. KRAUS, *J. Am. Chem. Soc.*, 76 (1954) 5916.
D. JENTZSCH *et al.*, *Z. Anal. Chem.*, 150 (1956) 241.
K. A. KRAUS, G. E. MOORE AND F. NELSON, *J. Am. Chem. Soc.*, 78 (1956) 2692.
E. W. BERG, *Anal. Chem.*, 30 (1958) 1827.

Zn(II)/Cd(II)

E. W. BERG AND J. T. TRUEMPER, *Anal. Chem.*, 30 (1958) 1827.
F. W. E. STRELOW, *Anal. Chem.*, 32 (1960) 1185.

(b) Cation exchangers. The fixed cations in the resin can be eluted, by means of a solution containing chlorides, by complex formation in the aqueous solution.

COMPARE

F. W. E. STRELOW, *Anal. Chem.*, 32 (1960) 1185.
C. K. MANN AND C. L. SWANSON, *Anal. Chem.*, 33 (1961) 459.

(iv) Separations in varied media. In an analogous way we can use the complex formation in SO_4^{2-}, F^-, Br^-, I^-, NO_3^-, $CH_3CO_2^-$, CO_3^{2-}, etc. media.

For example in a medium containing iodide Cd(II) is bound by an anionic resin and thus separated from Zn(II), Mn(II), Al(III), Ca(II), Mg(II), Ni(II) and Co(II).

U(VI) can be bound by an anionic resin in media containing NO_3^-, SO_4^{2-}, $CH_3CO_2^-$.

COMPARE

Cation exchanger, F^-

J. S. FRITZ, B. B. GARRALDA AND S. K. KARRAKER, *Anal. Chem.*, 33 (1961) 882.

Cation exchanger, Br^-

J. S. FRITZ AND B. B. GARRALDA, *Anal. Chem.*, 34 (1962) 102.

Anion exchanger, $Cl^- + F^-$

L. WISH, *Anal. Chem.*, 31 (1959) 326.
F. NELSON, R. M. RUSH AND K. A. KRAUS, *J. Amer. Chem. Soc.*, 82 (1960) 339.

Anion exchanger, F^-

U. SHINDEWOLF AND J. W. IRVINE, JR., *Anal. Chem.*, 30 (1958) 906.
J. P. PARIS, *Anal. Chem.*, 32 (1960) 520.

(v) Replacement of one ion by another for estimation purposes. If the reaction:

$$A^+_{(exchanger)} + B^+_{(solution)} \rightarrow B^+_{(exchanger)} + A^+_{(solution)}$$

is quantitative, then B^+ may be estimated indirectly by estimating A^+, which may be easier.

Similarly, an anion can be replaced by another which is easier to estimate—for example, SO_4^{2-} by SCN^-.

(vi) Estimation of trace concentrations. It is possible to retain traces of elements present in very dilute solutions (milk, urine, etc.) by passing a large volume of the solution through a small amount of resin. Elution may subsequently be carried out with a small volume of a suitable solution.

This method may, for example, be used to separate very small amounts of Fe(III) in HCl in the presence of much Al(III), with the aid of an anionic resin.

(3) Utilisation of ion exchangers in solvents

The use of a mixture of a solvent and water, or of a pure polar solvent, opens up many new possibilities, particularly in complexing media.

(4) Exchange of ions between precipitates

Equilibria analogous to those given above may be established between isomorphous crystals and solution.

$$\underset{}{M\downarrow} + \underset{\text{solution}}{M'} \rightleftarrows M'\downarrow + \underset{\text{solution}}{M} \quad \text{with} \quad \frac{[M'\downarrow] \cdot [M]_s}{[M\downarrow] \cdot [M']_s} = K$$

This exchange is very selective, since the ions M and M′ must exchange with one another in the crystalline lattice; the equilibria are however attained very slowly.

COMPARE

A. Knappwost, *Angew. Chem.*, 68 (1956) 371.

General bibliography

See Chromatography, Chapter 11

Chapter II

Separation of similar substances. Chromatography

Similar substances may only be separated by repeating the separation technique several times.

In most cases, fractionation procedures involve two phases. The following list presents such pairs of phases, together with the type of separation possible with each.

Liquid with solid	(a) precipitation or crystallization. (b) adsorption phenomena. (c) ion-exchange.
Two liquids	(d) extraction.
Liquid with gas	(e) distillation. (f) evolution of gases.
Solid with gas	(g) adsorption.

Finally, various separations of less importance may be carried out in a homogeneous medium: electromigration, diffusion, etc.

In early work, fractionation operations involving precipitation or crystallization were repeated one after another in separate stages and in some logical order. At present however, methods have been developed which enable the continuous separation of various substances by automatically displacing one of the phases with respect to the other. This may be achieved by sometimes very simple chromatographic techniques.

General principles. In chromatography, a series of exchanges between a solid and either a solution or a gas, are caused to occur successively and in logical order.

The greater the rates of exchange of the materials to be fractionated between the two phases, the easier is it to carry out a large number of exchanges under given conditions, so that a more efficient separation may be performed. Exchanges between a precipitate and a solution are of necessity slow; this is in general an inefficient method of fractionation. On the other hand, exchanges between two solvents are often quite fast, and those between gases and solvents or between resins and solutions may be very rapid indeed.

In all cases, attempts are made whenever possible to accelerate the exchanges by physical means—for example, by increasing the surface of contact.

The following types of chromatography will be considered successively:

partition chromatography (partition between two solvents or between a gas and a solvent);

adsorption chromatography (equilibrium between a solid adsorbent and a solution or a gas);

ion-exchange chromatography.

When a gas is involved, (liquid-gas or solid-gas), the operation is known as gas chromatography. For a treatment of this subject the reader is referred to the references on p. 146.

Chromatography may be performed by several methods such as elution, frontal analysis, and displacement, of which however, only the first is of interest here. A number of alternative techniques are available, such as column or paper chromatography.

Definition of the separation factor f. The separation factor is defined by the expression $f = \frac{C_1'/C_2'}{C_1/C_2}$, where C_1' and C_2' are the concentrations of the two materials to be separated in one of the phases, and C_1 and C_2 are their concentrations in the other.

(1) Partition chromatography

The elution method

(1a) *Partition of a substance between two solvents*

(i) Successive operations. In this method, the operation is carried out in separate steps. The substance A is distributed between two solvents S and S′ in a series of separating funnels. The successive operations are carried out in the following way (Fig. 52). A solution of A in solvent S is placed in the first funnel (no. 0) and it brought into equilibrium with the first fraction of the solvent S′;

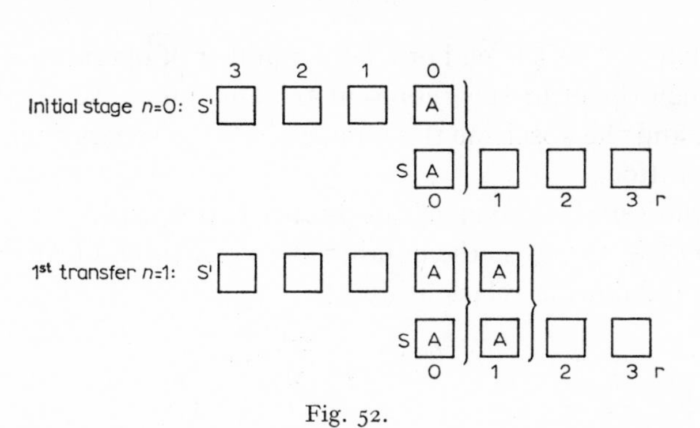

Fig. 52.

the two phases are then separated. In subsequent operations, fraction 0 of S′ is brought into equilibrium with fraction 1 of S and fraction 1 of S′ is brought into equilibrium with fraction 0 of S, and so on. The solvent S′ is continually displaced towards the right in the figure.

In the simple case where the volumes of the various fractions are equal and the partition coefficient D = 1, the results of the operations are shown in Table 10. The numbers occurring in the various columns of the table indicate the total concentrations of A in the two fractions when the latter are in contact.

TABLE 10

Serial numbers of funnels	0	1	2	3	4	...r
0	1					
1	0.5	0.5				
2	0.25	0.5	0.25			
3	0.125	0.375	0.375	0.125		
4	0.0625	0.25	0.375	0.25	0.0625	
.						
.						
.						
n						
Numbers of the transfers						

The fraction a of A present in the two separating funnels r at the moment of the nth series of equilibrium operations (nth transfer) can be calculated as follows:

$$a_{r,n} = \frac{n!}{r!\,(n-r)!}\left(\frac{1}{D\frac{V}{V'}+1}\right)^{n}\left(D\frac{V}{V'}\right)^{r}, \text{ for } D = \frac{[A]_s}{[A]_{s'}}$$

For example, in funnels no. 3 at the moment of the fourth transfer, assuming $D\frac{V}{V'} = 1$, we obtain:

$$a = \frac{4!}{3!}\left(\frac{1}{2}\right)^{4} = 0.25.$$

The distribution of A at the end of a large number of operations is indicated in Fig. 53, in which the ordinates represent the amounts of A as a fraction of the total amount, and the abscissae the numbers r of the separating funnels in each successive operation.

If r_{max} is the serial number of the funnel corresponding to the maximum concentration of A, and n_{max} the number of corresponding transfers, we obtain, in addition, the following relationship:

$$r_{max} = \frac{n_{max}}{1 + D\frac{V}{V'}}$$

NOTE. In cases when an apparent partition coefficient D′ must be used, this replaces D in the formulae.

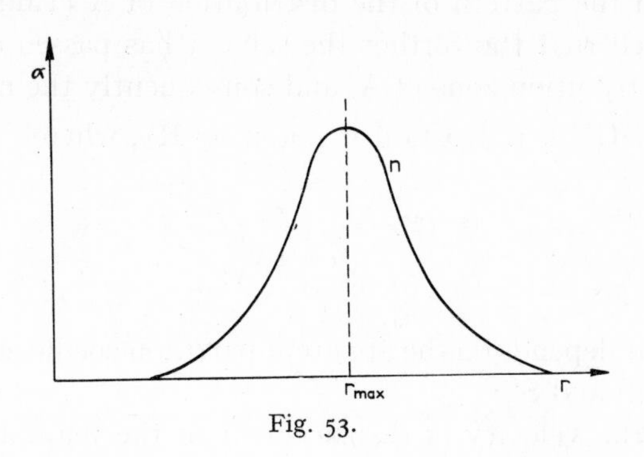

Fig. 53.

(ii) Operations on a column. The same series of operations can be carried out continuously on a column.

In this method, one of the solvents S, *e.g.* water, is immobilized by a solid support (cellulose, silica, etc.), and a little of the material A is introduced at the top of the column. Solvent S′ is then passed continuously down the column.

The experimental conditions in this method are slightly different to those prevailing in the previous case, since equilibrium is never quite established. It may however be assumed that the column is equivalent to a sequence in which the equilibria are in fact realized. Experimental evidence has proved this approximation to be adequate. If the concentration of A is measured at several points along the column, the same distribution is obtained as in the preceding case (Fig. 54).

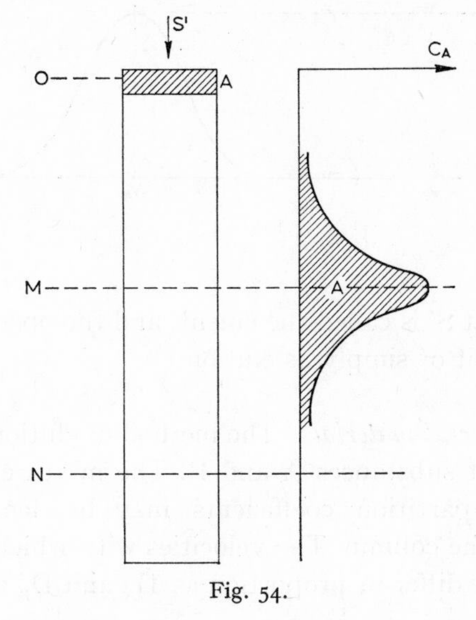

Fig. 54.

The solvent front is at N when the maximum concentration of A occurs at M.

NOTES. (1) From the pattern of the distribution of A (Table 10 and Fig. 53) it may be deduced that the further the solvent has passed down the column, the wider the distribution zone of A, and consequently the more dilute it is.
(2) The ratio $\overline{OM}/\overline{ON}$ is a constant denoted by R_F, where

$$R_F = \frac{1}{1 + D' \frac{V}{V'}}.$$

This constant depends on the apparent partition coefficient D′ of A between the two solvents S and S′.

If v_A denotes the velocity of displacement of the maximum of C_A and v_S' that of the solvent, it follows that $R_F = \frac{v_A}{v_S'}$.
(3) At the outlet of the column, the liquid which runs out may be collected in equal fractions, for example by means of an automatic apparatus called a fraction collector. The distribution curve of the concentration of A between the various fractions can then be plotted and the $C_A = f$ (volume of the solvent) or $C_A = f$ (time) diagram be derived (Fig. 55). In addition, the following relationships may be used:

$$\frac{V_S'}{V_A} = \frac{v_A}{v_S'} = R_F = \frac{1}{1 + D' \frac{V}{V'}}$$

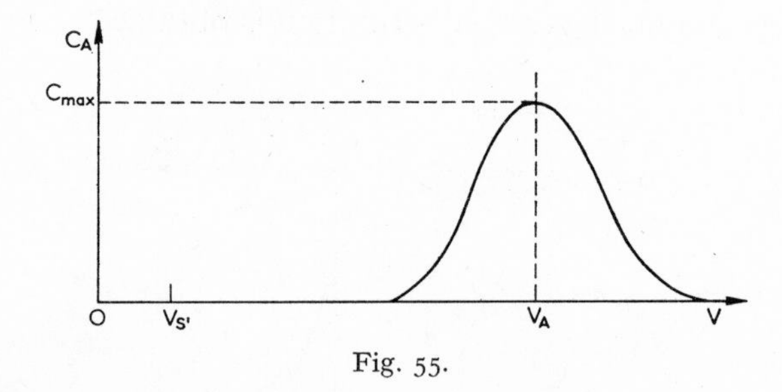

Fig. 55.

In this method solvent S′ is called the eluent, and the operation itself is known as elution development or simply as elution.

(iii) Separation of several materials. The method of elution may be adopted to separate two different substances A and B. The mixture of A and B, whose respective apparent partition coefficients may be denoted as D_A', D_B', is placed at the top of the column. The velocities with which the two substances move down the latter differ in proportion as D_A' and D_B' are different. At the

outlet of the column, a distribution like that of Fig. 56 is obtained with

$$\frac{V_B}{V_A} = \frac{1 + D'_B \dfrac{V}{V'}}{1 + D'_A \dfrac{V}{V'}}$$

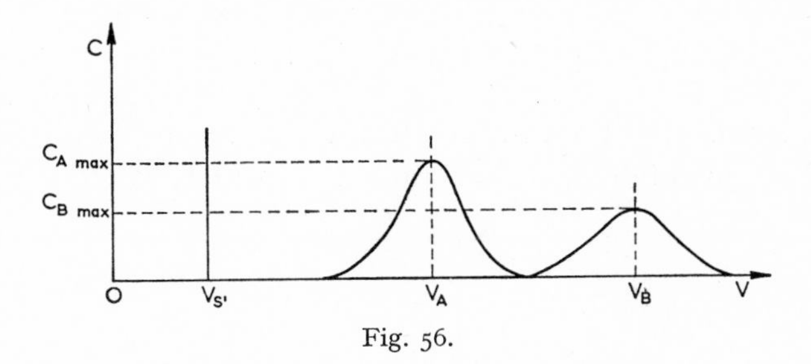

Fig. 56.

Typical applications. The positions of A and B are characterized by the corresponding values of R_F, which can be determined. Conversely, therefore, the substances may be characterized by their position. This is of very great importance in two cases: (a) where a mixture of unknown substances is analysed and (b) where the materials are too similar for them to be distinguished by means of various chemical reactions. In the latter case however, radioactive tracer techniques could also be used as an alternative method.

Variable factors. The efficiency of a separation is the greater the larger the number of successive equilibria established *i.e.* the greater the theoretically equivalent number of separate operations. For a column of a given length, the equivalent number of operations is determined by considering each of these to take place in a plate-like element of the column, and finding the height equivalent to a theoretical plate (H.E.T.P.). The smaller this height, the greater is the efficiency for a given column.

H.E.T.P. diminishes in proportion to increasing rate of exchange between the two solvents. The latter is governed by the nature of the substances in solution and that of the solvents, and is generally smaller the higher the mutual solubilities of the two solvents. The exchange surface can however be increased considerably by atomising the solvent S′ or increasing the surface of S by means of a subdivided support. In addition, it is possible to agitate the solutions by means of suitable mechanical attachments (pulsed columns).

The slower the rate of displacement of S′ with respect to S, the more time is available for the establishment of equilibrium within a given volume, and the smaller the H.E.T.P. Owing, however, to diffusion along the column, only low flow rates may be employed.

As is to be expected, the efficiency of separation of two substances depends on D'_A and D'_B. The separation factor may be expressed in the form:

$$\frac{1 + D'_A \dfrac{V}{V'}}{1 + D'_B \dfrac{V}{V'}}$$

(1b) *Paper Chromatography*

A strip of paper may be used to replace the column in the method described above. Cellulose immobilizes a large amount of water, which then acts as the

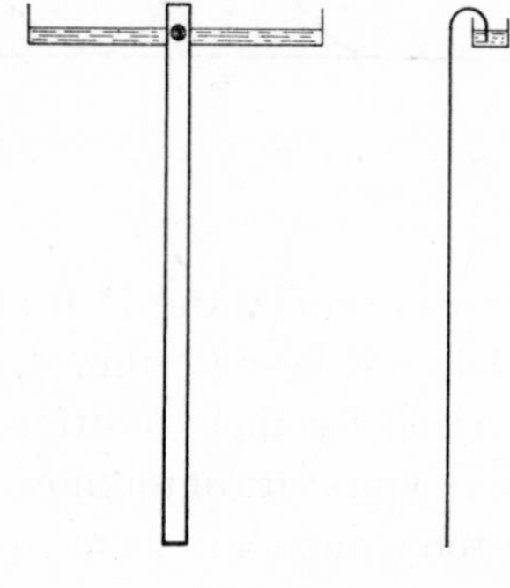

Fig. 57.

solvent S. A drop of the solution to be analyzed is deposited at the upper part of the strip (Fig. 57), and the latter is then immersed in the solvent S′; the constituents of the test solution are drawn down the strip of paper by capillarity and gravity.
The relevant formulae are identical to those used above.

In practice, a number of precautions are necessary. For example, the operation should be carried out in a closed vessel in an atmosphere saturated with the two solvents, to minimize evaporation.

(i) Two-dimensional chromatography. Chromatography of this type may be carried out using a sheet of paper, and performing two successive separations in directions at right angles to each other, if necessary with different solvents. The various substances present become localized in spots distributed over the whole surface of the sheet, and their position may then be compared with those of similarly treated pure materials (Fig. 58).

(ii) Application to quantitative analysis. If necessary, the spots may be rendered visible by staining them with a suitable reagent, which does not need to be specific since each material is identified by its position.

To a first approximation, the surface of a spot is proportional to the logarithm of the mass of the corresponding compound, *i.e.*

$$S = k \log m$$

Quantitative estimations may be carried out by comparison with calibration sheets, produced using known amounts of the various substances.

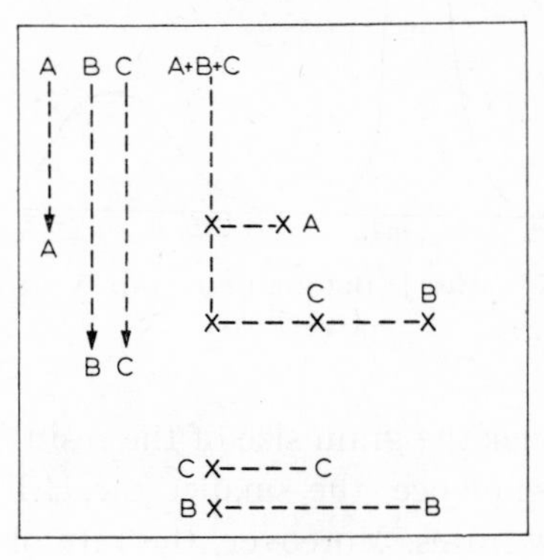

Fig. 58.

NOTE. The above method is extremely simple, and has many important applications, particularly in organic and biological chemistry.

(2) Ion-exchange chromatography

(i) The elution method. The principle of this method is similar to that of partition chromatography in that small quantities of the substances to be separated are placed at the top of the column, and are then eluted with a solution containing an ion possessing a lesser affinity for the resin than either of them. The latter may consist of the ion already present in the resin, as for example, H^+ for cationic resins. If the substances to be separated are denoted as A and B, equilibria of the following type are set up:

$$H_r^+ + A_s^+ \rightleftarrows H_s^+ + A_r^+, \text{ for which } \frac{[H^+]_s\,[A^+]_r}{[H^+]_r\,[A^+]_s} = K$$

Under normal conditions, A^+ is generally dilute in comparison with H^+, and $[H^+]_s$ and $[H^+]_r$ may therefore be regarded as essentially constant. It follows that $\frac{[A^+]_s}{[A^+]_r}$ = constant, as in the case of partition chromatography. The two processes are in fact equivalent.

The process of separation is illustrated in Fig. 59.

(ii) Variable factors. As in the case of partition chromatography, the operation

may be regarded as occurring in contiguous laminar elements of the column plates.

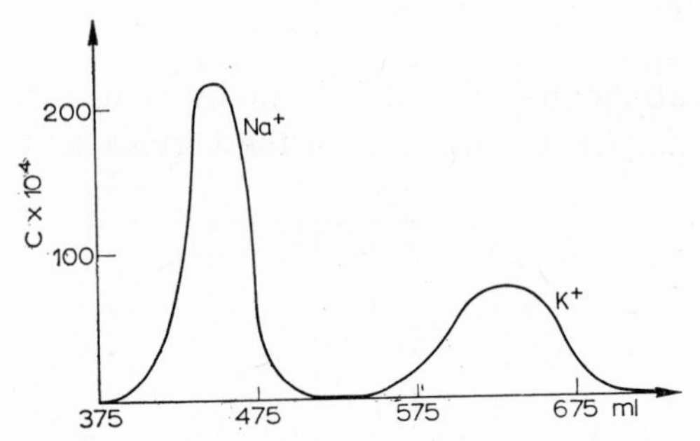

Fig. 59. Elution of Na^+ and K^+ (after J. Beukenkamp and W. Riemann, III, *Anal. Chem.*, 22 (1950) 582)

In this instance, the finer the grain size of the resin, the larger the exchange surface area and in consequence, the smaller the H.E.T.P. The latter is also reduced with smaller flow rates. Moreover, the rate of flow can be made very slow before longitudinal diffusion becomes an important factor.

All of the processes involved are accelerated at higher temperatures.

In the laboratory, an H.E.T.P. approximately the same size as the grains of resin (for example, 0.1 mm) may be attained. When this is so, a 1 m column corresponds to 10,000 equilibrium operations.

The various equilibrium constants may be altered in appropriate directions by any of the usual chemical means—for example, by the formation of complexes.

In view of the above characteristics, the method in question is one of the most efficient means of fractionation, and is generally much more efficient than partition chromatography between two solvents.

(3) Adsorption chromatography

It was shown above that in the case of dilute solutions, the law of adsorption leads to the definition of a partition coefficient of the dissolved substances between the solid adsorbent and the solution. As in the previous cases, the elution method can be used satisfactorily.

(4) Electrochromatography

(i) Electromigration. In this method, use is made of the fact that as ions in solution migrate in directions appropriate to their charges in an electric field between two electrodes, ions of the same size gradually become arranged in the

order of their mobilities, with corresponding variations in their local concentrations.

In consequence, with electrolytes contained in sufficiently long tubular cells, it has been possible to effect separations of very similar substances, including isotopes.

The electrolyte is sometimes immobilized by means of a gel.

(ii) Electrophoresis. The same phenomenon is used in the case of large charged molecules.

(iii) Electrochromatography. In this method an electrical means of separation is used in conjunction with one of the other methods of chromatography. The latter frequently consists of ordinary paper chromatography.

An electric field is created in the solution by means of two electrodes and at the same time a solid adsorbent is introduced. The ions then migrate as directed by the sign of their charge. In many cases, more efficient separations may be achieved in this manner than by simple chromatography.

Alternatively, two-dimensional chromatography can be carried out on paper by applying the electric field in a direction perpendicular to the flow of liquid.

Bibliography

CHROMATOGRAPHY

E. LEDERER AND M. LEDERER, *Progrès récents de la chromatographie.* I. Chimie organique et biologique, 1949; II. Chimie minérale, 1952, Hermann, Paris.

R. T. WILLIAMS AND R. L. M. SYNGE, *Partition Chromatography*, Cambridge University Press, 1951.

E. LEDERER AND M. LEDERER, *Chromatography*, Elsevier, 1953.

F. H. POLLARD AND J. F. W. MCOMIE, *Chromatographic Methods of Inorganic Analysis*, Butterworths, 1953.

O. C. SMITH, *Inorganic Chromatography*, Van Nostrand, 1953.

R. C. BRIMLEY AND F. C. BARRETT, *Practical Chromatography*, Chapman and Hall, 1958.

H. G. CASSIDY, *Fundamentals of Chromatography*, Interscience, 1957.

L. SAVIDAN, *La chromatographie*, Dunod, 1958.

E. BLASIUS, *Chromatographische Methoden in der analytischen und praparativen anorganischen Chemie*, F. Enke, Stuttgart, 1958.

E. LEDERER, *Chromatographie en chimie organique et biologique*, 1959.

I. SMITH, (Ed.), *Chromatographic and Electrophoretic Techniques*, vol. I, Heinemann, London, 1960.

M. LEDERER, H. MICHL, K. SCHLÖGL AND A. SIEGEL, *Chromatographische Methoden in der Anorganischen Analyse*, Springer Verlag, Wien, 1961.

Periodicals

Journal of Chromatography, Elsevier, Amsterdam.

Chromatographic Reviews, vols 1-6 and continuation, Elsevier, Amsterdam.

Reviews

H. H. STRAIN, T. R. SATO AND J. ENGELKE, *Anal. Chem.*, 26 (1954) 90.
H. H. STRAIN AND T. R. SATO, *Anal. Chem.*, 28 (1956) 687.
H. H. STRAIN, *Anal. Chem.*, 30 (1958) 624.

PAPER CHROMATOGRAPHY

R. J. BLOCK, E. L. DURRUM AND G. ZWEIG, *A Manual of Paper Chromatography and Paper Electrophoresis*, Academic Press, 1958.
F. CRAMER, *Papierchromatographie*, 5th ed., Verlag Chemie, Weinheim, 1962.
I. HAIS AND R. MACEK, *Handbuch der Papierchromatographie*, G. Fischer, Jena (1957–1960).
H. F. LINSKENS AND L. STANGE, *Praktikum der Papierchromatographie*, Springer Verlag, Berlin, 1961.
H. J. PAZDERA AND W. H. MCMULLEN, *Treatise on Analytical Chemistry*, Ed. I. M. KOLTHOFF AND P. J. ELVING, Part I, Vol. III, Interscience Publishers, New York, 1961.

THIN LAYER CHROMATOGRAPHY

E. STAHL, *Dünnschicht-Chromatographie*, Springer Verlag, 1962.
E. V. TRUTER, *Thin Film Chromatography*, Cleaver-Hume, London, 1963.
K. RANDERATH, *Thin Layer Chromatography*, 1963.
E. STAHL, *Thin Layer Chromatography*, 1963.
J. M. BOBBITT, *Thin Layer Cromatography*, Reinhold, 1963.

ION-EXCHANGERS

F. C. NACHOD AND J. SCHUBERT, *Ion Exchange Technology*, Academic Press, 1956.
J. A. KITCHENER, *Ion Exchange Resins*, Wiley, 1957.
C. CALMON AND T. R. E. KRESSMAN, *Ion Exchangers in Organic and Biochemistry*, Interscience 1957.
J. E. SALMON AND D. K. HALE, *Ion Exchange. A Laboratory Manual*, Butterworths, 1959.
R. KUNIN, *Elements of Ion Exchange*, Reinhold, 1960.
W. RIEMANN III AND A. C. BREYER, *Treatise on Analytical Chemistry*, Ed. I. M. KOLTHOFF AND P. J. ELVING, Part I, Vol. III, Interscience Publishers, New York, 1961.
F. HELFFERICH, *Ion Exchange*, McGraw Hill, 1962.
K. DORFNER, *Ionenaustauscher*, W. de Gruyter, Berlin, 1963.
J. SCHUBERT, *Principles of Ion Exchange*, 1963.
O. SAMUELSON, *Ion Exchangers in Analytical Chemistry*, 2nd. ed. Wiley, 1963.
J. E. SALMON, *Ion Exchange Chromatography*, Pergamon, 1961.

Reviews

R. KUNIN AND F. X. MCGARVEY, *Anal. Chem.*, 26 (1954) 104.
R. KUNIN, F. X. MCGARVEY AND A. FARREN, *Anal. Chem.*, 28 (1956) 729.
R. KUNIN, F. X. MCGARVEY AND D. ZOBIAN, *Anal. Chem.*, 30 (1958) 681.
D. K. HALE, *Analyst*, 83 (1958) 3.
Los Alamos Scientific Laboratory, Ion Exchange Resins, Bibliographie des rapports déclassés, 1947–1954.

Report

B. TRÉMILLON, *Displacement method. Bull. Soc. Chim.*, (1958) 1621.

GAS CHROMATOGRAPHY

A. T. JAMES, *Symposium on Vapour Phase Chromatography*, Butterworths, 1956.

C. PHILLIPS, *Gas Chromatography*, Butterworths, 1956.

A. I. M. KEULEMANS, *Gas Chromatography*, Reinhold, 1959.

V. J. COATES, H. J. NOEBELS AND I. S. FAGERSON, *Gas Chromatography*, New York, 1958.

D. H. DESTY, *Gas Chromatography*, Butterworths, 1958.

E. BAYER, *Gaschromatographie*, Berlin, 1959: *Gas Chromatography*, Elsevier, Amsterdam, 1961.

R. L. PECSOK, *Principles and Practice of Gas Chromatography*, Wiley, 1959.

Reviews and Reports

A. J. P. MARTIN, *XVth Congress of Pure and Applied Chemistry*, Lisbon, 1956, Birkhaüser, Basle.

P. CHOVIN, *Bull. Soc. Chim.*, (1957) 83.

C. J. HARDY AND F. H. POLLARD, *J. Chromatog.*, 2 (1959) 1–43.

ELECTROCHROMATOGRAPHY AND ELECTROPHORESIS

(see Chromatography and Paper Chromatography)

C. WUNDERLY, *Die Papier Elektrophorese*, 1955.

M. LEDERER, *Introduction to Paper Electrophoresis and Related Methods*, Elsevier, 1955.

R. AUDUBERT AND S. DE MENDE, *Les principes de l'électrophorèse*, Presses Universitaires Paris, 1957.

M. BIER, *Electrophoresis Theory and Applications*, 1959.

I. SMITH, *Zone Electrophoresis*, 1960.

L. P. RIBEIRO, E. MITIDIERI AND O. R. AFFONSO, *Paper Electrophoresis*, Elsevier, 1961.

J. R. CRAMER, *Treatise on Analytical Chemistry*, Ed. I. M. KOLTHOFF AND P. J. ELVING, Part I, Vol. III, Interscience Publishers, 1961.

C. WUNDERLY, *Principles and Applications of Paper Electrophoresis*, Elsevier, 1961.

R. J. WIEME, *Agar Gel Electrophoresis*, Elsevier, 1962.

J. G. FEINBERG AND I. SMITH, *Chromatography and Electrophoresis on Paper, London*, 1963.

H. BLOEMENDAL, *Zone Electrophoresis in Blocks aud Columns*, Elsevier, Amsterdam, 1963.

Reviews and Reports

R. MUNIER, *Chim. Anal.*, 37, (1955) 253, 283.

H. H. STRAIN AND T. R. SATO, *Anal. Chem.*, 28 (1956) 687.

H. H. STRAIN, *Anal. Chem.*, 30 (1958) 624.

Differential migration

Anal. Chem., 31 (1959) 818.

Chapter 12

Separations in the gaseous state

Once more, two phases are involved and, as in the previous cases (precipitation, extraction, ion-exchangers), separations may be carried out by suitably adjusting the pH, controlling the formation of complexes etc.

The quantitative relationships applicable to gas separations are of the same form as those appropriate to partition equilibria between two solvents (see p. 89). In the gaseous phase the substances present exist entirely in the undissociated molecular state, and their concentrations can be expressed in terms of partial pressures.

(i) Applications

(1) A certain number of materials may be separated in the gaseous state: CO_2, NH_3, AsH_3, H_2S, HCN, Cl_2, Br_2, I_2, Re_2O_7, RuO_4, OsO_4, etc.

(2) Certain chlorides can be distilled (see below).

(3) Azeotropic mixtures with water can be distilled: H_2SiF_6, HCl, etc.

Examples of the above applications are described in the second part of this book.

(4) Similar materials can be separated by chromatography (cf. p. 146, Gas chromatography).

Information concerning two further general cases is presented in Table 11.

The chlorides are usually distilled by one of the following methods (See Table 12):

1. Distillation of perchloric acid solutions: the solution is treated with 15 ml of concentrated perchloric acid, and is concentrated by boiling until the boiling point of the azeotropic mixture (200°) is attained. Concentrated hydrochloric acid (15 ml) is the added in small amounts over 20–30 minutes.

2. Distillation of sulphuric acid solutions: The procedure described above is repeated using concentrated sulphuric acid instead of perchloric acid.

The above table presents the yields of such distillations under such conditions. The yields are presented in milligrams assuming that 100 mg of the element being distilled were originally present.

A certain number of elements are bound into complexes by H_3PO_4 and do not distil over when it is present; these are Sn(II), Sn(IV), Bi(III), Mo(VI), Sb(V), and V(V).

The following are not volatilized: Ag(I), alkali metals, Al(III), Ba(II), Be(II),

TABLE 11

DISTILLATION OF CHLORIDES

Distillation or sublimation temperatures of some anhydrous chlorides

$GeCl_4$	86°	WCl_6	337°
$SnCl_4$	113°	$TeCl_4$	394°
$AsCl_3$	122°	$BiCl_3$	441°
$TiCl_4$	136°	$BeCl_2$	488°
VCl_4	164°	$SnCl_2$	600°
$SbCl_5$	172°	$ThCl_4$	922°
$AlCl_3$	180°	$ScCl_3$	967°
$SbCl_3$	219°	$MnCl_2$	1190°
$TaCl_5$	234°	$YbCl_3$	1200°
$NbCl_5$	243°	$CeCl_3$	1400°
$MoCl_5$	268°	$MgCl_2$	1410°
WCl_5	276°	KCl	1420°
$FeCl_3$	317°	NaCl	1440°
$HfCl_4$	317°	$CaCl_2$	1500°
$ZrCl_4$	331°		

TABLE 12

	Method 1 *mg*	*Method 2* *mg*
As(III)	30	100
As(V)	5	5
Au(III)	1	0.5
B(III)	20	50
Bi(III)	0.1	0
Cr(III)	99.7	0
Ge(IV)	50	90
Hg(I)	75	75
Hg(II)	75	75
Mn(II)	0.1	0.02
Mo(VI)	3	5
Os(VIII)	100	0
P(V)	1	1
Re(VII)	100	90
Ru	99.5	0
Sb(III)	2	33
Sb(V)	2	2
Se(IV)	4	30
Se(VI)	4	20
Sn(II)	99.8	1
Sn(IV)	100	30
Te(IV)	0.5	0.1
Te(VI)	0.1	0.1
Tl(I)	1	1
V(V)	0.5	0

Ca(II), Cd(II), Co(II), Cu(II), Fe(III), Ga(III), Hf(IV), In(III), Mg(II), Ni(II), Nb(V), Pb(II), Pt(IV), rare-earth metals, Si(IV), Ta(V), Th(IV), Ti(IV), U(VI), W(VI), Zn(II), and Zr(IV).

COMPARE

J. I. HOFFMAN AND G. E. F. LUNDELL. *J. Res. Natl. Bur. Std., A*, 22 (1939) 465.

(ii) Distillation of fluorides

Bi(III), Si(IV), and As(III) are volatilized completely by distillation from a mixture of concentrated perchloric and hydrofluoric acids. Ge(IV), Sb(III), Cr(III), Se(VI), Mn(VII), and Re(VII) are volatilized only partially.

Na(I), K(I), Cu(II), Ag(I), Au(III), Be(II), Mg(II), Ca(II), Sr(II), Ba(II), Zn(II), Cd(II), Hg(II), Sn(II), Ce(III), Ti(IV), Th(IV), Pb(II), V(V), Bi(III), Mo(IV), W(VI), Fe(III), Co(II), and Ni(II) are not volatilized.

COMPARE

F. W. CHAPMAN, G. G. MARVIN AND S. Y. TYREE, *Anal. Chem.*, 21 (1949) 700.

Bibliography

See Gas Chromatography. p. 146.

Chapter 13

Electrolytic separation

(1) Separation of trace amounts

It is seldom possible to separate trace amounts by the ordinary methods of electrodeposition, and even when this is possible, the operation is long and the conditions of electrolysis are unpredictable[1]. Small amounts of, for example, Cu, PbO_2, etc., may nevertheless be deposited on a platinum wire, redissolved, and estimated colorimetrically.

(i) Internal electrolysis. The method known as 'internal electrolysis' has often been found suitable for treating trace amounts. In this case, a suitable cell is set up and the two electrodes are short-circuited; the cell is arranged so that

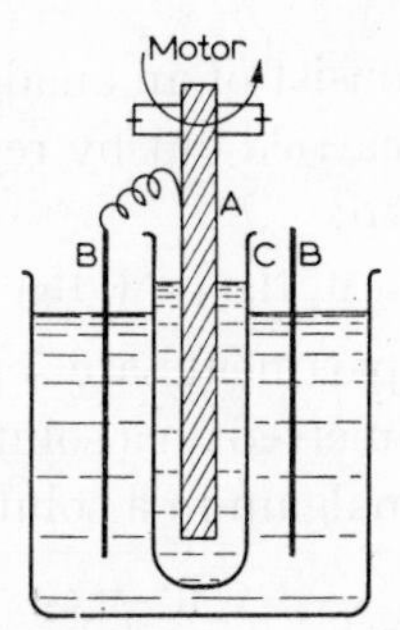

Fig. 60. Internal electrolysis apparatus. A: zinc rod, B: platinum gauze cylinder, C: porous vessel.

under these conditions an oxidation reaction takes place at one electrode, whilst reduction occurs at the other.

If, for example, a short-circuited platinum electrode and a zinc electrode are immersed in a solution containing Cu^{2+}, and the curves $i = f(E)$ (shown in Fig. 61) are plotted, it becomes evident that the two electrodes assume a common potential E_M at which the copper is deposited on the platinum, while the zinc is attacked. If zinc alone were to be used, this would become covered with copper and the reaction would stop, and in fact the copper may only be separated from the solution by being deposited on platinum. With similar apparatus, traces of antimony, cobalt, and bismuth can also be deposited.

EXAMPLES. Two electrodes, one of Pt and one of PbO_2, are immersed in a solution of Tl^+. The curves obtained are shown in Fig. 62. At the potential E_M, thallium deposits on platinum in the form of the oxide Tl_2O_3, whilst lead oxide is re-dissolved[2].

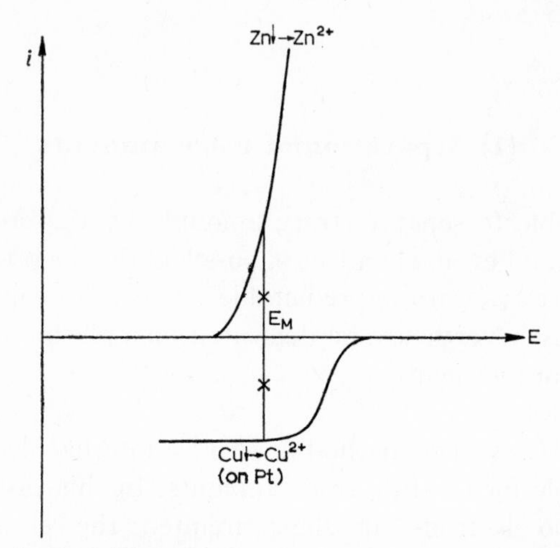

Fig. 61. Deposition of copper by internal electrolysis (on Pt)

One of the electrodes may consist of an amalgam. In this way, separations of an interfering element can be carried out by replacing it with another element from the amalgam, according to:

$$M_1^{n+} + M_2(Hg) \rightarrow M_1(Hg) + M_2^{n+}$$

If the anode and cathode compartments are separated, the anode may consist of an inert metal (platinum) immersed on a solution of a suitable reducing agent [V(II), Cr(II)], or of sodium amalgam in a solution of caustic soda[3]. The anode

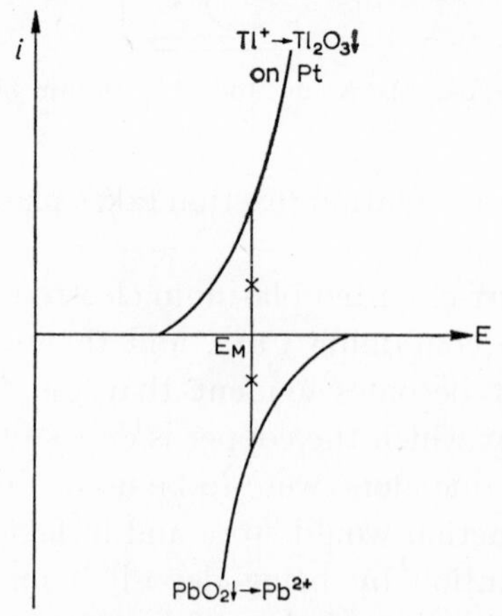

Fig. 62. Internal electrolysis. Deposition of Tl_2O_3 on Pb (after A. LIPCHINSKII, *Zh. Analit. Khim.*, 12 (1957) 83).

reaction then consists of the oxidation of V(II), Cr(II), or Na(Hg) (for example, Fig. 63).

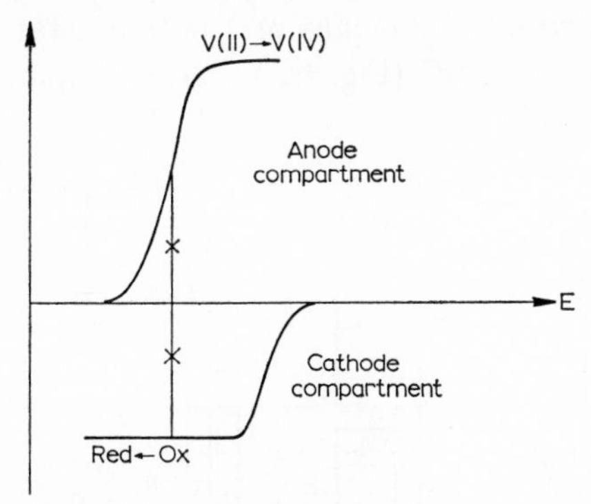

Fig. 63. Internal electrolysis with separate compartments.

(ii) Characteristics of the method. The above method has been known for a very long time, and in consequence has not yet been systematically employed in connection with the many recently available means of separating various elements. It is however capable of separating very small traces of elements, and may be very rapid, particularly when the electrolyte is efficiently stirred.

In addition, since the control of potential can nowadays be carried out automatically, the operating procedure is considerably simpler than that in other methods of separation.

(2) Separation of interfering materials

Substances present in a solution in large amounts can be separated by electrolysis, and the trace materials left behind in the solution, to be estimated.

(i) Determination of the conditions of electrolysis. With rapid systems (Ag, Zn, Hg, etc.), suitable conditions of electrolysis may be approximated quite satisfactorily by operating near the equilibrium potentials of the electrodes. A more precise estimate may however generally be obtained by plotting the $i = f(E)$ curves for the system with the same apparatus as is to be used, in the final operation. In this manner, it is possible to predict exactly the best conditions with regard to stirring, temperature, nature and surface of the electrodes, potential difference between the electrodes or intensity of the electrolysis current, and time of electrolysis.[4]

In the simplest cases practically no control is excercized over the conditions of electrolysis, and the operation is often carried out at an approximately constant current intensity. Thus, in the course of the separation of copper from a solution containing no other reducible metallic cations, the current intensity used is high enough to liberate hydrogen at the cathode. When however substances which are themselves reducible are to be separated, the cathodic potential must be controlled by means of a potentiostat. The latter maintains the potential between B and C (Fig. 64) constant, operating on the potential difference between A and C.

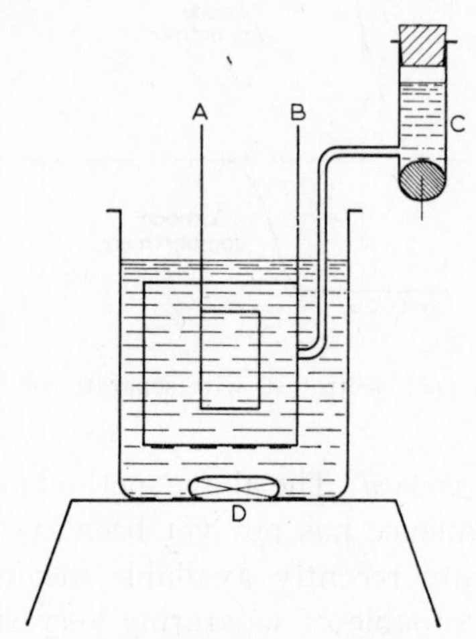

Fig. 64. Separations by electrolysis at a controlled potential. A, B electrodes, C: reference electrode, D: magnetic stirrer.
(after J. J. LINGANE, *Anal. Chim. Acta*, 2 (1948) 592.

(ii) Electrolysis with a mercury cathode. The most frequent application of this method is the extraction of large amounts of interfering elements from solutions, leaving the trace elements behind, ready for estimations. For example, 4 g of cadmium can be separated from 0.3 g of zinc in this way.

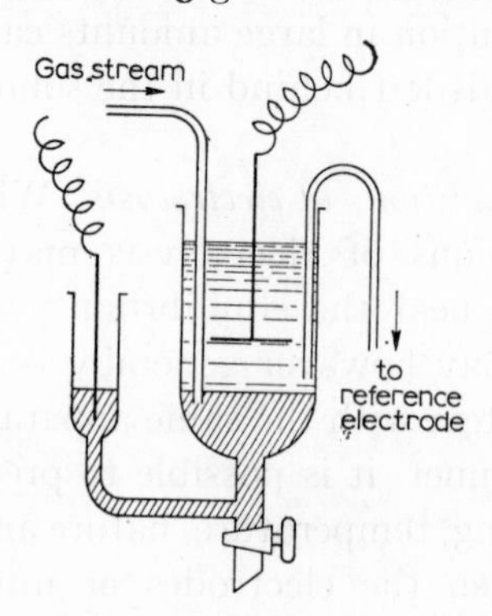

Fig. 65. Electrolysis with a mercury cathode.

In addition, the elements to be determined are occasionally deposited on to the mercury and are then recovered by distilling the latter, or redissolved again by a reoxidation process which may take place at a controlled potential.

Since the overvoltage of hydrogen on mercury is high, quite difficult reductions may be performed without the copious evolution of hydrogen. In addition, the formation of the amalgam enables certain metals, which would not otherwise be deposited, to come out. The separations are rapid, as is illustrated by the fact that with a conventional apparatus and a current of 4 amperes, 1 g of copper or ion is deposited in about an hour.

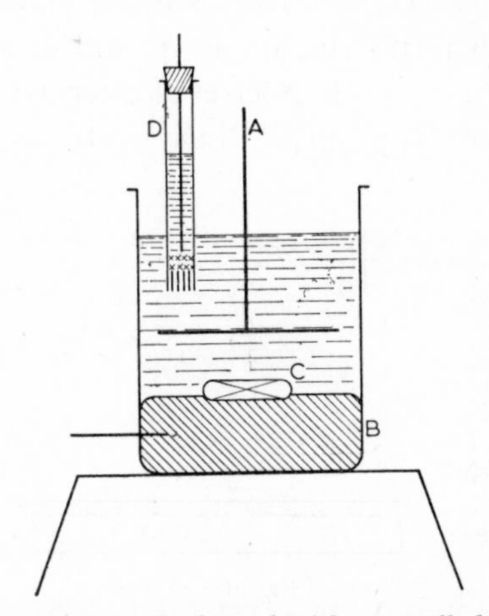

Fig. 66. Electrolysis with a mercury cathode and with controlled potential. A: platinum plate, B: mercury electrode, C: magnetic stirrer, D: reference electrode.
(after J. J. LINGANE, *Anal. Chim. Acta*, 2 (1948) 595.

The following metals may be deposited from 0.1 N H_2SO_4, with no control over the potential: Fe, Cu, Ni, Co, Zn, Ge, Ag, Cd, In, Sn, Cr, Mo, Pb, Bi, Se, Te, Os, Tl, Hg, Au, Pt, Ir, Rh, and Pd; Mn is difficult to deposit. Ru, As, and Sb are not deposited quantitatively. Al, B, Be, Ta, Nb, W, the rare-earth metals, Ti, Zr, Th, U, V, and Pu remain in solution. Certain of the latter elements are brought to the lower state of oxidation, for example, Ti(IV) → Ti(III), U(VI) → U(III).

Selectivity of the method may be increased by controlling the potential. It is possible to predict the course of a separation with the aid of polarographic curves obtained with a dropping mercury electrode. When two elements possess half-wave potentials which differ by at least 250 to 300 mV, they may be separated quantitatively by selecting an appropriate potential on the diffusion plateau of the more reducible element.

(iii) Chemical means used to effect a separation. Highly selective separations may be performed by chemical means, incorporating the variation of the pH, formation of complexes and so on.

(3) Electrography

In electrography, a metal or a conducting mineral is attacked anodically, and the ions liberated are identified with the aid of appropriate reagents. Although the method is basically qualitative, semiquantitative results may be obtained if the procedure is carried out under well-defined conditions.

(i) Principles of operation. The cell used is made up of the solid to be studied (metal or mineral) which forms the anode A, and of a cathode C, consisting, for example, of an aluminium block. A sheet of filter paper, B, impregnated with a solution of an electrolyte (KCl, NH_4Cl, $NaNO_3$, etc.) is interposed between the two electrodes (Fig. 67).

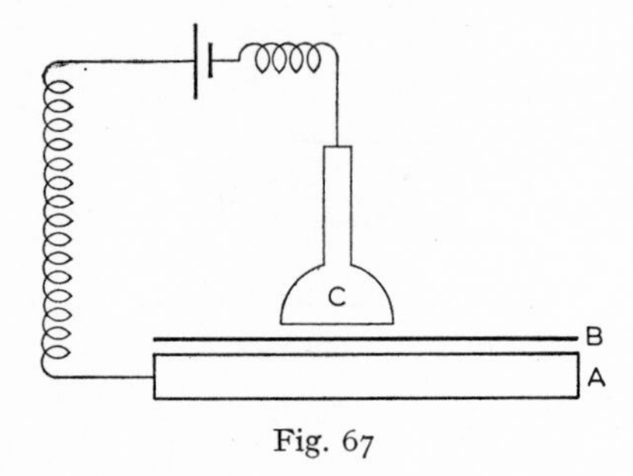

Fig. 67

An appreciably high continuous voltage is applied between A and C for several seconds. The anode A is attacked and the ions formed pass into the electrolyte solution with which the paper is impregnated. The paper is then 'developed' with the aid of suitable reagents to identify the ions formed.

For a quantitative determination, the experimental conditions (time, etc.) must be fixed exactly, and the intensity of the spot obtained must be compared with those of a series of standards prepared under identical conditions.

(ii) Apparatus. The apparatus used may be extremely simple; the source of continuous current may even be a pocket torch battery.

Commercial instruments—'electrographs'—incorporate voltage and current intensity controls, and may include a chronometer which is used to time the electrolysis. The solid plate anode is brought into contact with the cathode by a suitable mechanical arrangement.

(iii) Field of application. The method is very simple, and very rapid, and moreover allows an object to be tested with no apparent deterioration. It can be used to determine the distribution of impurities, for which purpose it is preferable to replace the filter paper by a paper coated with gelatin. In this way, inclusions, discontinuities in a coat of paint or an electrolyte coating, etc., may be studied.

The sensitivity of the method may be extremely high; for example, 0.1 p.p.m. of copper may be detected in antimony.

References and bibliography

Books

(1) A. SCHLEICHER, *Elektroanalytische Schnellmethoden*, F. Enke, Suttgart, 1947.
J. J. LINGANE, *Electroanalytical Chemistry*, 2nd. ed., Interscience, New York, 1958.
(4) G. CHARLOT, J. BADOZ-LAMBLING AND B. TRÉMILLON, *Les réactions électrochimiques. Les méthodes électrochimiques d'analyse*, Masson et Cie, Paris, 1959; idem, *Electrochemical Reactions. The Electrochemical Methods of Analysis*, Elsevier, Amsterdam, New York, 1962.

Reviews

S. E. Q. ASHLEY, *Anal. Chem.*, 21 (1949) 70; 22 (1950) 1379; 24 (1952) 91.
D. D. DE FORD, *Anal. Chem.*, 26 (1954) 135; 28 (1956) 660.

Ultramicroelectrolysis

L. B. ROGERS, *Anal. Chem.*, 22 (1950) 1386.
M. HAÏSSINSKY, *Électrochimie des substances radioactives et des solutions extrêmement diluées*, Hermann, Paris, 1946.

MERCURY CATHODE

F. T. RABBITS, *Anal. Chem.*, 20 (1948) 181.
J. CORIOU, J. GUÉRON, H. HÉRING AND P. LÉVÊQUE, *J. Chim. Phys.*, 48 (1951) 55.
H. AND S. BOZON, *Bull. Soc. Chim.*, 18 (1951) 917.
J. J. LINGANE, *Electroanalytical Chemistry*, 2nd. ed., Interscience, New York, 1958.

Reviews

W. BÖTTGER, *Physikalische Methoden der analytischen Chemie*, Akad. Verlag, Leipzig, vol. II, p. 142.
J. A. MAXWELL AND R. P. GRAHAM, *Chem. Rev.*, 46 (1950) 4171.

INTERNAL ELECTROLYSIS

A. SCHLEICHER, *Z. Anal. Chem.*, 26 (1944) 412.
A. SCHLEICHER AND T. TODOROFF, *Z. Elektrochem.*, 50 (1944) 2
A. SCHLEICHER AND O. SCHLÖSSER, *Z. Anal. Chem.*, 130 (1949) 1.

Applications

J. J. LINGANE, *Electroanalytical Chemistry*, 2nd. ed., Interscience, New York, 1958, p. 443.

Review

S. E. Q. ASHLEY, *Anal. Chem.*, 24 (1951) 91.
(2) A. LIPSCHINSKY, *Zh. Analit. Khim.*, 12 (1957) 83.
(3) S. ISHIMARU AND S. MIZOGUCHI, *J. Chem. Soc. Japan*, 73 (1952) 267.

ELECTROGRAPHY

H. W. HERMANCE AND H. V. WADLOW, *Electrography and Electro-spot testing. Physical Methods in Chemical Analysis*, vol. II, Academic Press, New York, 1951.
H. W. HERMANCE AND H. V. WADLOW, A.S.T.M. Special Report no. 98. *Symposium on Rapid Methods for the Idetiufication of Metals*, A.S.T.M., Philadelphia, 1949.

Reviews

S. E. Q. ASHLEY, *Anal. Chem.*, 24 (1952) 93.
P. R. MONK, *Analyst*, 78 (1953) 141.
R. A. IRKOVSKY, *Zavodsk. Lab.*, 22 (1956) 898.

Part II

Estimation of the Principal Elements

(Selected Methods)

Aluminium

Al = 26.97

(1) Separation of trace amounts

(i) The dry method

After fusion with carbonates or caustic soda and extraction with water, even traces of Al(III) pass into solution. The metal may in this way be separated from Fe(III) and Ti(IV).

cf. T. D. PARKS AND L. LYKKEN, *Anal. Chem.*, 20 (1948) 1102.

(ii) Precipitation of the hydroxide

This can be precipitated by ammonia. Entraining agent: pure Ti(IV). Fe^{3+} can be complexed by thioglycollic acid.

(iii) Precipitation of the phosphate

Entraining agent: Fe(III). Al may be separated from the alkaline-earth metals in this way.

cf. J. CHOLAK, D. M. HUBBARD AND R. V. STORY, *Ind. Eng. Chem., Anal. Ed.*, 15 (1943) 57.

(iv) Extraction with cupferron

Fe(III), Ti(IV), Zr(IV) and other elements, may be extracted from 4 *N* (1/9) sulphuric acid by chloroform while Al(III) remains in the aqueous phase.

cf. N. STRAFFORD AND P. F. WYATT, *Analyst*, 72 (1947) 54.
H. G. SHORT, *Analyst*, 75 (1950) 420.
H. THALER AND F. H. MÜHLBERGER, *Z. Anal. Chem.*, 144 (1955) 241.

Extraction of Ti(IV)
J. A. CORBETT, *Analyst*, 78 (1953) 20.

(v) Precipitation and extraction of the oxinate

This is one of the most useful methods. Under suitable conditions, the separation is practically specific. A particularly effective separation from Be(II) may be effected at about pH 5.

In an ammoniacal medium, Al may be separated from P(V), As(V), B(III), Cr(VI), W(VI), etc., and if CN^- is added, from Zn(II), Cu(II), etc. as well.

In the presence of hydrogen peroxide in an ammoniacal medium, Al may be separated from V(V) and V(IV), Mo(VI), Ti(IV), Ta(V), and Nb(V), but not from Zr(IV), which is colloidal.

In this manner, Al may be separated from Cu(II), Zn(II), Sn(IV), Fe(III), Mn(II), Pb(II), Ni(II), and Co(II) in the presence of CN^-, NH_3, tartrates and EDTA.

cf. D. A. DETMAR AND H. C. VAN ALLER. *Rec. Trav. Chim.*, 75 (1956) 1429.
C. REUTEL, *Metall u. Erz*, 38 (1951) 170.
R. C. CHIRNSIDE, C. F. PRITCHARD AND H. P. ROOKSBY, *Analyst*, 66 (1941) 399.
J. L. KASSNER AND M. A. OZIER, *Anal. Chem.*, 23 (1951) 1453.
R. H. LINNELL AND F. H. RAAB, *Anal. Chem.*, 33 (1961) 1955.

(vi) Ion-exchange separations

(a) *Anion-exchanger* Al^{3+} does not give anionic complexes with SCN^-, I^- or Cl^- and may thus be separated from a number of elements which themselves form complex anions which may be exchanged (see p. 133).

In 9 *N* HCl, the following are readily exchanged in addition to the usual anions: MnO_4^-, $Cr_2O_7^{2-}$, vanadates, and the anionic chloride complexes of Fe(III), Sn(IV), and Sb(V): this occurs to a lesser extent with: U(VI), Zr(IV), Mo(VI), Zn(II), Sn(II), Sb(III), Fe(II), Cu(II), Cd(II), Co(II), and W(VI). The following are not absorbed: Al(III), Mn(II), Cr(III), Ni(II), Ti(IV), Th(IV), rare earths(III), Be(II), Sb(III), Pb(II), V(IV), and Mg(II).

In hydrofluoric acid, Zr(IV) is extracted, but not Al(III)[1].

Phosphates interfere.

COMPARE

Fe(III) + SCN⁻

H. TEICHER AND L. GORDON, *Anal. Chem.*, 23 (1951) 930; *Anal. Chim. Acta*, 9 (1953) 507.

9 N HCl

A. D. HORTON AND P. F. THOMASON, *Anal. Chem.*, 28 (1956) 1326.
(1) H. FREUND AND F. J. MINER, *Anal. Chem.*, 25 (1953) 564, 567.

(b) *Cation-exchangers.* In the presence of a slight excess of EDTA, any substance which is not complexed or insufficiently complexed is extracted: Ti(IV), Mn(II), Ca^{2+} and Mg^{2+}. AlY^- passes through the column.

cf. C. CIMERMAN, A. ALON AND J. MASHALL, *Talanta*, 1 (1958) 314.

(vii) Electrolysis with a mercury cathode

This is the most general method of separation. Under the usual conditions, 10–100 micrograms of Al(III) are present. The separation is carried out in 8 *N* 1/4) sulphuric acid at about 3–5 amperes, using the apparatus described on

p. 156. In this way, 1 g of Cu or Fe may be separated in 1 hour, and 1 g of Sn, Sb, Pb, or Zn in 2–3 hours. Al(III), Be(II), V, the rare-earth metals, the alkali metals, the alkaline-earth metals, etc. remain in solution, together with a little manganese.

About 2 micrograms of iron may remain in solution[1].

cf. P. ROCQUET, *Rev. Met. (Paris)*, 40 (1943) 276.
F. P. PETERS, *Chemist-Analyst*, 24 (1935) 4.
(1) N. STRAFFORD AND P. F. WYATT, *Analyst*, 68 (1943) 319.

(viii) Standard solutions

1 g of pure metal is dissolved in a suitable acid. If during solution the metal is put in contact with Pt, the reaction is accelerated. The solution is finally made up to 1 litre.

Alternatively, 1 g of potash alum of guaranteed purity ($K_2SO_4.Al_2(SO_4)_3.24\,H_2O$) may be dissolved in 1 litre of water.

$$Al_2O_3/Al_2(SO_4)_3 \cdot K_2SO_4 \cdot 24H_2O = 0.1076$$

$$2Al/Al_2(SO_4)_3 \cdot K_2SO_4 \cdot 24H_2O = 0.0569$$

$$Al_2O_3/2Al = 1.890$$

(2) Colorimetry

Provided a suitable pH is employed, a large number of colouring agents, among which aluminon may be particularly distinguished, combine with Al(III) to give colorations which are due not only to definite compounds but often also to adsorption compounds (lakes). The corresponding colorimetric determinations are not very precise since the results vary with the time, the grain size, etc. The pH is a very important factor. Many other substances, in particular Fe(III), give analogous colorations and in view of this, they should be separated out or complexed.

There are two other compounds which do not possess these disadvantages:

(1) Aluminium oxinate in chloroform solution: this also enables the performance of a very good separation of Al(III) in a complexing medium.

(2) Eriochrome cyanine: this is very sensitive.

NOTE. In many experiments, the results are spoiled by contamination with aluminium derived from the attack of glass vessels. Further, numerous reagents contain Al(III).

(i) Oxinate

Aluminium oxinate is yellow when dissolved in chloroform, carbon tetrachloride, or benzene. The coloration is not very appreciable to the eye, but the absorption is high in violet and ultraviolet light.

Sensitivity: $\varepsilon = 6{,}700$ at 390 mμ and 80,000 at 260 mμ.

(ii) Interfering ions

Separation by extraction of the oxinate. Aluminium oxinate can be extracted quantitatively at pH 4.5–5.0 (acetic buffer). In this way, aluminium may be separated from a number of elements, including Be(II), Th(IV) etc.

More specific separation may be obtained by operating in an ammoniacal medium at pH 9 in the presence of tartrate, cyanide, and hydrogen peroxide. Separation from the following may thus be achieved: Cu(II), Co(II), Ni(II), Zn(II), Cd(II), Fe(III), Ti(IV), V(V), V(IV), U(VI), Mn(II), Cr(III), Mo(VI) Sn(IV), and Ag(I). Less than 10 mg of Zr(IV) or Nb(V) does not interfere.

F^- interferes, but small amounts may be complexed by the addition of Be(II), provided that extraction is carried out at a pH of 4.5–5.0.

It is also possible to work in the presence of EDTA and CN^- at pH 8.5–9.0.[1] Under these conditions, the alkaline-earth and rare-earth metals do not interfere. On the other hand, Bi(III), Ga(III), In(III), Nb(V), Ta(V), Sb(III), (V), Ti(IV), U(VI), V(VI), Zr(IV), and a little Be(II), are extracted.

REAGENTS

Oxine, 2 g in 100 ml of chloroform;
Potassium cyanide, 13% in water;
Sodium sulphite, 20% in water;
Tartaric acid, 10% in water;
Hydrogen peroxide, 10-volume;
Standard solution of Al(III): 1 g of 99.9% aluminium dissolved in hydrochloric acid.

OPERATING PROCEDURE

2 ml of tartaric acid and 1 ml of hydrogen peroxide are added to 25 ml of a weakly acidic solution containing 10–150 μg of Al(III). The solution is allowed to stand for 5 minutes, and 5 ml of sulphite are then added. After standing for a further 3 minutes, 10 ml of cyanide are added, and the solution is heated to about 70–80° and then cooled to 25–30°C. Finally, 2 g of ammonium nitrate are added, and the pH is adjusted to 8.9 ± 0.3 with ammonia or HCl. The solution is then transferred to a separating funnel, 5 ml of oxine are added, and the funnel is shaken for 2 minutes. The solution is allowed to settle, and the organic phase is collected in a measuring flask. The extraction is repeated three times, the solution being made up to 50 ml with chloroform.

Finally, a colorimetric estimation is performed at 389 mμ in comparison with a blank.

NOTES

(1) In the presence of large amounts of Nb(V), U(VI) or Ti(IV) (> 10 mg), an excess of hydrogen peroxide should be added.

(2) Fluorescence of the oxinates may also be used. This method is more specific and more sensitive, but less precise.

COMPARE

J. L. KASSNER AND M. A. OZIER, *Anal. Chem.*, 23 (1951) 1453.
W. SPRAIN AND C. V. BANKS, *Anal. Chim. Acta*, 6 (1952) 363.
C. H. R. GENTRY AND L. G. SHERRINGTON, *Analyst*, 71 (1946) 432.
K. MOTOJIMA AND H. HASHITANI, *Bull. Chem. Soc. Japan*, 29 (1956) 458.

In steels

A. CLAASSEN, L. BASTINGS AND J. VISSER, *Anal. Chim. Acta*, 10 (1954) 373.
J. L. KASSNER AND M. A. OZIER, *Anal. Chem.*, 23 (1951) 1453.
H. SPECKER, M. KUCHTNER AND H. HARTKAMP, *Z. Anal. Chem.*, 142 (1954) 166.

In thorium

D. W. MARGERUM, W. SPRAIN AND C. V. BANKS, *Anal. Chem.*, 25 (1953) 249.
G. GOLDSTEIN, D. L. MANNING AND O. MENIS, *Talanta*, 2 (1959) 52.

In phosphates

F. S. GRIMALDI AND H. LEVINE, *U.S. Geol. Survey.*, Bull no. 992 (1953) 39.

In rocks

J. P. RILEY, *Anal. Chim. Acta*, 19 (1958) 413.
(1) A. CLAASSEN, L. BASTINGS AND J. VISSER, *Anal. Chim. Acta*, 10 (1954) 373.

(iii) With eriochrome cyanine R

A red-violet compound is obtained.

Sensitivity. Molar extinction coefficient at pH 5.4, at 530 mμ: $\varepsilon \sim 40{,}000$.

Accuracy. $\pm 2\%$ for 10 μg.

Interfering ions. A large number of elements give the same reaction, but the choice of the concentration of the reagent, the fixing of the pH at about 6.0, and the addition of thioglycollic acid as a reducing and complexing agent permit 5 μg of Al(III) to be estimated in the presence of at least 750 μg of: W, Th, Ni, Co, Ce, Sb, Cu, Mo, Cr, Sn, Cd, Bi, Pb, Zn, Nb, Ta, Ti, Fe, U, and Mn.

Only V, Zr, and Be interfere. If these elements are present in small amounts, corrections can be made: only vanadium gives a coloration in the presence of F^-, and only Zr and Be give a coloration in the presence of EDTA. It is therefore possible to determine the absorption due to these elements in separate tests.

REAGENTS

Eriochrome cyanine R: 750 mg of the reagent are dissolved in 200 ml of water, and 25 g of sodium chloride, 25 g of ammonium nitrate, and finally

2 ml of concentrated nitric acid are added. The solution is made up to 1 litre.
Acetic buffer: 320 g of ammonium acetate and 5 ml of acetic acid are made up to 1 litre. The pH of the solution is checked with a pH-meter and should be 5.9–6.0.
Sodium thioglycollate: 1.5 g of thioglycollate are dissolved in 450 ml of water, and 50 ml of alcohol added. This solution should be renewed every 3 days.
EDTA: 1 g of the disodium salt of ethylenediaminetetraacetic acid is dissolved in 950 ml of water and 50 ml of alcohol added.
Sodium fluoride: 24 g of sodium fluoride are dissolved in 1 litre of water.
Methyl red: 0.1% in alcohol.
Standard solution of aluminium: 1 g of 99.9% aluminium is dissolved in 25 ml of 6 *N* (1/2) hydrochloric acid and the solution made up to 1000 ml. By successive dilutions, a solution of 1 μg of aluminium per ml is prepared.
3 N (1/4) ammonia.

OPERATING PROCEDURE. Two 10 ml portions of a solution containing 5–30 μg of aluminium are each transferred into a 50 ml flask: 1 drop of methyl red is added, and the solution neutralized with ammonia. The solution is then made just acid with dilute (0.5 *N*) hydrochloric acid.

25 ml of thioglycollate are added to each of the two samples. 1 ml of EDTA is introduced into one of them, and then 5 ml of Eriochrome cyanine and 5 ml of acetate buffer are added to both. The solutions are made up to 50 ml and their optical density determined at 535 mμ. The reference cell contains a solution in which the aluminium is masked by the EDTA. The difference between the two readings is the optical density due to the aluminium.

If necessary, the vanadium present may be estimated in another sample, to which 2 ml of sodium fluoride have been added.

COMPARE

W. F. THRUN, *Anal. Chem.*, 20 (1948) 1117.
B. J. MACNULTY, G. J. HUNTER AND D. G. BARRETT, *Anal. Chim. Acta*, 14 (1956) 368.
U. T. HILL, *Anal. Chem.*, 28 (1956) 1419.
S. HENRI AND P. HANISET, *Ind. Chim. Belge*, 27 (1962) 24.

In ferrous alloys

K. STUDLAR AND V. EICHLER, *Chemist-Analyst*, 51 (1962) 68.

In copper alloys

H. POHL, *Z. Anal. Chem.*, 133 (1951) 322.
M. C. STEELE AND L. G. ENGLAND, *Anal. Chim. Acta*, 16 (1957) 148.

In zinc

L. C. IKENBERRY AND A. THOMAS, *Anal. Chem.*, 23 (1951) 1806.

(iv) Nephelometry of the cupferrate

cf. H. OSTERTAG AND Y. CAPPELLIEZ, *Compt. Rend.*, 246 (1958) 1550.

Antimony

Sb = 121.8

(1) Separation of trace amounts

(i) Precipitation of the sulphide

This is still sometimes performed in a tartaric medium, using Cu(II) as the entraining agent.

cf. H. ONISHI AND E. B. SANDELL, *Anal. Chim. Acta*, 11 (1954) 444.

(ii) Precipitation of the hydroxide

The medium used is slightly acidic: permanganate and a manganese salt are added to the solution. The hydrated manganese dioxide formed entrains the antimony.

cf. C. L. LUKE, *Ind. Eng. Chem., Anal. Ed.*, 15 (1943) 526.

(iii) Extraction of Sb(V) from a chloride medium

In sufficiently concentrated hydrochloric acid, Sb(V) is extracted by numerous solvents (ethers, esters) in the form of a complex (see Colorimetry, p. 169).

(iv) Extraction of the iodide

Sb(III) is extracted in the form of SbI_3 by benzene (see Colorimetry, p. 170).

(v) Internal electrolysis

cf. Y. Y. LURE, E. N. TAL AND L. R. FLIGELMAN, *Zavodsk. Lab.*, 8 (1939) 1222.

(vi) Standard solutions

0.1 g of '99.9 powdered' antimony is dissolved in 25 ml of hot, concentrated sulphuric acid. The solution is made up to 1000 ml with 3 *N* (1/12) H_2SO_4. More dilute solutions are also prepared in 4 *N* H_2SO_4.

For colorimetry, it is also possible to dissolve 99–100% Sb_2O_3.

$$2Sb/Sb_2O_3 = 0.836$$

(2) Colorimetry

(i) Using Rhodamine B

(1) In fairly concentrated hydrochloric acid in the presence of Rhodamine B, Sb(V) gives a violet-red compound RH $SbCl_6$ which can be extracted, in particular, by *iso*-propyl ether and benzene.

Sensitivity. Coefficient of molar extinction $\varepsilon \sim 40{,}000$ at 545 mμ in *iso*-propyl ether.

Interfering ions. Oxidizing agents, such as the nitrate ion, can destroy the dye.

Au(III), Tl(III), Fe(III), and Ga(III) give analogous reactions. W(VI) is precipitated.

(2) Alternatively, Sb(V) is first separated by extraction with *iso*-propyl ether from 1–2 *N* HCl. Colorimetry is then carried out directly in the solvent after the addition of reagent, the excess of which remains in aqueous solution[1]. The operating procedure for this method will be described below.

Fe(III), which interferes, may be reduced by hydroxylamine.

In this way it is possible to estimate 2 μg of antimony in the presence of 30 mg of iron. Tl(III), As(III), and Au(III) interfere when present in amounts exceeding 250 μg.

Oxidation of Sb(III) to Sb(V). In general, Sb is initially present in its trivalent form, which may be oxidized by ceric salts.

Oxidation can only take place in sufficiently concentrated hydrochloric acid – 6 *M* HCl. The solutions often contain Sb(IV) and since this is only oxidized very slowly under these conditions, it should be reduced to Sb(III) by sulphurous acid before oxidation. The excess of oxidizing agent is subsequently removed by reaction with additions of hydroxylamine hydrochloride.

The compositions of Sb(V) solutions change very rapidly on standing, doubtless due to condensation.

REAGENTS

Sodium sulphite, 1%;

M hydrochloric acid: 10 ml of conc. HCl + 100 ml of water;

Ceric sulphate: 3.3 g of anhydrous ceric sulphate are dissolved in 100 ml of 0.5 *M* (3/100) sulphuric acid;

Hydroxylamine hydrochloride, 1%;

Washing solution: 1 g of hydroxylamine hydrochloride in 100 ml of *M* hydrochloric acid;

Rhodamine B: 200 mg in 100 ml of *M* hydrochloric acid;

Iso-propyl ether saturated with acid by shaking with M HCl;

Standard solution of antimony: A quantity of Sb_2O_3 of guaranteed purity is weighed out and dissolved in 100 ml of 6 *N* (1/2) hydrochloric acid. The solution is made up to 1000 ml with *N* hydrochloric acid. Under these conditions, 1.197 g of Sb_2O_3 corresponds to 1000 μg of antimony per ml. This is diluted with *N* HCl to obtain 2 μg/ml.

OPERATING PROCEDURE

Oxidation of Sb(III). 10 ml of concentrated hydrochloric acid and 2 ml of sulphite are added to 10 ml of a solution containing 5–50 μg of antimony, and the mixture is shaken with 3 ml of a ceric salt. 10 drops of hydroxylamine solution are then added, the solution is further agitated, and allowed to stand for one minute.

Separation of Sb(V). The solution as prepared above is transferred to a separating funnel with 60 ml of water and 5 ml of *iso*-propyl ether, and is shaken for 30 seconds. The aqueous phase is separated, 2 ml of the washing solution are added to the organic phase, and the funnel shaken for 1–2 seconds: the aqueous phase is then drawn off. 2 ml of *M* hydrochloric acid are added to the solvent, the funnel is shaken for a further few seconds and the aqueous phase separated.

Colorimetry. 2 ml of Rhodamine are added to the solvent and shaken for 10 seconds. The solvent phase is then transferred into a 25 ml measuring flask, made up to volume with the solvent, and estimated colorimetrically at 550 mμ.

COMPARE

W. Nielsch and G. Boltz, *Z. Anal. Chem.*, 143 (1954) 264.
R. W. Ramette and E. B. Sandell, *Anal. Chim. Acta*, 13 (1955) 455.
B. J. Macnulty and L. D. Wollard, *Anal. Chim. Acta*, 13 (1955) 64.
(1) F. N. Ward and H. W. Lakin, *Anal. Chem.*, 26 (1954) 1168.

Extraction of Sb(V)

G. K. Schweitzer and L. E. Storms, *Anal. Chim. Acta*, 19 (1958) 154.
S. S. M. A. Khorasani and M. H. Khundkar, *Anal. Chim. Acta*, 21 (1959) 24.

In lead

C. L. Luke, *Anal. Chem.*, 31 (1959) 1680.

In rocks

F. N. Ward and H. W. Lakin, *Anal. Chem.*, 26 (1954) 1168.
H. Onishi and E. B. Sandell, *Anal. Chim. Acta*, 11 (1954) 444.

In tin and alloys

W. C. Coppins and J. W. Price, *Metallurgia*, 53 (1956) 183.

In biochemistry

T. H. Maren, *Anal. Chem.*, 19 (1947) 487.

With methyl violet (Reaction analogous to the preceding one)

$\varepsilon \sim$ 50,000 at 600 mμ.

H. Gotô and Y. Kakita, *Sci. Rept. Res. Inst. Tohoku, Univ. Ser., A*. 4, (1952) 589; 8 (1956) 243.

In white alloys

M. Jean, *Anal. Chim. Acta*, 7 (1952) 462; 11 (1954) 82.

In Zn and ZnO

R. E. VAN AMAN, F. D. HOLLIBAUGH AND J. H. KANZELMEYER, *Anal. Chem.*, 31 (1959) 1783.

(ii) From the iodide. Acid solutions of Sb(III) or Sb(V) give yellow complexes with I^-. The procedure is carried out in 2.2–3.6 N H_2SO_4. The concentration of the iodide should also be fixed at 6–8% of KI. The compound may be extracted with benzene.

Accuracy of ± 2% may be attained.

Sensitivity. Molar extinction coefficient: $\varepsilon \sim 4{,}000$ at 425 mμ. There is a second maximum at 330 mμ with $\varepsilon = 32{,}000$. At 425 mμ however, the interference due to Hg(II) and Pb(II) is much less pronounced.

Bi(III) gives an analogous coloration, and the two elements may therefore be estimated simultaneously (see Bismuth).

Tl(I) and W(VI) interfere. Oxidizing agents and even atmospheric oxygen produce free iodine in the solution; this may be removed by the addition of a reducing agent, for example ascorbic acid. Sulphites give a pale yellow coloration. Cu(II) can be complexed by thiourea.

10 to 20 ppm of As(III), Hg(II), Fe(III), Cu(II), Pb(II), and W(VI) do not interfere at 425 mμ.

REAGENTS

Potassium iodide and ascorbic acid: 140 g of potassium iodide and 10 g of ascorbic acid are dissolved in 1 l of water.

9 N (1/4) sulphuric acid.

OPERATING PROCEDURE

25 ml of the iodide solution are added to 10 ml of a solution containing 10–100 μg of antimony (when working at 330 mμ) or 200 to 1000 μg (when working at 425 mμ). The solution is made up to 50 ml with sulphuric acid, and measured colorimetrically at the desired wavelength.

COMPARE

E. W. McCHESNEY, *Ind. Eng. Chem., Anal. Ed.*, 18 (1946) 146.

Extraction

R. W. RAMETTE, *Anal. Chem.*, 30 (1958) 1158.

At 330 mμ

A. ELKIND, K. H. GAYER AND D. F. BOLTZ, *Anal. Chem.*, 25 (1953) 1744.

In aluminium

A. L. A. R. *Modern Methods for the Analysis of Aluminium Alloys*, Chapman and Hall, London, 1949.

In Cu-Sn alloys

A. C. HOLLER, *Anal. Chem.*, 19 (1947) 353.

In cast iron

R. C. ROONEY, *Analyst*, 82 (1957) 619.

Arsenic

As = 74.9

(1) Separation of trace amounts

NOTE. In several instances, arsenic may be lost by volatilization, as for example during the treatment of arsenic-bearing compounds with fused sodium carbonate, or during the boiling of hydrochloric acid solutions of As(III) or even As(V).

(i) Distillation of $AsCl_3$

$AsCl_3$ boils at 130°C, but begins to pass over appreciably at 110°. The distillation is carried out from concentrated hydrochloric acid, in the presence of a reducing agent.

$AsBr_3$ exhibits similar properties, and hydrobromic acid may sometimes be used instead of hydrochloric.

As little as 1 μg of arsenic can be recovered to within 5%.

Interfering compounds. Distillation is generally carried out in the presence of hydrazine; As(V) is then reduced.

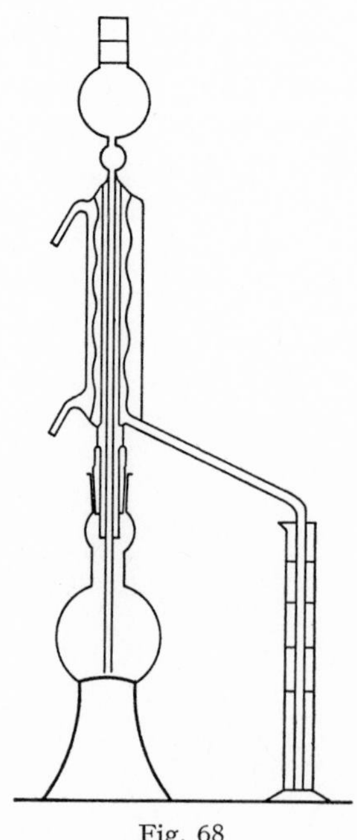

Fig. 68

$GeCl_4$, which boils at 86°, comes over at the same time.

$SnCl_4$, which boils at 115°, can be kept back by carrying out the distillation in the presence of phosphoric acid. The addition of phosphoric acid is also recommended when W(VI) is present.

$SbCl_3$ boils at 220° but begins to pass over appreciably at 107°. When this compound is present, the distillation temperature should therefore be controlled carefully; it should not generally exceed 107°. When large amounts of phosphoric acid are present, traces of phosphoric acid pass over into the distillate, and a second distillation should therefore be carried out.

Oxidizing agents, and in particular the nitrate ion, interfere with the distillation. Se(IV) distils over.

OPERATING PROCEDURE

The process is carried out in a fractionating apparatus with ground-glass joints.

2 ml of hydrobromic acid, 10 ml of concentrated hydrochloric acid, and 0.3 g of hydrazine sulphate are added to 10–15 ml of the solution. A slow current of carbon dioxide or nitrogen is passed through the apparatus and the mixture is heated to a temperature of 107°, which should not be exceeded. Under these conditions, about 10 ml of distillate are recovered.

COMPARE

E. Schaaf, *Z. Anal. Chem.*, 126 (1943) 298.
C. J. Rodden, *J. Res. Natl. Bur. Std.*, 24 (1940) 7.
D. M. Hubbard, *Ind. Eng. Chem., Anal. Ed.*, 13 (1941) 915.
H. J. Magnuson and E. B. Watson, *Ind. Eng. Chem., Anal. Ed.*, 16 (1944) 339.
J. C. Bartlet, M. Wood and R. A. Chapman, *Anal. Chem.*, 24 (1952) 1821.

(ii) In the form of AsH_3

Arsine, AsH_3, is a gas the liberation of which enables trace amounts of arsenic to be separated (see Colorimetry).

(iii) Extraction of $AsCl_3$

$AsCl_3$ may be extracted from 8–9 *N* HCl with carbon tetrachloride and then reclaimed into a dilute HCl medium. Although Ge(IV) is extracted at the same time, the two may be separated by oxidizing the arsenic to the pentavalent state, when only Ge(IV) is extracted; the arsenic may subsequently be extracted in the form of As(III). Se(IV) also comes through.

COMPARE

W. Fischer, W. Harre, W. Freese and K. G. Hackstein, *Angew. Chem.*, 66 (1954) 165.
M. C. Beard and L. A. Lyerly, *Anal. Chem.*, 33 (1961) 1780.

Theory

M. Green and J. A. Kafalas, *J. Chem. Phys.*, 22 (1954) 760.
G. O. Brink, P. Kafalas, R. A. Sharp, F. L. Weiss and J. W. Irvine, *J. Am. Chem. Soc.*, 79 (1957) 1303.

(iv) Extraction with xanthate

A complex which can be extracted with carbon tetrachloride is formed.

cf. A. K. Klein and F. A. Vorhes, Jr., *J. Assoc. Offic. Agr. Chem.*, 22 (1939) 121.
K. Sugawara, M. Tanaka and S. Kanomori, *Bull. Chem. Soc., Japan*, 29 (1956) 670.

(v) Extraction with diethylammonium diethyldithiocarbamate

With As(III), a compound is formed which can be extracted by chloroform from 1–10 *N* sulphuric acid. As(V) does not come through.

Sb(III), Sn(II), Cu(II), Hg(II), and Bi(III) are also extracted. Pb(II), Zn(II), Ge(IV) etc. may be rejected. Se precipitates.

cf. N. Strafford, P. F. Wyatt and F. G. Kershaw, *Analyst*, 78 (1953) 624.
S. T. Payne, *Analyst*, 77 (1952) 278.
C. L. Luke and M. E. Campbell, *Anal. Chem.*, 25 (1953) 1588.
P. F. Wyatt, *Analyst*, 80 (1955) 368.
H. Gotô and Y. Kakita, *J. Chem. Soc. Japan, Pure Chem. Sect.*, 77 (1956) 739.

(vi) Extraction of the arsenimolybdate

cf. M. Daniels, *Analyst*, 82 (1957) 133.
S. Yokosuka, *Japan Analyst*, 5 (1956) 395.

(vii) Separation of As(V) by means of ion-exchangers.

As(V), which is present in the state of H_3AsO_4, may be separated from most cations in this manner.

EXAMPLE

Separation from Fe^{3+} and Cu^{2+}.

The arsenic present is first converted to its higher oxidation state if necessary. In this case, a cation-exchanger will retain Fe^{3+} and Cu^{2+}, while H_3AsO_4 passes through.

REAGENT

Amberlite IR 100 AG.

OPERATING PROCEDURE

The exchange occurs between 10 g of resin in the H^+ form, and 40 ml of a solution which should not contain more than 100 mg of the substances to be collected. The HCl acidity should be 0.3 N. After the solution has been passed through the column at the rate of 5–10 ml per minute, the resin should be washed with 60 ml of 0.1 N hydrochloric acid.

The column may be regenerated with 350 ml of 2 N hydrochloric acid and is finally washed with 200 ml of water.

(viii) Standard solutions

About 1 g of As_2O_3 of guaranteed purity are weighed out and dissolved in 25 ml of 2 N (1/5) caustic soda. The solution is neutralized either by bubbling through CO_2 to saturation, or with dilute sulphuric acid until phenolphtalein is just decolorized, and is made up to 1000 ml. The diluted solutions should be prepared afresh every two or three days.

If a solution of As(V) is required, arsenic is oxidized with permanganate before making up to 1000 ml.

$$2As/As_2O_3 = 0.757$$

A 0.1000 N arsenic solution contains 3.75 mg of arsenic per millilitre.

98–100% sodium arsenate $Na_2HAsO_4 \cdot 7H_2O$ is also available.

(2) Colorimetry

It is important to be able to estimate trace amounts of arsenic present in many substances. Among the latter are blood, the skin of fruit, gelatin, tobacco, food colouring agents, phosphoric acid, copper, bone, malt, sugar, water, pyrites, vegetables, fuels and ashes, oils, etc. Although colorimetric estimations of arsenic are sometimes not very accurate, they do however permit the detection of very minute traces of this element.

Paper strip colorimetry is a laborious process, requiring the preparation of standards which are difficult to keep. It is however still frequently used when the method of separation involves AsH_3.

Colorimetry of a preparation of molybdenum blue is more rapid, but numerous ions interfere.

(i) Paper strip colorimetry (Gutzeit's method)

Arsine produced by reduction with zinc and an acid, or alternatively by electrolysis, is passed over a paper impregnated with mercuric chloride, or, better still, mercuric bromide. A yellow-brown spot is produced, which becomes larger and more intense as the amount of arsine is increased. This coloration is due to the formation of the compounds $H(HgBr)_2As$ (yellow), $(HgBr)_3As$ (brown), and Hg_3As_2 (black). A comparison is carried out against a series of standard papers prepared under identical conditions with titrated solutions of arsenious acid.

When protected from light and (in the case of $HgBr_2$) moisture, the standard papers may be kept for several months.

Precautions. The reagents should be particularly pure ('for toxicology') and should therefore be checked by means of a blank test.

The reduction to AsH_3 is only quantitative if a certain number of precautions are observed, in particular with respect to the dimensions of the zinc granules.

A constant temperature should be maintained during all the tests.

The zinc may be activated by the addition of a little stannous chloride.

Limit of sensitivity and accuracy. It is usually possible to estimate 10 to 30 μg to within 5 to 10%. 0.1 to 0.01 μg can be estimated with appropriate instruments.

Interfering compounds. Certain metals which precipitate, *e.g.* Cu, Ni, Co, will interfere; Hg(II), Bi(III), Fe(III), and Se(IV) diminish the sensitivity. 10 mg of Ag(I) and 50 mg of Pb(II) do not interfere if the test is carried out with heating. A small amount of F^- does not interfere.

Phosphorous and hypophosphorous acids give rise to PH_3, which accompanies the $AsH_3 \cdot SO_4^{2-}$ and P(V) do not interfere.

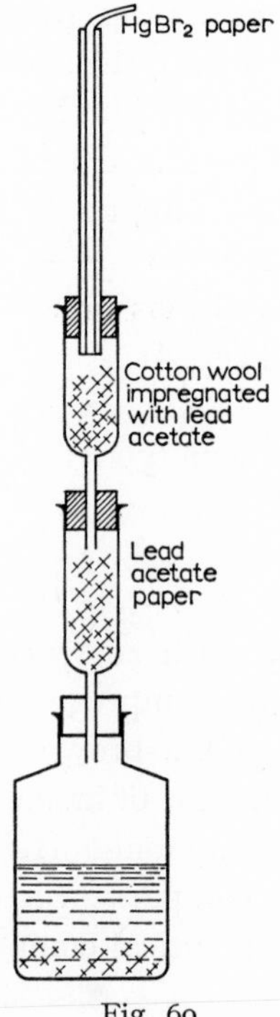

Fig. 69

H_2S, PH_3, SbH_3, and GeH_4, which all appear during the reduction, will interfere. H_2S may be blocked by means of a cotton plug impregnated with a 10% solution of lead acetate. The following solution, which keeps back PH_3, SbH_3 and H_2S at least partially, has also been recommended: 50 g of cuprous chloride in 100 ml of 6 *N* (1/2) hydrochloric acid.

As(V) should be pre-reduced to As(III), for example with stannous chloride in the presence of iodide, in the hot.

Paper. A fine-grained paper is dried at 105°; before use, strips 2.5 mm × 120 mm are impregnated with a 5% solution of mercuric bromide in alcohol. These are dried for 10 minutes and should be used within 2 hours.

Apparatus. A large variety of instruments have been described. The simplest device is illustrated in Figure 69 (A.F.N.O.R. information sheet).

OPERATING PROCEDURE

30 μg of arsenic are dissolved in 20 ml of 8 *N* (1/5) sulphuric acid or 3 *N* (1/4) hydrochloric acid and made up to 40 ml. 4 ml of 20% potassium iodide solution and 3 drops of 40% stannous chloride in concentrated hydrochloric acid are then added, and the solution is heated for 10 minutes. After cooling to 5°C in ice water, 15 g of zinc are added in the form of needles; the solution is left in ice water for 15 minutes and then for 1 hour at room temperature. A blank test is also performed.

cf. Of the many available references, the following may be consulted:
H. GRIFFON AND J. THURET, *Bull. Soc. Chim.*, 5 (1938) 1129.
N. I. GOLDSTONE, *Ind. Eng. Chem., Anal. Ed.*, 18 (1946) 797.
N. C. MARANOVSKI, R. E. SNYDER AND R. O. CLARK, *Anal. Chem.*, 29 (1957) 353.
A. E. HOW, *Ind. Eng. Chem., Anal. Ed.*, 10 (1938) 226.
H. S. SATTERLEE AND G. BLODGETT, *Ind. Eng. Chem., Anal. Ed.*, 16 (1944) 400.

Paper
A. BERTON, *Bull. Soc. Chim.*, 12 (1945) 296.

Electrolytic reduction equipment. In such cases, the need for very pure zinc is obviated.

As little as 0.1 μg may be estimated to within 10%.

cf. A. E. OSTERBERG AND W. S. GREEN, *J. Biol. Chem.*, 155 (1944) 513.
H. GRIFFON AND J. THURET, *Bull. Soc. Chim.*, 5 (1938) 1129.
G. VAN DER KELEN, *Ind. Chim. Belge*, 17 (1952) 119.

(ii) Reduction of the arsenimolybdate

Under similar conditions to the phosphates, arsenates also combine with molybdate to give a complex which can easily be reduced to give a blue colour.

The pH is very critical. If the medium is too acidic the reaction is slow, whereas if the pH is too high the excess of molybdate is reduced. 0.5 *N* sulphuric acid is used. Concentration of the molybdate is an important factor, and should have a fixed value.

The temperature and the duration of heating have little influence after the maximum depth of coloration is attained.

Sensitivity. The molar extinction coefficient, $\varepsilon \sim 25{,}000$ at 840 mμ.

Interfering ions. Phosphates give the same reaction. Interference is also obtained from the ions listed for the instance of P(V) (see Phosphorus); in particular Ge(IV) and Si(IV) give colorations which are, respectively, 10 times and 50 times less sensitive. Zr(IV) precipitates As(V); W(VI), Nb(V), and Ta(V) partially complex the molybdate.

As(III) can be separated by the distillation of $AsCl_3$ or by extracting the phosphomolybdate.

Sb, and Se do not interfere.

REAGENTS

Ammonium molybdate: 1 g in 100 ml of 5 *N* (1/7) sulphuric acid;
Hydrazine sulphate: 0.15 g in 100 ml of water.

OPERATING PROCEDURE

1 ml of molybdate and then 0.4 ml of hydrazine sulphate are added to 10 ml of a solution containing from 10 to 40 μg of As(V). The latter is left for 15 min on the water bath, cooled, and made up to 25 ml. Colorimetry is performed at 840 mμ.

COMPARE

G. R. Kingsley and R. R. Schaffert, *Anal. Chem.*, 23 (1951) 914.
R. J. Evans and S. L. Bandemer, *Anal. Chem.*, 26 (1954) 595.

In rocks

H. Onishi and B. Sandell, *Mikrochim. Acta*, 34 (1953).

In biochemistry

G. R. Kingsley and R. R. Schaffert, *Anal. Chem.*, 23 (1951) 914.
H. J. Magnuson and E. B. Watson, *Ind. Eng. Chem., Anal. Ed.*, 16 (1944) 339.
R. J. Evans and S. L. Bandemer, *Anal. Chem.*, 26 (1954) 595.
A. Dyfverman and R. Bonnischen, *Anal. Chim. Acta*, 23 (1960) 491.

Metals and alloys

C. J. Rodden, *J. Res. Natl. Bur. Std.*, 24 (1940) 7.

In steels

M. JEAN, *Anal. Chim. Acta*, 14 (1956) 172.

In Se

J. F. REED, *Anal. Chem.*, 30 (1958) 1122.

Colorimetry of arsenomolybdic acid after extraction by n-butanol, at 370 mμ. ε = 5,000.

Y. C. WADELIN AND M. G. MELLON, *Analyst*, 77 (1952) 708.

Separation in the state of AsH_3 and then As (Marsh)

A. DYFVERMAN AND R. BONNISCHEN, *Anal. Chim. Acta*, 23 (1960) 491.

(iii) Turbidimetry of As

In copper alloys. Méthodes des Industries de la Fonderie [Methods of the Foundry Industry] (As-I).

Barium

$Ba = 137.4$

Barium is usually estimated by means of flame spectrophotometry, after the interfering materials have been removed.

(i) Separation of trace amounts

Barium is usually separated in the form of the sulphate or chromate, sometimes in the presence of alcohol or acetone to diminish their solubility. The precipitation of $BaSO_4$ in a very dilute acid is almost specific: only Pb(II) and Sr(II) come down as well.
Nephelometry or turbidimetry is then used.

Precipitation of the chromate in a weakly acidic medium may often however require previous separations, since Pb(II) also precipitates. In both cases, traces of Pb(II) can be extracted with dithizone.

(ii) Standard solutions

$BaCl_2 \cdot 2H_2O$ and $Ba(NO_3)_2$ of guaranteed purity are available.

$$Ba/BaCl_2 \cdot 2H_2O = 0.562.$$
$$Ba/Ba(NO_3)_2 = 0.526.$$

Alternatively, barium carbonate may be used.

$$Ba/BaCO_3 = 0.696.$$

Colorimetry

(i) Chromate

If the concentration of barium is high (greater than 20 ppm) barium chromate can be separated. The precipitate is then redissolved and chromium (VI) is estimated by colorimetry: see p. 227.

cf. J. Agterbendos, *Z. Anal. Chem.*, 159 (1957) 202.

(ii) Using o-cresolphthalein complexone. $\varepsilon = 30{,}000$ at 575 mμ.
The other alkaline-earth metals interfere.

cf. F. H. Pollard and J. V. Martin, *Analyst*, 81 (1956) 348.

(iii) Nephelometry or turbidimetry of barium sulphate

Very minute traces cannot be detected by this method. Although not very accurate, the method is however very selective in an acid medium, and in the majority of cases avoids the the necessity for carrying out separations (see Sulphate).

Beryllium

Be = 9.01

(1) Separation of trace amounts

(i) Precipitation of the hydroxide

This is effected by ammonia. Even traces can be separated at about pH 8.5, using Al(III) as the entraining agent. If Fe(III) is present, it may be complexed with thioglycollic acid. Less than 4 μg of Be(II) per litre are left. Ti(IV) also precipitates. Cr(III) can be oxidized beforehand.

The separation can be completed by digestion with fused caustic soda. Traces of Mg(II), Ca(II), Ni(II) and Co(II) may be separated in this way.

cf. W. FISCHER AND J. WERNET, *Angew. Chem.*, 60 (1948) 129.

(ii) Extraction of the oxinates

Many oxinates can be extracted in a slightly acid or neutral media: most of the Be(II) present remains in solution. Fe(III), Ti(IV), Zr(IV), Cu(II), Ni(II), and Zn(II) are extracted, together with Al(III) (see p. 111).

(iii) Extraction with acetylacetone

In the presence of EDTA the method is very selective (see Colorimetry).

(iv) Ion-exchangers in the presence of EDTA

A cation exchanger will collect Be^{2+} from a medium of pH 3–4: most of the other materials present are in the form of anionic complexes: FeY^-, NiY^{2-}, etc.

Interfering compounds. U(VI), Ti(IV), P(V), As(V), and F^-; Ti(IV) can however be complexed by hydrogen peroxide[1]. The resin may initially be in the Na^+ form. Be(II) can be eluted with 3 *N* HCl.

COMPARE

S. BANERJEE, *Thesis*.

(1) M. N. NADKARNI, M. S. VARDE AND V. T. ATHAVALE, *Anal. Chim. Acta*, 16 (1957) 421.

T. Y. TORIBARA AND R. E. SHERMAN, *Anal. Chem.*, 25 (1953) 1596.

(v) Electrolysis with a mercury cathode

This permits the separation of a number of elements: Be(II) remains in solution.

cf. F. W. KLEMPERER AND A. P. MARTIN, *Anal. Chem.*, 22 (1950) 828.

(vi) Standard solutions

These can be prepared from the pure metal, which is dissolved in a dilute acid.

(2) Colorimetry

Numerous colouring agents give rise to lakes with Be(II). Colorimetry with these compounds possesses all the disadvantages mentioned earlier.

Fluorimetry with pentahydroxyflavone is specific: it is very sensitive, but also very inaccurate.

The absorption of the sulphosalicylate using ultraviolet radiation is an accurate method and quite specific in the presence of EDTA.

(i) Spectrophotometry by absorption in the ultraviolet

This is carried out with the aid of sulphosalicylic acid in the presence of EDTA, which strongly complexes numerous ions, including Al(III).

Interfering compounds. Cu(II) (which can be separated electrolytically), Ce (IV), Zr(IV), Fe(III), nitrate, and phosphate.

REAGENTS

Sulphosalicylic acid, 2 g per litre;
Disodium salt of ethylenediaminetetraacetic acid, 16%.

OPERATING PROCEDURE

Exactly 10 ml of sulphosalicylic acid and about 50 ml of EDTA are added to a slightly acid solution containing 0.2 to 1 mg of Be(II) and less than 200 mg of Al(III). The pH is adjusted to about 9.2–10.5 with ammonia and the solution is made up to 200 ml. Absorptiometry is performed at 317 mμ against a blank sample.

cf. H. W. MECK AND C. V. BANKS *Anal. Chem.*, 22 (1950) 1512.

(ii) p-Nitrobenzene-azo-orcinol

Under alkaline conditions, the dye is yellow: in the presence of Be(II) a red-brown lake is formed.

The conditions of alkalinity, concentrations, and time are important. The operations are carried out in a borate buffer solution.

Sensitivity: molar extinction coefficient, $\varepsilon \approx 2{,}000$ at 525 mμ.

Interfering ions. A number of ions give interfering colorations. Hydroxides precipitate and adsorb the reagent. Al(III) does not interfere below 10 mg per millilitre. 1 mg of Zn(II) is equivalent to 13 μg of Be(II). Cu(II) must be separated. Ti(IV) can be complexed with hydrogen peroxide. Small amounts of Cu(II), Ni(II), Fe(III), and Ca(II), can be complexed with EDTA[1].

REAGENTS

p-Nitrobenzene-azo-orcinol: 25 mg in 100 ml of 0.4% caustic soda are shaken for several hours, filtered, and kept in a brown bottle.

Buffer solution: 116 g of citric acid, 58.7 g of anhydrous sodium borate, and 216 g of caustic soda are dissolved and made up to 1 litre.

Complexing solution: EDTA, disodium salt of ethylenediaminetetraacetic acid, 15%.

Hydrogen peroxide, 3% (10 volumes).

OPERATING PROCEDURE

15 ml of solution containing 50 to 600 μg of Be(II) are treated with 15 ml of hydrogen peroxide and 5 ml of EDTA solution. The pH is adjusted to 5.5, and the solution is allowed to stand for 5 minutes. 10 ml of the buffer solution are then added, and the mixture left for a further 5 minutes. Following this, an accurately measured 10 ml portion of the reagent is added, and the volume adjusted to 100 ml. The measurement is performed at 515 mμ.

COMPARE

W. STROSS AND G. H. OSBORN, *J. Soc. Chem. Ind.*, 63 (1944) 249.
A.L.A.R., *Modern Methods for Analysis of Aluminium Alloys*, Chapman and Hall, London, 1949.

Precise colorimetry

J. C. WHITE, A. S. MEYER, JR. AND D. L. MANNING, *Anal. Chem.*, 28 (1956) 956.
(1) F. A. VINCI, *Anal. Chem.*, 25 (1953) 1583.

In beryl

G. H. OSBORN AND W. STROSS, *Metallurgia*, 30 (1944) 3.
J. B. POLLACK, *Analyst*, 81 (1956) 45.

In titanium

L. C. COVINGTON AND M. J. MILES, *Anal. Chem.*, 28 (1956) 1728.

(iii) Eriochrome Cyanine R in the presence of EDTA and CN^-. 512 mμ.

The constituents of steels, Al(III), Cu(II), Ti(IV) etc., do not interfere.

cf. U. T. HILL, *Anal. Chem.*, 30 (1958) 521.

(iv) Fluorimetry with pentahydroxyflavone in an alkaline medium

A very sensitive but very imprecise method.

C. W. SILL, C. P. WILLIS AND J. K. FLYGARE, JR., *Anal. Chem.*, 33 (1961) 1671.

In rocks

E. B. SANDELL, *Anal. Chim. Acta*, 3 (1949) 89.
H. A. LAITINEN AND P. KIVALO, *Anal. Chem.*, 24 (1952) 1467.

In biochemistry

F. W. Klemperer and A. P. Martin, *Anal. Chem.*, 22 (1950) 828.

⩾ 0,0004 μg

C. W. Sill and C. P. Willis, *Anal. Chem.*, 31 (1959) 598.

(v) *Spectrophotometry of the acetylacetonate in the ultraviolet*

Acetylacetone is used after extraction with chloroform. The presence of EDTA eliminates the interfering ions. At 295 mμ, $\varepsilon \sim 32{,}000$.

COMPARE

J. A. Adam, E. Both and J. D. H. Strickland, *Anal. Chim. Acta*, 6 (1952) 462.
I. P. Alimarin and I. M. Gibalo, *Zh. Analit. Khim.*, 11 (1956) 389.

In water

J. R. Merrill, M. Honda and J. R. Arnold, *Anal. Chem.*, 32 (1960) 1420.

Bismuth

Bi = 209.0

(1) Separation of trace amounts

Of the more familiar methods, the following can be used in the case of trace amounts.

(i) Precipitation of the sulphide

Bismuth sulphide may be precipitated with hydrogen sulphide in 0.1–0.3 *N* hydrochloric acid; the solution should stand for 12 hours. Cu(II) is used as the entraining agent.

Alternatively, ammonium sulphide may be used, in the presence of cyanide if necessary.

(ii) Precipitation of the hydroxide

Bi(III) can be precipitated in a weakly acidic media; the hydroxide is entrained with $Mn(OH)_4$.

The solution is neutralized until methyl orange just changes colour, and is then acidified with nitric acid up to 0.02 *N* (about 1/500). After boiling, 5 ml of 20% potassium bromide and 3 ml of 3% potassium permanganate are added. The solution is then boiled until the colour of the permanganate disappears. The operation is repeated on the filtrate.

cf. Y. C. Yao, *Ind. Eng. Chem., Anal. Ed.*, 17 (1945) 114.

$Fe(OH)_3$ entraining agent

A. J. G. Smout and J. L. Smith, *Analyst*, 58 (1933) 475.

(iii) Precipitation by cupferron

This may be used to separate bismuth from Co(II), Pb(II), Zn(II), As(III), Sb(V), and Ag(I).

cf. L. Silverman and M. Shideler, *Anal. Chem.*, 26 (1954) 911.

(iv) Precipitation by phenylarsonic acid

The method is suitable for trace amounts.

A. K. Majumdar and R. N. S. Sarmra, *J. Indian Chem. Soc.*, 26 (1949) 477.

(v) Extraction of Bi(III) from a hydrochloric acid medium

Separation from Pb(II) may be effected with butanol by the use of this method.

cf. S. KALLMANN, *Anal. Chem.*, 23 (1951) 1291.

(vi) Extraction with diethyldithiocarbamate

See Colorimetry.

J. H. THOMPSON AND B. W. PETERS, *Analyst*, 84 (1959) 180.

(vii) Extraction from an iodide medium

See Colorimetry.

H. A. MOTTOLA AND E. B. SANDELL, *Anal. Chim. Acta*, 24 (1961) 301.

(viii) Extraction with dithizone

See Colorimetry.

(ix) Extraction with cupferron

The cupferrate of Bi(III) can be extracted with chloroform at pH 1. Bi(III) passes back into aqueous 2 *N* sulphuric acid. The method is used to separate bismuth from Sb(III); in addition, as little as 1.3 μg of Bi(III) can be separated from 1 g of Pb(II). Sn(IV) is distributed between the two phases.

cf. H. BODE AND G. HENRICH, *Z. Anal. Chem.*, 135 (1952) 98.

(x) Internal electrolysis

Bi(III) can be separated from Pb(II) and from Sn(IV).

cf. *Chemical Analysis of Metals*, American Society for Testing Materials, Philadelphia, 1956.

(xi) Standard solutions

In the absence of the pure metal, $Bi(NO_3)_3 \cdot 5H_2O$, of guaranteed purity, may be used. 0.2–0.3 g are weighed out and dissolved in 50 ml of *N* nitric acid (1/10); the solution is made up to 1000 ml.

$$1.150 \text{ g of } Bi_2O_3 = 1.000 \text{ g of Bi}$$

$$Bi/Bi(NO_3)_3 \cdot 5H_2O = 0.431$$

$$2Bi/Bi_2O_3 = 0.897$$

(2) Colorimetry

One of four important methods may be used. The dithizone method and the iodide method are very sensitive. The thiourea method is also frequently used. The diethyldithiocarbamate method is however the most selective. The choice of one or another of these methods is governed by the particular circumstances.

(i) Iodide complexes

The iodide complexes of Bi(III) are orange coloured, and Beer's law is obeyed in the presence of an excess of I^- ions (concentration of potassium iodide greater than 1%).

The complex is soluble in alcohols, esters, and ketones.

Sensitivity. The molar coefficient $\varepsilon \sim 34{,}000$ at 337 mμ in water.

The coloration is stable for three to four hours. The acidity should be fixed between the limits 1–2 *N* sulphuric acid.

Interfering ions. Oxidizing agents liberate iodine and should be reduced with *e.g.* sulphurous acid. CuI and AgI can be separated by precipitation without loss of Bi(III), but PbI_2 retains Bi(III) and interferes, as does TlI. Large amounts of Cd(II) consume I^- by the formation of complexes. Hg(II) does so to an even greater extent. 1,000 ppm of Fe(III), 100 ppm of Pb(II), 20 ppm of Cu(II) and 400 ppm of As, F^-, and tartrate ions do not interfere.

Pt(IV), Pd(II), Sn(IV) and Sb(III) produce interfering colours, but Sb(III) only interferes above 200 ppm, and the same doubtless applies to Sn(IV).

Sb(III) and Bi(III) can be estimated simultaneously[1].

Cl^- and F^- weaken the coloration.

REAGENTS

Potassium iodide, 10% in water;
Sulphurous acid solution, 5%, freshly prepared;
Hypophosporous acid, 30% in water.

OPERATING PROCEDURE

The initial solution should consist of 10–20 ml, containing from 5 to 50 μg of bismuth (III). The acidity should be adjusted to 1–2 *N* H_2SO_4. Then, 0.1 ml of sulphurous acid, 1 ml of hypophosphorous acid, and 3 ml of iodide are added, and the volume made up to 25 ml. Colorimetry is performed at 465 mμ, or alternatively in the ultraviolet at 337 mμ[2].

The blank tests or the calibration curve should be determined under identical conditions with respect to acidity and the concentration of salts and iodide.

COMPARE

R. C. Sproull and A. O. Gettler, *Ind. Eng. Chem., Anal. Ed.*, 13 (1941) 462.
C. J. W. Wiegand, G. H. Lann and F. V. Kalich, *Ind. Eng. Chem., Anal. Ed.*, 13 (1941) 912.

(1) E. W. Mc Chesney, *Ind. Eng. Chem., Anal. Ed.*, 18 (1946) 146.
(2) N. M. Lisicki and D. F. Boltz, *Anal. Chem.*, 27 (1955) 1722.

In lead

C. J. Hall, *Analyst*, 77 (1952) 318.

In copper

D. T. Englis and B. B. Burnett, *Anal. Chim. Acta*, 13 (1955) 574.

In biochemistry

N. J. Giacomino, *Ind. Eng. Chem., Anal. Ed.*, 17 (1945) 456.

(ii) Absorptiometry in hydrochloric acid in the ultraviolet, in the presence of Pb(II) and Tl(I).

$\varepsilon = 14{,}900$ at 327 mμ.

C. Merritt, Jr., H. M. Hershenson and L. B. Rogers, *Anal. Chem.*, 25 (1953) 572.

In hydrobromic acid

$\varepsilon = 19{,}000$ at 375 mμ.

I. A. Stolyarova, *Zh. Analit. Khim.*, 8 (1953) 270.
N. W. Fletcher and R. Wardle, *Analyst*, 82 (1957) 747.

(iii) Thiourea

With thiourea in a diluted acid medium, Bi(III) gives a yellow complex.

Concentration of the reagent is very critical.

Sensitivity; $\varepsilon = 35{,}000$ at 322 mμ.

The coloration is stable for 1 hour.

Interfering ions. Large amounts of Ag(I), Hg(II), Pb(II), Cu(II), Cd(II) and Sn(IV), precipitate. Sb(III), which gives a coloration, should be complexed with F^-, and Fe(III) must be reduced.

The medium may contain citric and tartaric acids[2].

Large amounts of F^-, Cl^- and Br^- interfere.

REAGENTS

Thiourea, 12% in water;
Perchloric acid, 5 N (1/2).

OPERATING PROCEDURE

10 ml of perchloric acid are added to 10 ml of solution containing 50 to 500 μg of bismuth (III). If necessary Fe(III) may be reduced with a little hydrazine sulphate at the boil. Antimony can be complexed by sodium fluoride (not more

than 50 mg). 25 ml of thiourea are then added, and colorimetry is performed at 322 mμ or at 470 mμ.

COMPARE

W. NIELSCH AND G. BÖLTZ, *Z. Anal. Chem.*, 142 (1954) 321.
(1) N. M. LISICKI AND D. F. BOLTZ, *Anal. Chem.*, 27 (1955) 1722.

Lead alloys
(2) E. ASMUS, *Anal. Chem.*, 142 (1954) 255.
J. H. THOMPSON AND B. W. PETERS, *Analyst*, 84 (1959) 180.

In aluminium
J. H. BARTRAM AND P. J. C. KENT, Light Metals 9 (1946) 229.
A. L. A. R., *Modern Methods for the Analysis of Aluminium Alloys*, Chapman and Hall, London, 1949.
G. NORWITZ, *Metallurgia*, 43 (1951) 56.

In white metals
B. B. BENDIGO, R. K. BELL AND H. A. BRIGHT, *J. Res. Natl. Bur. Std.*, 47 (1951) 252
J. H. THOMPSON AND B. W. PETERS, *Analyst*, 84 (1959) 180.

(iv) Dithizone (separation and estimation)

Bismuth dithizonate, orange in colour, is stable from pH 2 to pH 11, and, can be separated from Zn(II), Cd(II), Cu(II), etc. by extraction in the presence of cyanide at pH 9. Only Pb(II) and Tl(I) interfere. Bi(III) can be separated from Pb(II) by extraction of the bismuth dithizonate at pH 2–3.

As little as 1 μg of Bi can be estimated in this way.

H. A. MOTTOLA AND E. B. SANDELL, *Anal. Chim. Acta*, 24 (1961) 301.

Separation of Bi(III) and Pb(II) at pH 3.
H. FISCHER AND G. LEOPOLDI, *Z. Anal., Chem.*, 119 (1940) 161.

In biochemistry
D. M. HUBBARD, *Anal. Chem.*, 20 (1948) 363.

In copper
YU LIN YAO, *Ind. Eng. Chem., Anal. Ed.*, 17 (1945) 114.

(v) Diethyldithiocarbamate.

In the presence of EDTA and cyanide, the extraction into carbon tetrachloride from ammoniacal solutions is almost specific. A yellow compound is obtained.

Interfering ions. Only Hg(II) and Pb(II) interfere. The absorption maximum for the bismuth compound is located at 370 mμ; by operating at 400 mμ however, the determination of 20 μg of Bi(III) remains unaffected by the presence of 100 mg of Pb(II).

Relatively large amounts of other extractable materials may generally be present without interference, since the Bi(III) complex is very stable.

The coloration is stable for 30 min.

Sensitivity. At 370 mμ, $\varepsilon \sim 9{,}000$, and at 400 mμ, $\varepsilon \sim 5{,}000$.

REAGENTS

Complexing mixture; 50 g of EDTA (ethylenediaminetetraacetic acid) with 50 g of potassium cyanide, in 1 litre of 1.5 *M* (1/10) ammonia.

Sodium or ammonium diethyldithiocarbamate; 200 mg in 100 ml of water, stored in a brown bottle.

Carbon tetrachloride.

Standard solution of bismuth (III).

OPERATING PROCEDURE

10 ml of complexing solution followed by 1 ml of reagent are added to 10 ml of solution containing 50 to 200 μg of bismuth. The solution is then agitated with exactly 10 ml of carbon tetrachloride for 30 seconds, and allowed to settle. The organic phase is filtered, and colorimetry is performed at 370 or 400 mμ.

COMPARE

K. L. Cheng, R. H. Bray and S. W. Melsted, *Anal. Chem.*, 27 (1955) 24.
G. W. C. Milner and J. W. Edwards, *Anal. Chim. Acta*, 18 (1958) 513.

In metallurgical products

J. Kinnunen and B. Wennerstrand, *Chemist-Analyst*, 43 (1954) 88.

Boron

B = 10.82

(1) Separation of trace amounts

(i) General remarks

(1) Losses of boron in the form of boric acid occur by evaporation from even weakly acidic media, since steam behaves as an entrainer.

(2) BF_3 and HBF_4 are volatile.

(3) Analyses performed with Pyrex equipment give results which are too high, since boron is extracted from the pyrex.

C. FELDMAN, *Anal. Chem.*, 33 (1961) 1916.

(ii) Using oxine

Many oxinates, for example those of Fe(III), Al(III), Zn(II), Ni(II), etc. may be extracted from ammoniacal and other media, leaving the boron in aqueous solution.

cf. H. SCHÄFER AND S. SIEVERTS, *Z. Anal. Chem.*, 121 (1941) 161.

(iii) Separation by electrolysis with a mercury cathode

cf. N. CHISHEVSKII, *Ind. Eng. Chem.*, 18 (1926) 607.

In steels

H. A. KAR, *Metals and alloys*, 9 (1938) 175.

(iv) Distillation of methyl borate

This method, which is the most general, may be the only one suitable for the estimation even of trace amounts. The operation is best carried out in fused silica apparatus.

The interfering ions include F^-, which also distills over in the form of fluorine compounds of borine, and interferes in the colorimetric estimations. This may be avoided by adding aluminium chloride to the flask, when F^- is complexed with Al(III) in the methanol solution[1].

Gelatinous silica retains boron. V(V) is entrained during the distillation[2].

Large amounts of Al(III), Fe(III), and Ge(IV) interfere[3].

The boron content of metallic specimens may be estimated by dissolving the latter in bromine, in methanolic solution, and then distilling the methyl borate formed[4].

COMPARE

C. E. White, A. Weissler and D. Büsker, *Anal. Chem.*, 29 (1947) 802.

H. A. Bewick, F. E. Beamish and J. C. Bartlet, *Natl. Res. Counc. Can., Techn. Rept. MC*, (1948) 228.

(1) *Appreciable (several μg) quantities of boron in fluorides*

C. Gaestel and J. Huré, *Bull. Soc. Chim.*, 16 (1949) 830.

(2) G. Weiss and P. Blum, *Bull. Soc. Chim.*, 14 (1947) 1077.

(3) J. R. Martin and J. R. Hayes, *Anal. Chem.*, 24 (1952) 182.

(4) A. R. Eberle and M. W. Lerner, *Anal. Chem.*, 32 (1960) 146.

1956 Book of A.S.T.M. Methods of Chemical Analysis of Metals, American Society for Testing Materials, Philadelphia.

In plants

L. V. Wilcox, *Ind. Eng. Chem., Anal. Ed.*, 2 (1930) 358.

W. W. Scott and S. K. Webb, *Ind. Eng. Chem., Anal. Ed.*, 4 (1932) 180.

Distillation of traces from a neutral medium. 0.01–0.3 ppm.

U.K. Atomic Energy Authority IGO-AM/S-124

J. J. Russel, *Canada Energy Project, Bull. MC-47*.

(v) *Extraction in the form of tetraphenylarsonium fluoborate*

Similarly to ClO_4^-, ReO_4^-, etc., BF_4^- reacts with tetraphenylarsonium $(C_6H_5)_4As^+$ ions to give a compound which can be extracted with chloroform.

The medium should be adjusted to pH 2 to 3 with a HF/F^- buffer. The presence of a large excess of fluoride ensures that the formation of BF_4^- is quantitative. Since, however, the formation of the complex is slow, the solution should be allowed to stand; thus, very minute amounts of boron (10^{-6} g) are fully complexed only after 18 hours.

OPERATING PROCEDURE

See Curcumin, p. 196.

J. J. Coursier, J. Huré, and K. Platzer, *Anal, Chim. Acta.*, 13 (1955) 379.

In Si

L. Ducret and P. Seguin, *Actas do Congresso, 1956, Lisboa,* 1957, p. 88.

(vi) *Extraction of the fluoborate of methylene blue*

See Colorimetry.

(vii) *Separation by ion-exchangers*

Boron (III) may be separated from cations and from anions such as phosphate, arsenate, nitrate etc. with the aid of a mixture of a strong acidic resin and a weak basic resin. Small amounts of oxalates and acetates do not interfere.

Ti(IV) can be complexed with H_2O_2; the cation formed is then collected.

Fluorides interfere by the formation of BF_4^- which is collected by the anionic resin.

This method may be used to separate even traces of boric acid.

REAGENTS

Strong acid resin: Dowex 50, 50–100 mesh, in the H^+ form;

Weak base resin: Amberlite IR 45, 60–100 mesh, in the basic HO^- form;

The operation is performed with a 1 cm diameter column containing 50 ml of the mixture of the two resins, present in proportions giving equivalent exchange capacities.

OPERATING PROCEDURE

100 ml of weakly acid solution containing 20 to 30 mg of boron (III) are passed through the resin at the rate of 2.5 ml per minute. The solution is collected in a 250 ml measuring flask after washing.

COMPARE

G. Brunisholz and J. Bonnet, *Helv. Chim. Acta*, 34 (1951) 2074.
J. R. Martin and J. R. Hayes, *Anal. Chem.*, 24 (1952) 182.
H. Kramer, *Anal. Chem.*, 27 (1955) 144.
J. D. Wolszon, J. R. Hayes and W. H. Hill, *Anal. Chem.*, 29 (1957) 829.

In Ti

R. C. Calkins and V. A. Stenger, *Anal. Chem.*, 28 (1956) 399.
E. G. Newstead and J. E. Gulbierz, *Anal. Chem.*, 29 (1957) 1673.

(viii) Standard solutions

Boric acid $B(OH)_3$ of guaranteed purity and borax $Na_2B_4O_7 \cdot 10\ H_2O$ are available.

$$B/B(OH)_3 = 0.1748$$

$$B/Na_2B_4O_7 \cdot 10H_2O = 0.1165.$$

(2) Colorimetry

Colorimetric methods permit the detection of very minute traces, but are somewhat inaccurate and are thus only suitable for trace amounts. Above 10 μg of boron acidimetry may be used, with suitable precautions.

A number of dyes give colorations with boric acid in very concentrated sulphuric acid. The coloration obtained depends graetly on the concentration of H_2SO_4. Carminic acid may be used with more ordinary concentrations of sulphuric acid, when more water is present.

Very minute traces of boron, such as 1 μg or below, may be determined by one of two methods, either of which however requires a large number of precautions: the curcumin method (possessing the disadvantage that the reagent is unstable), and the methylene blue method.

(i) Carminic acid

The red colour of the reagent changes to blue in the presence of boric acid. The concentration of sulphuric acid is also an important factor. However, 1 ml of water per 10 ml of concentrated sulphuric acid can be tolerated.

REAGENTS

Concentrated sulphuric acid;

Carminic acid: 50 mg of carminic acid are dissolved in 100 ml of concentrated sulphuric acid.

OPERATING PROCEDURE

10 ml of concentrated sulphuric acid followed by 10 ml of carminic acid solution are slowly added to 2 ml of a solution containing 1 to 10 μg of boron (III), with cooling. The solution is allowed to stand for 30–45 min and colorimetry is performed at 585 mμ.

COMPARE

J. T. Hatcher and L. V. Wilcox, *Anal. Chem.*, 22 (1950) 567.
D. L. Callicoat and J. D. Wolszon, *Anal. Chem.*, 31 (1959) 1434.

In titanium

R. C. Calkins and V. A. Stenger, *Anal. Chem.*, 28 (1956) 399.

In biological materials

W. C. Smith, A. J. Goudie and J. N. Sievertsen, *Anal. Chem.*, 27 (1955) 295.

In uranium-aluminium alloys

K. W. Puphal, J. A. Merrill, G. L. Booman and J. E. Rein, *Anal. Chem.*, 30 (1958) 1612.

In nitrates

W. J. Ross and J. C. White, *Talanta*, 3 (1960) 311.

(ii) Quinalizarin

COMPARE

B. A. Ripley-Duggan, *Analyst*, 78 (1953) 183.

In alloys

A. H. Jones, *Anal. Chem.*, 29 (1957) 1101.

In steels, without separation

G. A. Rudolph and L. C. Flickinger, *Steel*, 112 (1943) 114.

Other dyes

J. A. RADLEY, *Analyst*, 69 (1944) 47.

G. H. ELLIS, E. G. ZOOK AND O. BAUDISCH, *Anal. Chem*., 21 (1949) 1345.

N. TRINDER, *Analyst*, 73 (1948) 494.

C. M. AUSTIN AND J. S. MCHARGUE, *J. Assoc. Offic. Agr. Chem.*, 31 (1948) 284, 427.

Tetrabromochrysazine (Limit 0.02 μg)

J. H. YOE AND R. L. GROB, *Anal. Chem.*, 26 (1954) 1465.

(iii) Curcumin

When a solution of curcumin is evaporated to dryness in the presence of boric acid, pink rosocyanine is obtained.

This method is more sensitive than the preceding one. $\varepsilon = 40{,}000$ at 540 mμ. However, numerous ions interfere, and the method is lengthy and not very accurate.

Moreover, solutions of curcumin are rarely stable and the reproducibility of the results is poor.

The procedure described is effective for estimating traces of boron by separating it in the form of tetraphenylarsonium fluoborate so as to eliminate the interfering ions.

REAGENTS

Solution of tetraphenylarsonium chloride, 10^{-2} *M*, obtained by dissolving 1.046 g of the crystalline salt in 250 ml of chloroform.

Phenolphthalein solution, 0.1%, in alcohol.

Curcumin solution, 0.125%, obtained by dissolving the solid product in alcohol.

OPERATING PROCEDURE

For the separation and estimation of 0.05–1 μg of boron in 5 ml of solution.

5 ml of an aqueous solution of pH less than 3.2 and containing a total concentration of fluoride at least equal to 0.8 *M* are left to stand for 18 hours. 5 ml of a chloroform solution of tetraphenylarsonium chloride are then added, the mixture is stirred mechanically for 30 minutes, and then centrifuged. The aqueous solution is drawn off with a polythene pipette, and 5 ml of distilled water are added and then separated. A further 5 ml of distilled water are added and the mixture is stirred for some moments and centrifuged; the aqueous phase is drawn off. This last operation is repeated twice. The chloroform solution is placed in a platinum capsule, the polythene tube in which the extraction was carried out is rinsed with 1 ml of chloroform, which is then added to the solution in the capsule. 15 drops of 0.1 *N* sodium hydroxide and 1 drop of phenolphthalein are added to the latter and the solution is evaporated to dryness by placing the capsule beneath a heating rack consisting of a silica tube 40 cm long carrying a heating resistance on the inside.

The solid residue is taken up in 1 ml of 1 *N* trichloroacetic acid, left to stand for 10 minutes and then thoroughly mixed with 1 ml of 95% alcohol. 1 ml of curcumin is now added, and the mixture is placed in the oven at 106 ± 1° for 60 ± 3 minutes. The residue is taken up in alcohol and made to 25 ml, and colorimetry is performed, in a 1 cm cell, at 540 mμ.

COMPARE

T. PHILIPSON, *Lantbruks-Högskol. Ann.*, 12 (1944–1945) 251.

B. V. ZALETEL, *Rec. Trav. Inst. Rech. Struc. Mat. (Belgrade)*, 2 (1953) 31.

In steels

H. W. WINSOR, *Anal. Chem.*, 20 (1948) 176.

G. S. SPICER AND J. D. H. STRICKLAND, *Anal. Chim. Acta*, 18 (1958) 231, 523.

J. BORROWDALE, R. H. JENKINS AND C. E. A. SHANAHAN, *Analyst*, 84 (1959) 426.

In U

L. SILVERMAN AND K. TREGO, *Anal. Chim. Acta*, 15 (1956) 439.

In Na

J. RYNASIEWICZ, M. P. SLEEPER AND J. W. RYAN, *Anal. Chem.*, 26 (1954) 935.

In Si and Ge

C. L. LUKE, *Anal. Chem.*, 27 (1955) 1150.

In plants and soils

W. T. DIBLE, E. TRUOG AND K. C. BERGER, *Anal. Chem.*, 26 (1954) 418.

In Ni

R. C. CHIRNSIDE, H. J. CLULEY AND P. M. C. PROFFITT, *Analyst*, 82 (1957)

In Cd

C. L. LUKE, *Anal. Chem.*, 30 (1958) 1405.

Separation via tetraphenylarsonium fluoborate

J. COURSIER, J. HURÉ AND R. PLATZER, *Anal. Chim. Acta*, 13 (1955) 379.

In silicon

L. DUCRET AND P. SEGUIN, *Actas do Congresso 1956, Lisboa* (1957) 88; *Anal. Chim. Acta*, 17 (1957) 207.

(iv) Methylene blue fluoborate in 1,2-dichloroethane.

As in the preceding method, the reaction is

$$\underset{\text{water}}{HR^+ + BF_4^-} \rightleftarrows \underset{\text{solvent}}{RHBF_4}.$$

Here HR^+ represents a cation of methylene blue in an acid medium. The limit of sensitivity is 0.01 μg of boron.

REAGENTS

Ammonium difluoride solution, 7.5 M;
Methylene blue, medicinal, 0.02 M solution;
1,2-Dichloroethane, rectified (83 ± 1°);
Standard solutions of boric acid, 0.5 mg of boron per litre.

Equipment. The HF solutions should be contained and handled in polythene vessels.

The latter should be free from fatty acids, which may be extracted in the presence of methylene blue.

Microburettes.

OPERATING PROCEDURE

10 ml of the solution to be analyzed are placed in each of 6 polythene bottles. 1.00 ml of water is added to bottles 1 and 2, 0.50 ml of water and 0.5 ml of boric acid are added to bottles 3 and 4, and 1.00 ml of boric acid is added to bottles 5 and 6. 1 ml of the difluoride solution is added to each bottle; the latter are then stoppered and left to stand for 18 hours at a constant temperature. After this period, 3.00 ml of the methylene blue solution and 15.0 ml of the solvent are added to each bottle, and the latter are shaken for 5 minutes and allowed to settle. The aqueous layers are then withdrawn by means of a polythene pipette. 10 ml of water are then shaken with the solvent phases for 1 minute and the aqueous layers removed as before. Colorimetry is then performed in each case, at 645 mμ.

The results should be grouped two by two, and are found to fall on a straight line if the optical densities are plotted as a function of the amounts of boron added. From each result, the value of a blank test carried out under the same conditions with all the reagents should be substracted.

In silicon and silica

L. DUCRET, *Anal. Chim. Acta,* 17 (1957) 213.

In steels

L. PASZTOR, J. D. BODE AND Q. FERNANDO, *Anal. Chem.,* 32 (1960) 277.

Bromine and bromides

Br = 79.9

(i) Standard solutions. KBr of guaranteed purity is available. Br/KBr = 0.671

The corresponding amount of bromine may be liberated by oxidation (see below).

Alternatively, $KBrO_3$ and $NaBrO_3$ may be used.

(ii) Colorimetry of bromine. Colorimetry is occasionally performed directly on an aqueous solution of bromine: alternatively a known carbon tetrachloride solution may be used, ($\varepsilon = 200$ at 415 mμ). These methods do not permit the detection of very minute traces.

1. *Formation of eosin.* Bromine reacts with fluorescein to give a red brominated derivative, eosin, as a result of which a colour change from green to pink occurs. This may be compared with solutions containing a known amount of bromine.

The method is not very accurate: Cl_2 and I_2 interfere.

REAGENTS

Solution of fluorescein: 0.3 g of fluorescein in 5 ml of $N(^1/_{10})$ caustic soda, diluted to 1 litre;

Buffer solution: 10 ml of *M* (8%) sodium acetate and 1 ml of *N* (1/18) acetic acid: pH 5.5–5.6;

Reducing solution: 10 g of caustic soda, 1 g of crystalline sodium thiosulphate, and 200 ml of water.

OPERATING PROCEDURE

1 drop of fluorescein and 3 drops of buffer solution are added to 1 ml of the neutralized solution to be estimated. The mixture is shaken and the reaction allowed to proceed for 60–90 seconds. Reaction is then interrupted by the addition of one drop of the thiosulphate solution and the colorimetric determination is carried out.

0.1 to 0.2 μg of bromine may be detected in this manner. If more bromine is present, the amount of fluorescein should be increased.

cf. F. A. POHL, *Z. Anal. Chem.*, 149 (1956) 68.

2. *Using fuchsin.* In sulphuric acid, the reagent assumes a violet colour, and may be extracted with organic solvents. The limit of sensitivity is 7 μg per ml of solvent.

REAGENT

100 ml of 2 N (1/20) sulphuric acid and 10 ml of fuchsin in 1% aqueous solution. The solution is shaken and left for 1 hour, after which it becomes colourless.

OPERATING PROCEDURE

20 ml of reagent are added to 20 ml of the solution to be estimated and the mixture shaken vigorously. The extraction is performed with 5 ml of *iso*-amyl alcohol or chloroform. Colorimetry is then performed directly.

cf. R. C. LOPEZ, *Ann. fals. et fraudes*, 28 (1935) 115.
C. LEPIERRE, *XVe Cong. de Ch. ind.*, (1936) 725.

3. *Other methods*

With phenol red

1 to 18 μg of Br^- can be estimated.

cf. V. A. STENGER AND I. M. KOLTHOFF, *J. Am. Chem. Soc.*, 57 (1935) 831.

With pentabromorosaniline (ε = 60,000 at 585 mμ in benzyl alcohol)

W. M. J. TURNER, *Ind. Eng. Chem., Anal. Ed.*, 14 (1942) 599.

(iii) Colorimetry of bromides. Bromides are oxidized to bromine which is then estimated colorimetrically (p. 199).

1. *Oxidation with Chloramine T.* (Colorimetry by the formation of eosin)

The solution is brought to pH 5.5 and is then oxidized with one drop of 1% solution of Chloramine T. Colorimetry is carried out as on p. 199. Chlorides are not oxidized: iodides should first be removed by oxidation with Fe(III) and the iodine formed should be expelled by boiling.

cf. F. A. POHL, *Z. Anal. Chem.*, 149 (1956) 68.

2. *Oxidation with chlorine water.* The bromine formed can be estimated by direct colorimetry or by the fuchsin method (p. 199). Iodides may be converted to iodates via iodine itself, with an excess of chlorine water.

OPERATING PROCEDURE

Assuming for example, that there are 5 mg of bromide present in 100 ml of solution, 10 ml of 18 *N* (1/2) sulphuric acid are added to the latter, followed by saturated chlorine water, drop by drop, until the coloration is deepest (about 5 ml).

3. *Oxidation by permanganate.* The bromine formed is expelled and can then be estimated, for example, with the aid of a fluorescein paper (p. 199).

4. *The formation of complexes.* Compare the method used for chlorides: ferric chloride complexes, decoloration of ferric thiocyanate, Hg(II)-diphenylcarbazide, etc. (p. 222).

Cadmium

Cd = 112.4

(1) Separation of trace amounts

(i) Extraction by dithizone. This is one of the best methods of separation; see Colorimetry.

(ii) Extraction by amines. The cadmium iodide complexes are extracted by xylene in the presence of long-chain secondary amines.

Interfering ions. Large quantities of Fe(III) interfere in chloride solutions, being extracted and preventing the extraction of cadmium. Interference is also encountered with Cu(II) and Tl(III).

As little as 0.00005% of cadmium may thus be separated from zinc, but chloride ions interfere.

REAGENTS

Amine LA-2, 1% solution: (N-lauryltrialkylmethylamine; Amberlite LA-2) in xylene.

Potassium iodide, 166 g per litre.

Washing solution: 8 g of potassium iodide are dissolved in 500 g of water, and 1 ml of concentrated sulphuric acid is added.

OPERATING PROCEDURE

5 ml of iodide and 10 ml of the LA-2 solution are added to 50 ml of solution, acidified to $pH < 3$, and the mixture is shaken for 20 seconds. The aqueous phase is rejected, 50 ml of the washing solution are added, and the mixture is again shaken for 20 seconds.

The cadmium is transferred into the aqueous phase by shaking with two 10 ml portions of 10% sodium carbonate.

cf. J. R. KNAPP, R. E. VAN AMAN AND J. H. KANZELMEYER, *Anal. Chem.*, 34 (1962) 1374.

(iii) By anion exchange

(a) Cadmium is held back as the iodide complexes, whilst the considerably less stable zinc complexes are not retained.

100 μg of Cd(II) may thus be separated from 3 g of Zn(II). Mn(II), Al(III), Ca(II), Mg(II), Ni(II), and Co(II) do not interfere.

cf. E. R. BAGGOTT AND R. G. WILLCOKS, *Analyst*, 80 (1955) 53.
J. A. HUNTER AND C. C. MILLER, *Analyst*, 81 (1956) 79.
S. KALLMANN, H. OBERTHIN AND R. LIU, *Anal. Chem.*, 30 (1958) 1846.

(b) The chloride complexes of Cd(II) are more stable than those of Zn(II), Ni(II), Co(II) etc., and are therefore more strongly retained. The various elements may be separated off by HCl of different concentrations.

cf. E. W. BERG AND J. T. TRUEMPER, *Anal. Chem.*, 30 (1958) 1827.
S. L. JONES, *Anal. Chim. Acta*, 21 (1959) 532.

(iv) By cation exchange

(a) Cd^{2+} is eluted by iodide, in the form of anionic complexes, and may thus be separated from Zn^{2+}, Fe^{3+}, Mn^{2+}, Al^{3+}, Ni^{2+}, Co^{2+}, In^{3+}, and Cr^{3+}.

cf. S. KALLMANN, H. OBERTHIN AND R. LIU, *Anal. Chem.*, 32 (1960) 58.

(b) In 0.5 *N* HCl Cd^{2+} exists in the form of anionic chloride complexes. It is not retained, and can be separated from U(VI), Co(II), Mn(II), Zn(II), Ni(II), Cu(II), Ti(IV), and Fe(III).

Sn(IV) accompanies Cd(II).

cf. F. W. E. STRELOW, *Anal. Chem.*, 32 (1960) 363.
J. S. FRITZ AND T. A. RETTIG, *Anal. Chem.*, 34 (1962) 1562.
J. KALLMANN, H. OBERTHIN AND R. LIU, *Anal. Chem.*, 30 (1958) 1846.

(v) Internal electrolysis. Separation of Zn(II).

cf. J. G. FIFE, *Analyst*, 65 (1940) 562.

(2) Colorimetry

Dithizone. Cadmium dithizonate, which is red, may be extracted from media of pH greater than 10 (quantitatively above pH 12) and also from 2.5 N caustic soda with carbon tetrachloride as the solvent. The precision is poor, since the compound is unstable to light, particularly in the tetrachloride solution.

Sensitivity: ε = 80,000 at 518 mμ.

Interfering ions. In principle, Zn(II) is not extracted from 0.5 *N* caustic soda; one part of Cd(II) can be separated from 1000 parts of Zn(II). This may no longer remain true when larger amounts of Zn(II) are present.

In this way, Pb(II) and Bi(III) may be removed (except in very large amounts). Only the following are extracted, at least to some extent: Cu(II), Ag(I), Au(III), Hg(II), Pd(II), Ni(II), and Co(II). It is however possible to eliminate a certain number of them previously – Ag(I), Cu(II), Hg(II), and most of the Co(II) and Ni(II) – by extraction from an acid medium at pH 1–2. (The inverse operation – agitation of the dithizonates with dilute acid to cause Cd(II) to pass back into aqueous solution – is less to be recommended because it is slow).

The operation may be performed in the presence of cyanide[1].

The precipitation of hydroxides under alkaline conditions may be avoided by the addition of citrate or tartrate.

REAGENTS

Potassium sodium tartrate: 25 g in 100 ml of water.

Potassium cyanide, 1%: 400 g of caustic soda and 1.0 g of potassium cyanide are dissolved in 1 l. of water.

Potassium cyanide, 0.05%: 400 g of caustic soda and 0.5 g of potassium cyanide are dissolved in 1 l. of water.

Hydroxylamine hydrochloride, in 20% solution.

Solution I of dithizone: 80 mg of dithizone are dissolved in 1 l. of chloroform.

Solution II of dithizone: 8 mg of dithizone are dissolved in 1 l. of chloroform.

Tartaric acid, 20 g per l.

Thymol blue, 0.1% in alcohol.

OPERATING PROCEDURE

The initial solution should be neutralized to the yellow colour of thymol blue and should contain not more than 10 μg of cadmium, in 25 ml. To a portion of this volume, the following are added in the order given, with stirring: 1 ml of tartrate, 5 ml of 1% potassium cyanide, and 1 ml of hydroxylamine hydrochloride. The mixture is shaken with 15 ml of solution I of dithizone for 1 minute, and the organic phase is then transferred to a separating funnel containing 25 ml of tartaric acid. A further 10 ml of chloroform are added to the alkaline solution; after shaking for 1 minute, the chloroform phase is added to the first extract. The extracts are shaken with the tartaric acid solution for 2 minutes: the cadmium passes into the aqueous solution. After rinsing with 5 ml of chloroform, 0.25 ml of hydroxylamine hydrochloride and 15 ml (accurately measured) of solution II of dithizone are added, followed by 5 ml of 0.05% potassium cyanide. The mixture is shaken for 1 minute, and the chloroform solution filtered. Colorimetry is performed at 520 mμ.

NOTES. (1) No contact between the chloroform phase and the strongly alkaline phase should occur, beyond that strictly necessary for the extraction and the separation of the two layers.

(2) Under these conditions, 5 to 10 mg of Zn, Co, Ni, Pb, Bi, Ag, and Cu do not interfere; Tl interferes.

COMPARE

(1) B. E. Saltzman, *Anal. Chem.*, 25 (1953) 493.
A. Petzold and I. Lange, *Z. Anal. Chem.*, 146 (1955) 1.

Traces of cadmium in zinc
H. Fischer and G. Leopoldi, *Mikrochim. Acta*, 1 (1937) 30; *Metall u. Erz.*, 35 (1938) 119.

In lead
L. Silverman and K. Trego, *Analyst*, 77 (1952) 143.

In rocks
E. B. Sandell, *Ind. Eng. Chem., Anal. Ed.*, 11 (1939) 364.

In biochemistry
R. L. Shirley, E. J. Benne and E. J. Miller, *Anal. Chem.*, 21 (1940) 300.
J. C. Smith, J. E. Kench and R. E. Lane, *Biochem. J.*, 61 (1955) 698.

In tin
W. C. Coppins and J. W. Price, *Metallurgia*, 48 (1953) 149.

Calcium

Ca = 40.1

(1) Separation of trace amounts

Calcium must frequently only be estimated after the removal of other elements apart from the alkali and alkaline-earth metals. Methods of separation are therefore presented below; those quoted are also suitable as preliminaries in determinations of Ba(II), Sr(II), and Mg(II).

(i) Electrolysis at a mercury cathode. A large number of elements may be separated by this technique, if the potential is controlled to leave Ca(II) in solution.

cf. T. D. PARKS, H. O. JOHNSON AND L. LYKKEN, *Anal. Chem.*, 29 (1948) 148.

(ii) Extraction of the oxinate. Calcium oxinate can be extracted from a medium of pH 13. The pH of an aqueous solution containing less than 80 μg of Ca(II) is brought to this value by means of dilute caustic soda. 5.0 ml of butyl cellosolve are added and the solution is made up to 40–50 ml. The extraction is performed with 20 ml of a 3% solution of oxine in chloroform.

cf. C. JANKOWOSKI AND H. FREISER, in G. H. MORRISON AND H. FREISER: *Solvent Extraction in Analytical Chemistry*, Wiley, 1957.

(iii) Separation by ion-exchangers. (1) *Cation-exchangers.* Ca^{2+} can be collected by a cation-exchange resin, and may thus be separated from H_3PO_4, H_3BO_3, etc. The Ca^{2+} is subsequently eluted by means of an acid. The method is applicable to as little as 10 μg of Ca.

Other cations can also be complexed in a citric acid medium[1].

(2) *Anion-exchangers.* Separation of phosphates. The slightly acidic solution is passed through a weakly basic anion-exchanger; the phosphate ions are bound by the resin and the eluate is analyzed for calcium.

PREPARATION OF THE COLUMN

Resin: Amberlite IR 4B in the chloride form (grain size: 0.3–0,4 mm). Capacity: 75 milliequiv./g.

Column height: 30 cm.

The resin is converted to the chloride form by passing 2 *N* (1/16) hydrochloric acid through the column, which is subsequently washed with water to neutrality.

Methyl yellow (at pH 2.9, yellow; at pH 4.0 red).

OPERATING PROCEDURE

100–150 ml of solution are neutralized with approximately 2 *N* (1/7) ammonia until the indicator assumes an orange shade (not as far as the pure yellow coloration). The solution is then passed through the resin at the rate of 10 ml per minute. After washing, the eluate is analyzed for calcium.

COMPARE

A. S. MASON, *Analyst*, 77 (1952) 529.
G. BRUNISHOLZ, M. GENTON AND E. PLATTNER, *Helv. Chim. Acta*, 36 (1953) 782.
(1) O. SAMUELSON, L. LUNDEN AND K. SCHRAMM, *Z. Anal. Chem.*, 140 (1953) 330.
R. G. HEMINGWAY, *Analyst*, 81 (1956) 164.

Separation of Ca^{2+} — Sr^{2+} — Ba^{2+} — Mg^{2+}

M. LERNER AND W. RIEMAN, III, *Anal. Chem.*, 26 (1954) 610.
F. NELSON AND K. A. KRAUS, *J. Am. Chem. Soc.*, 77 (1955) 801.

Ca^{2+}/Mg^{2+}

R. L. GRISWOLD AND N. PACE, *Anal. Chem.*, 28 (1956) 1035.
M. HONDA, *Japan Analyst*, 3 (1953) 132.

(iv) Standard solutions. The starting material may be $CaCO_3$ of guaranteed purity, pre-dried at 110°. This is dissolved in 0.2 *N* (1/50) HCl, and the CO_2 is driven off by boiling.

Alternatively, $CaCl_2 \cdot 4H_2O$ of guaranteed purity may be used.

$$Ca/CaCO_3 = 0.401$$

$$Ca/CaCl_2 \cdot 4H_2O = 0.219$$

(2) Colorimetry

Although direct colorimetry can be performed in several ways, the available methods are very unselective. To estimate small amounts of Ca(II), it is thus often preferable to use microvolumetric estimations, spectrography, or flame spectrophotometry.

(i) With ammonium purpurate (murexide). Under suitable pH conditions, murexide gives a yellow-red coloration with Ca(II). The absorption maximum is located at 506 mμ; the pure reagent shows an absorption maximum at 537 mμ.

The coloration disappears slowly.

Sensitivity: molar extinction coefficient, $\varepsilon = \sim 10{,}000$ at 506 mμ.

Interfering ions. Mg(II) gives a yellow coloration, but the Mg(II) murexide compound has, at 500 mμ, approximately the same extinction coefficient as the pure reagent. Consequently, Mg(II) does not interfere, unless present in an amount exceeding 10 ppm when precipitation takes place.

The majority of ions interfere and give colorations or precipitates with the reagent.

REAGENTS

Saturated solution of ammonium purpurate in water (about 0.5%) to which $2\,^1/_2$ times its own volume of alcohol has been added. The reagent is stable for one day.

Caustic soda, 0.1 N (1/100), or 4 g/l.

Standard solution of Ca(II) at 10 ppm, prepared from pure calcium carbonate.

OPERATING PROCEDURE

The solution is neutralized if necessary. For the best results, it should contain 30 μg of Ca(II). 1 ml of reagent, accurately measured by means of a microburette is added and the volume is made up to about 45 ml. After a further addition of 2 ml of 0.1 *N* caustic soda, the volume is adjusted to 50 ml, and the solution is estimated colorimetrically at 506 mμ. The calibration curve should be checked daily.

COMPARE

H. OSTERTAG AND E. RINCK, *Compt. Rend. (Paris)*, 232 (1951) 629.
L. E. TAMMELIN AND S. MOGENSEN, *Acta Chem. Scand.*, 6 (1952) 988.
M. B. WILLIAMS AND J. H. MOSER, *Anal. Chem.*, 25 (1953) 1414.
F. H. POLLARD AND J. V. MARTIN, *Analyst*, 81 (1956) 348.

(ii) Other methods.

Colorimetry of Ca(II) and Mg(II) using eriochrome black T at pH 2.
A. YOUNG, T. R. SWEET AND B. B. BAKER, *Anal. Chem.*, 27 (1955) 356.

With o-cresolphthalein complexone
$\varepsilon \sim$ 30,000 at 575 mμ.
F. H. POLLARD AND J. V. MARTIN, *Analyst*, 81 (1956) 348.

Turbidimetry of the oxalate
50 to 500 μg.
J. G. HUNTER AND A. HALL, *Analyst*, 87 (1953) 106.

With 'Calcichrome' after extraction, in the presence of Ba(II), Sr(II) and Mg(II)
R. A. CLOSE AND T. S. WEST, *Talanta*, 5 (1960) 221.

Carbon and its compounds

C = 12.01

(1) Carbon

Free carbon is generally estimated by combustion in a current of oxygen. The apparatus used to determine the carbon contents of steels may be employed. The carbon dioxide formed can then be determined by colorimetry with pH indicators, or by the turbidimetry of barium carbonate – see CO_2.

(2) Carbon monoxide CO

(2a) Separation

Most methods of estimating carbon monoxide colorimetrically utilize its reducing power with respect to catalysts, or in the presence of a catalyst. It follows that the presence of other reducing gases, such as H_2, H_2S, hydrocarbons etc., will give rise to erroneous results. These should therefore be separated. It is possible, for example, to oxidize the most powerful reducing agents – aldehydes and unsaturated hydrocarbons – with chromic acid.

(2b) Colorimetry

Reduction of palladous salts

Carbon monoxide reduces Pd(II) to palladium black. Since the reaction is slow, sufficient time should be allowed for it to proceed to completion.

One of the simplest methods consists in introducing a sample of the gas to be analyzed into a previously evacuated receptacle containing only a standardized solution of palladous chloride. The palladium formed is flocculated with aluminium sulphate. The excess of Pd(II) is determined colorimetrically (see Palladium).

Limit of sensitivity: 0.005% by volume.

Accuracy 5 to 10%.

REAGENTS

Palladium chloride, crystalline: 0.5 g in 2.5 ml of concentrated hydrochloric acid are made up to 500 ml; Aluminium sulphate, 10% in water.

OPERATING PROCEDURE

500 ml of the gas to be analyzed are shaken with 0.3 ml of reagent and 0.2 ml of aluminium sulphate for 2 hours. The container is then left to stand for at

least 4 hours. Air is permitted to enter only at the conclusion of this operation. The solution is filtered through sintered glass, and washed. The residual Pd(II) is determined colorimetrically (see Palladium).

cf. C. H. GRAY AND M. SANDIFORD, *Analyst*, 71 (1946) 107.
A. A. CHRISTMAN, W. D. BLOCK AND J. SCHULTZ, *Ind. Eng. Chem., Anal. Ed.*, 9 (1937) 153.

NOTE. The reaction is occasionally carried out on paper strip.

cf. A. LAMBRECHTS AND R. ROSEMAN, *Compt. Rend. Soc. Biol.*, 140 (1946) 801.

Reduction of iodine pentoxide. At 150° the following reaction takes place:

$$I_2O_5 + 5CO \rightarrow I_2 + 5CO_2.$$

The iodine liberated may be determined by colorimetric methods.

Many other substances also reduce iodine pentoxide.

The method is sensitive and accurate, but presents experimental difficulties. It may be used in place of titrimetry of the iodine, for contents lower than 0.005% by volume. A litre of gas should be used. The accuracy is of the order of 5–10%. A blank test should be performed by heating I_2O_5 with nitrogen.

cf. B. SMALLER AND J. F. HALL, JR., *Ind. Eng. Chem., Anal. Ed.*, 16 (1944) 64.
K. H. NELSON, M. D. GRIMES, D. E. SMITH AND B. J. HEINRICH, *Anal. Chem.*, 29 (1957) 180.

If I_2O_5 is dispersed on silica gel, the surface area of contact is increased, and the reaction may proceed in the cold. In this case, either I_2 or CO_2[1] can be titrated, or advantage may be taken of the coloration produced. By this method as little as 0.001–1% CO in air may be estimated.

COMPARE

(1) H. G. GLOVER, *Mikrochim. Acta*, 5 (1955).

By spectrophotometry
G. A. GRANT, M. KATZ AND R. L. HAINES, *Can. J. Technol.*, 29 (1951) 43.

Review of methods with I_2O_5
E. G. ADAMS AND N. T. SIMMONS, *J. Applied Chem. (Lond.)*, 1 (1951) 820.

Formation of molybdenum blue. Carbon monoxide reduces phosphomolybdate and silicomolybdate ions in the presence of catalysts such as Pd(II).

0.06% by volume may be estimated to within 10%.

Limit: 0.001% by volume.

Many other reducing gases interfere.

COMPARE

M. Shepherd, *Anal. Chem.*, 19 (1947) 77.
M. Shepherd, S. Schuhmann and M. V. Kilday, *Anal. Chem.*, 27 (1955) 380.

Automatic apparatus

J. W. Cole, J. M. Salsbury and J. H. Yoe, *Anal. Chim. Acta*, 2 (1948) 115.

Reduction of HgO. Red HgO is reduced in the hot by CO in a convenient portable apparatus:

$$CO + HgO\ (red) \rightarrow CO_2 + Hg\uparrow$$

$$3Hg\uparrow + SeS_2 \rightarrow 2HgS + HgSe$$

A paper is blackened by the formation of HgS.
The method may be used to estimate 0.0003 to 3% of CO to within 1%.

cf. A. O. Beckman, J. D. McCullough and R. A. Crane, *Anal. Chem.*, 20 (1948) 674.

Infrared absorptiometry. Simple instruments are available.

cf. A. Jäger and W. Grebe, *Glückauf*, 85 (1949) 294.
H. Moureu, P. Chovin, L. Truffert and J. Lebbe, *Chim. Anal.*, 39 (1957) 3.

(3) CO_2 and carbonates

(3a) Separation

Decomposition of the carbonates

CO_2 is liberated by the action of acids on carbonates. Other gases accompany it, in particular SO_2 and H_2S, but these may be retained by bubbling the gaseous mixture through an oxidizing medium (H_2O_2, chromic acid mixture, etc.) or by the addition of oxidizing agents to the acid.

cf. P. S. Roller and G. Ervin, Jr., *Ind. Eng. Chem., Anal. Ed.*, 11 (1939) 150.
H. A. J. Pieters, *Anal. Chim. Acta*, 2 (1948) 263.

The most reliable method employs an arrangement similar to that used for the determination of carbon in steels, whereby the substance to be analyzed is calcined in a boat placed in a current of gas.

(3b) Colorimetry of CO_2

Colorimetric methods are not, in general, the best methods of determining CO_2. Nevertheless, they possess the advantage of simplicity and may consequently be of interest.

(i) Colorimetric determination of the pH. The most generally used method of this type consists in bubbling the gases containing CO_2 into a solution containing very dilute sodium bicarbonate. After a certain time, equilibrium is set up between the gas and the solution: the pH of the latter then depends only on the partial pressure of the carbon dioxide. The estimation of carbon dioxide is thus reduced to a measurement of the pH, which is best carried out by means of a pH-meter incorporating a glass electrode. Alternatively, the colour of a pH indicator may be noted. The method may be used for mixtures containing 0.03 to 100% of CO_2 by volume.

The colorimetric method can be adapted to estimations and to automatic control by means of photoelectric cells.

cf. R. J. WINZLER AND J. P. BAUMBERGER, *Ind. Eng. Chem., Anal. Ed.*, 11 (1939) 371.
N. A. SPECTOR AND B. F. DODGE, *Anal. Chem.*, 19 (1947) 55.
W. T. SUMERFORD, D. DALTON AND R. JOHNSON, *Ind. Eng. Chem., Anal. Ed.*, 15 (1943) 38.
W. D. MAXON AND M. J. JOHNSON, *Anal. Chem.*, 24 (1952) 1541.

(ii) Turbidimetry of barium carbonate. The gases are bubbled into a solution of baryta. Turbidimetry is then performed. The method is not accurate, but is rapid and simple. Up to 10 mg of CO_2 can be estimated in this way. Instruments known as opacimeters are available commercially. Such instruments consist of two cells placed in front of photoelectric cells: in one the gas to be estimated is bubbled into baryta water. The stream of gas is cut off when a certain opacity, which corresponds to a predetermined amount of carbon dioxide, has been obtained. The volume of gas passed through the cell is measured by a suitable arrangement.

COMPARE

0.5 — 10 mg of CO_2 to within ± 2.5%
P. S. ROLLER AND G. ERVIN, JR., *Ind. Eng. Chem., Anal. Ed.*, 11 (1939) 150
N. IVANOV, *Ann. fals. et fraudes*, 29 (1936) 488.

Carbon in steels
P. AGASSANT AND J. L. ANDRIEUX, *Bull. Soc. Chim.*, 17 (1950) 253.

(4) Cyanides CN^-

(4a) Separation

HCN is liberated from solutions of pH < 9. If a cyanide is present in solution, the latter may be acidified, for example with tartaric acid, and HCN may then be distilled off[1] or displaced by a current of air.[2]

If complexing agents such as Zn(II), Fe(II), Ni(II), etc. are present, HCN can only be distilled from a strongly acid solution such as *e.g.* phosphoric acid[3]. Certain other substances which give rise to more stable complexes, Cu, Co, Hg, interfere.

HCN present in other gases in quantities up to 1 μg per litre of gas can be absorbed by caustic soda. Separation from SCN^- may be effected by extraction of the HCN with *iso*-propyl ether[3].

COMPARE

(1) A. E. CHILDS AND W. C. BALL, *Analyst*, 60 (1935) 294.
R. A. FULTON AND M. J. VAN DYKE, *Anal. Chem.*, 19 (1947) 922.
(2) W. O. WINKLER, *J. Assoc. Offic. Agr. Chem.*, 22 (1939) 349.
(3) J. M. KRUSE AND M. G. MELLON, *Anal. Chem.*, 25 (1953) 446.

(i) Destruction of complex cyanides. (1) If complex cyanide solutions are boiled with mercuric acetate in a weakly alkaline medium, soluble $Hg(CN)_2$ is formed. The metallic constituent of the complex is precipitated as a hydroxide, and may be filtered off. $Hg(CN)_2$ can be decomposed by hydrogen sulphide, with the formation of HgS: the cyanide remains in solution. All cyanide complexes can be destroyed in this way, with the exception of the cobalticyanides and palladous cyanide.

cf. R. HÜNERBEIN, *Angew. Chem.*, 51 (1938) 539.

(2) Hot concentrated sulphuric acid destroys all cyanide complexes.

(3) Formaldehyde destroys the less stable complexes, *i.e.* those of Cd(II), Zn(II), etc.

(ii) Standard solutions. The potassium and sodium cyanides available in France contain 95–96% of KCN or NaCN, and may be used for the preparation of standard colorimetric solutions. They may be titrated against acids.

$$HCN/KCN = 0.415$$

$$CN^-/KCN = 0.399$$

A 0.1000 *M* solution of cyanide contains 2.60 mg of CN^- per millilitre.

(4b) Colorimetry

(i) As NH_3 using Nessler's reagent. In an acid or alkaline medium, CN^- is hydrolyzed as follows:

$$CN^- + 2H_2O \rightarrow NH_3 + HCO_2^-$$

The reaction is performed in an autoclave at 140–150° in hydrochloric acid, and is completed in 30 minutes. The NH_3 produced is then determined colorimetrically.

(ii) Formation of Prussian blue. CN^- is converted to ferrocyanide. Fe(III) is then added to form Prussian blue. As little as 0.1 μg of HCN may be estimated in this way.

COMPARE

A. O. Gettler and L. Goldbaum, *Anal. Chem.*, 19 (1947) 270.
R. A. Fulton and M. J. van Dyke, *Anal. Chem.*, 19 (1947) 922.

On reagent paper
B. E. Dixon, G. C. Hands and A. F. F. Bartlett, *Analyst*, 83 (1958) 199.

(iii) With picric acid. Red isopurpurate is formed. It is possible to estimate 80 μg in 25 ml to within 2%. 1 ppm can be detected. The colour, which develops rapidly in the hot, is very stable.

Interfering compounds. Aldehydes, acetone, and hydrogen sulphide give the same coloration.

COMPARE

F. B. Fischer and J. S. Brown, *Anal. Chem.*, 24 (1952) 1440.

On paper
M. T. Francois and N. Laffitte, *Bull. Soc. Chim. Biol.*, 17 (1935) 1088.
P. G. Hogg and H. L. Ahlgren, *J. Am. Soc. Agron.*, 34 (1942) 199.

(iv) Oxidizing power of Cu(II) in the presence of CN^-. In the presence of CN^-, Cu(II) acquires powerful oxidizing characteristics, owing to the formation of extremely stable cuprocyanide complexes. Under these conditions, many redox indicators are oxidized. A particularly suitable indicator is phenolphtaline, which is oxidized to phenolphthalein (red in an alkaline medium). Other sufficiently energetic oxidizing agents give the same reaction.

COMPARE

W. O. Winkler, *J. Assoc. Offic. Agr. Chem.*, 24 (1941) 380.
W. A. Robbie, *Arch. Biochem.*, 5 (1944) 49.

Oxidation to CNCl and formation of a dye (Down to 0.2 μg to within ± 4%)
G. Epstein, *Anal. Chem.*, 19 (1947) 272.
J. M. Kruse and M. G. Mellon, *Anal. Chem.*, 25 (1953) 446.

With reagent Hg(II) diphenylcarbazone
J. Trtilek, *Coll. Trav. Chim. Tchec.*, 10 (1938) 242.

(5) Cyanates CNO^-

(5a) Separation

The cyanate of Cu(II) and pyridine can be extracted with chloroform. CN^- interferes (see Colorimetry).

COMPARE

S. TAKEI *Japan Analyst*, 4 (1955) 479.

(5b) Colorimetry

By means of the Cu(II) – pyridine–CNO^- complexes.

E. L. MARTIN AND J. MACCLELLAND, *Anal. Chem.*, 23 (1951) 1519.

(6) Thiocyanates SCN^-

(6a) Separation

AgSCN can be precipitated in dilute nitric acid. The thiocyanate radical may then be separated from the silver by treating the precipitate with hydrogen sulphide: SCN^- passes back into solution and Ag_2S remains insoluble. The method is also suitable for trace amounts.

CuSCN is sparingly soluble: it may be precipitated by the addition of copper sulphate and bisulphite to a thiocyanate solution.

Separation of CN^- by volatilisation of HCN (p. 212).

Separation of ferrocyanide by precipitating thorium ferrocyanide from slightly acid solutions.

Separation of ferricyanide by the precipitation of cadmium ferricyanide.

Standard solutions. Potassium thiocyanate with a guaranteed purity of 99.5–100% is available.

$$SCN^-/KSCN = 0.597.$$

Alternatively, a standardized solution of thiocyanate may be used. 1 ml of 0.1000 *N* solution contains 5.82 mg of SCN^-.

(6b) Colorimetry

(i) Ferric thiocyanate. In the presence of an excess of Fe(III) the red $FeSCN^{2+}$ complex is formed (cf. p. 272).

The colour fades in light, which catalyzes the oxidation of SCN^- by Fe(III).

The following interfere: Hg(II), Ni(II), CN^-, and CNO^-. The last two of these can be driven off in an acid medium. $Fe(CN)_6{}^{3-}$ and $Fe(CN)_6{}^{4-}$ interfere.

In biochemistry

W. N. POWELL, *J. Lab. Clin. Med.*, 30 (1945) 1071.
P. DE FRANCISCIS, *Bull. Soc. Ital. Biol. Sper.*, 22 (1946) 779.

(ii) Thiocyanate of Cu(II) and pyridine. A blue to green complex, which can be extracted by chloroform, is formed.

It is usual to estimate 20 to 200 μg of SCN^- in this way. The accuracy is of the order of 4%.

REAGENTS

Copper sulphate, crystalline, 10%;
Pyridine;
Chloroform.

OPERATING PROCEDURE

2 ml of copper sulphate and 4 ml of pyridine are added to 10 ml of solution brought to a pH of about 2.5–4 and containing 20–200 μg of thiocyanate. The complex is extracted with 25 ml of chloroform, and colorimetry is then performed at 410 mμ.

cf. R. LANG, *Biochem. Z.*, 262 (1933) 14.
J. M. KRUSE AND M. G. MELLON, *Anal. Chem.*, 25 (1953) 446.

With the reagent Hg(II) - diphenylcarbazone

J. TRTILEK, *Coll. Trav. Chim. Tchec.*, 10 (1938) 242.

(7) Ferricyanides $Fe(CN)_6^{3-}$

Colorimetry

By means of redox indicators.

o-Dianisidine

B. BUSCARONS AND Y. ARTIGAS, *Anal. Chim. Acta*, 19 (1958) 434.

Cerium

Ce = 140.1

(1) Separation of trace elements

For Ce(III), see the Rare-Earth Elements. Traces of Ce(III) may be entrained by La(III).

(i) Separation of the rare-earths

Precipitation of the hydroxide of Ce(IV). Cerium may be converted to its tetravalent state by a vigorous oxidizing agent and be precipitated as $Ce(OH)_4$ from acid solutions (pH 2). The method is not very satisfactory; $Th(OH)_4$ also precipitates.

Precipitation of Ce(IV) iodate. The iodates of Ce(IV) and Th(IV) are precipitated from about 0.4 *N* nitric acid. Ce(III) may be oxidized to Ce(IV) with bromate or persulphate.

The precipitation is quantitative if suitable conditions of heating, stirring, and time of precipitation are observed.

In a homogeneous medium

H. H. Willard and S. T. Yu, *Anal. Chem.*, 25 (1953) 1754.

Z. Hagiwara, *Technol. Rept. Tohoku Univ.*, 17 (1952) 70, 77.

Extraction of Ce(IV). Ce(IV) may be extracted from an HNO_3 medium by ether, ketones, or esters. It is not separated from U(VI), Th(IV), or Zr(IV).

cf. A. W. Wylie, *J. Chem. Soc.*, (1951) 1474.

L. E. Glendemin, K. F. Flynn, R. F. Buchanan and E. P. Steinberg, *Anal. Chem.*, 27 (1955) 59.

B. D. Blaustein and J. W. Gryder, *J. Am. Chem. Soc.*, 79 (1957) 540.

(ii) Standard solutions. These can be made from pure CeO_2. 1 g of CeO_2 is heated with 100 ml of 12 *N* (1/3) sulphuric acid to complete dissolution. The solution is then made up to 1000 ml.

$$Ce/CeO_2 = 0.814$$

1 ml of a standardized 0.1000 *N* solution of Ce(IV) corresponds to 14.0 mg of cerium.

(2) Colorimetry

Ce(IV). Ceric ions are yellow, with an absorption maximum at 320 mμ.

Sensitivity. $\varepsilon = 6000$ at 320 mμ in $\geqslant 0.1\ N\ H_2SO_4$.

Interfering ions. If Ce(III) has been oxidized to Ce(IV) with persulphate, most of the excess of the latter must be destroyed, since it absorbs light at about 320 mμ. Since NO_3^- also absorbs, ammonium persulphate, which gives traces of nitrate by oxidation, cannot be employed.

Many substances interfere: Fe(III), Mn(II), Cr(VI), V(V), U(VI), and Th(IV). A blank test can be carried out after the selective reduction of Ce(IV) by hydrogen peroxide or by oxalic acid. Cl^-, F^-, and P(V) interfere. The interference due to Fe(III), U(VI), Th(IV), and V(V) is much less at 350 mμ.

OPERATING PROCEDURE

A 10 ml of a solution containing 0.01 to 0.2 mg of Ce, and about 0.1 *N* with respect to sulphuric acid, should be used. This is boiled with 0.5 mg of silver nitrate and 25 mg of sodium or potassium persulphate for 5 minutes. The volume is made up to 10 ml after cooling. Colorimetry is performed at 320 or 350 mμ.

cf. A. J. FREEDMAN AND D. N. HUME, *Anal. Chem.*, 22 (1950) 932.
A. I. MEDALIA AND B. J. BYRNE, *Anal. Chem.*, 23 (1951) 453.

In uranium

J. HURÉ AND R. SAINT JAMES-SCHONBERG, *Anal. Chim. Acta*, 9 (1953) 415.

Ce(III) in H_2SO_4 absorbs in the ultraviolet at 253.6 mμ.

H. L. GREENHAUS, A. M. FEIBUSH AND L. GORDON, *Anal. Chem.*, 29 (1957) 1531.

Ce(III) oxinate dissolves in chloroform to give a red-brown solution at pH 9.9–10.5 with an absorption maximum at 505 mμ.

In steels

W. WESTWOOD AND A. MAYER, *Analyst*, 75 (1948) 275.

Cesium

See **Rubidium** and **Cesium,** p. 361.

Chlorine and its compounds

Cl = 35.46

Chlorine

Standard solution. A solution of chlorine in water may be standardized immediately before use by an iodometric titration.

Colorimetry

Using o-tolidine. o-Tolidine is a redox indicator, which is yellow in the oxidized form; this coloration is unstable.

Limit of sensitivity. 0.01 ppm.

Interfering ions. Many strong oxidizing agents give the same reaction. It is sometimes possible to calculate the chlorine content from the difference between two measurements – one on the initial solution and the other on a solution from which Cl_2 has been expelled by boiling (blank test).

Fe^{3+} and particularly NO_2^- interfere.

REAGENT

1 g of *o*-tolidine is dissolved in 5 ml of 2 *N* (1/6) hydrochloric acid and the volume is made up to 1000 ml with 2 *N* hydrochloric acid.

Standards. Chlorine water is titrated immediately after preparation, diluted at once, and used immediately.

OPERATING PROCEDURE

The solution, which should contain less than 1 ppm of chlorine, is treated with 1 ml of the reagent per 100 ml. Colorimetry is carried out at 438 mμ, after 5 minutes.

Artificial standards consisting of mixtures of copper sulphate and dichromate are used.

cf. H. C. MARKS AND R. R. JOINER, *Anal. Chem.*, 20 (1948) 1197.
J. VERBESTEL AND A. BERGER, *Bull. Centre Belge Etude Doc. Eaux*, no. III (1950) 545.

VARIOUS METHODS

With 3,3'-dimethylnaphthidine

R. BELCHER, A. J. NUTTEN AND W. I. STEPHEN, *Anal. Chem.*, 26 (1954) 772.

Decoloration of helianthin. The reaction is more specific than the above; for example, chloramine T does not interfere.

M. TARUS, *Anal. Chem.*, 19 (1947) 342.

By extraction by CCl_4 and absorption in the ultraviolet

M. I. SHERMAN AND J. D. STRICKLAND, *Anal. Chem.*, 27 (1955) 1778.

Chlorides

Standard solutions. These can be made from a chloride of guaranteed purity, such as KCl, or from a standardized hydrochloric acid.

$$Cl/KCl = 0.476$$

1 ml of 1.000 *N* hydrochloric acid contains 35.5 mg of Cl^-.

Colorimetry

(i) As elementary chlorine. Chlorine is liberated by oxidation in an acid medium. The Cl_2 evolved is estimated colorimetrically (p. 219).

REAGENTS

Dilute sulphuric acid, 14 *N* (4/10).

Potassium periodate.

Standard solution of chloride: 2.1 g of potassium chloride in 1000 ml. The solution should be diluted 1000-fold in order to obtain 1 μg of Cl^- per ml during the test.

Distillation apparatus. A suitable apparatus consists of a 500 ml flask with a 10 cm Vigreux column; a glass wool plug is placed at the top of the column in order to prevent mechanical entrainment.

o-Tolidine solution: 1 g of reagent suspended in 5 ml of 2 *N* (1/5) hydrochloric acid. 500 ml of water are added, followed by 500 ml of 2 *N* hydrochloric acid.

OPERATING PROCEDURE

50 ml of sulphuric acid and 2 g of periodate are placed in the flask; 10 ml are distilled over and collected in a mixture of 1 ml of *o*-tolidine and 9 ml of water. (The tube leading from the condenser should dip into the reagent). Distillation should take place in about 10 minutes. The object of this procedure is to free the reagents from any chlorides present in them. The reagent solution becomes yellow. 10 ml of water are added to the flask and the distillation is repeated in order to check that no more chlorine passes over; when this is so, it may be assumed that there is no mechanical entrainment, that the distilled water does not contain chlorides, and that the reagents have been well purified.

The substance (solution or solid) to be analyzed is then added, and the volume made up to that present before the preliminary distillation, with chlorine-free water. This solution is distilled, and the volume of the distillate is made up to 25 ml; colorimetry is then performed at 438 mμ.

Comparison can be made with artificial standards consisting of mixtures of copper sulphate and dichromate (see Colorimetry of chlorine).

(ii) Iodometry. The following reaction takes place when the solution is shaken with silver iodate.

$$AgIO_3\downarrow + Cl^- \rightarrow AgCl\downarrow + IO_3^-$$

This is followed by:

$$IO_3^- + 6H^+ + 5I^- \rightarrow 3I_2 + 3H_2O$$

Finally, the iodine is determined colorimetrically.

REAGENTS

Silver iodate;

Potassium iodide, 20%.

OPERATING PROCEDURE

60 mg of silver iodate are added to 5–10 ml of solution and the mixture is shaken vigorously for 1 minute. After centrifuging, the liquid is made up to a known volume. An aliquot portion is made up to 5 ml, mixed with 1 ml of the iodide solution and several drops of acid, and allowed to stand for 10 minutes. Iodine is then determined colorimetrically.

cf. H. A. STIFF, JR., *J. Biol. Chem.*, 172 (1948) 695.

(iii) Ferrichloride complexes. Fe^{3+}, which is colourless, combines with Cl^- to give yellow complexes such as $FeCl^{2+}$, which absorb in the ultraviolet.

The optimum conditions correspond to 5 ppm of chlorine.

Interfering ions. The large number of interfering ions in existence may be classified as follows: ions which give complexes with Cl^-, *e.g.* Hg^{2+}; a large class of ions which give coloured complexes, with Fe^{3+}-sulphate, thiocyanate, etc. In contrast, ions which give colourless complexes, with Fe^{3+} – for example, fluorides, do not interfere, and in such cases an excess of ferric ions should be added.

REAGENTS

Solid ferric perchlorate, purified by repeated agitation with conc. $HClO_4$ until yellow colour in the acid disappears, and filtered on to sintered glass. 120 g of the product is dissolved in a mixture of 540 ml of 60% $HClO_4$ and 460 ml of water.

OPERATING PROCEDURE

4.9 ml of the reagent are mixed with 5 ml of the solution to be determined, (which should not contain more than 100 ppm of Cl^-) and the volume is made up to 10 ml. The absorption is determined at 348 mμ, and compared with that of a blank sample. The temperature should be taken into account.

cf. P. W. WEST AND H. COLL, *Anal. Chem.*, 28 (1956) 1834.

(iv) Coloration of ferrithiocyanate. SCN^- is displaced from mercuric thiocyanate by the formation of a mercuric chloride complex. The SCN^- is then determined by measuring the amount of ferric thiocyanate colorimetrically.

As little as 0.05 ppm can be determined in this way.

Interfering ions. Interference is obtained with a large class of ions which give complexes with divalent mercury. They can be estimated in the same way.

REAGENTS

Mercuric thiocyanate, saturated solution, about 0.07%. Ferric perchlorate. 6% solution in 4 *N* (4/10) $HClO_4$.

The reagent can also be prepared as follows: 14.0 g of iron are dissolved in dilute HNO_3. After dissolution, 120 ml of $HClO_4$ are added, and the mixture is heated until white fumes of perchloric acid are produced. Heating is continued until the solution becomes purple, and the volume is made up to 1000 ml after cooling.

OPERATING PROCEDURE

10 ml of the solution, which should contain 0.05 to 5 ppm are placed in a 50 ml measuring flask. 5 ml of 60% perchoric acid and 1 ml of mercuric thiocyanate are added, followed by 2 ml of ferric perchlorate. The volume is made up to 50 ml, and the solution is stirred and left for 10 minutes. Colorimetry is performed at 480 mμ (see p. 215).

COMPARE

D. M. ZALL, D. FISHER AND M. Q. GARNER, *Anal. Chem.*, 28 (1956) 1665.
J. S. SWAIN, *Chem. Ind., London* (1956) 418.
I. IWASAKI, S. UTSUMI, K. HAGINO AND T, OZAWA, *Bull. Chem. Soc. Japan*, 29 (1956) 860

In H_2O_2
I. GELD AND I. STERNMAN, *Anal. Chem.*, 31 (1959) 1662.

(v) With Hg(II) diphenylcarbabazide. A complex is formed between Hg^{2+} and Cl^-. The loss of coloration of an unstable coloured complex of Hg(II), for exemple that formed with diphenylcarbazide, is determined.

Interfering ions. NH_4^+, Co^{2+}, Cu^{2+}, Cr(VI), Fe^{3+}, Pb^{2+}, Zn^{2+}, SO_4^{2-}, etc.

REAGENTS

Mercuric nitrate, 0.025% in 0,008 *N* nitric acid.

Diphenylcarbazide, 1% in methanol.

Standard solutions of chloride containing 0.020 mg per ml prepared from pure NaCl.

OPERATING PROCEDURE

10 ml of the mercuric solution are placed in a 100 ml measuring flask. The chloride solution (neutralized to pH 5–6) is added, followed by 5 ml of reagent; the total volume is made up to 100 ml.

The colorimetric measurement should be performed 20 minutes later, at 525 mμ.

COMPARE

D. F. BOLTZ, *Colorimetric Determination of Non-Metals*, Interscience, 1958.

VARIOUS METHODS

By means of Hg(II) and diphenylcarbazone. Limit < 0.1 ppm.

F. E. CLARKE, *Anal. Chem.*, 22 (1950) 553.

J. L. GERBACH AND R. G. FRAZIER, *Anal. Chem.*, 30 (1958) 1142.

With mercuric chloranilate

$HgR^{2-} + 2Cl^- + H^+ \rightarrow HgCl_2 + HR^-$. Limit of sensitivity: 0.2 ppm.

J. E. BARNEY AND R. J. BERTOLACINI, *Anal. Chem.*, 29 (1957) 1187.

(vi) Nephelometry of silver chloride. This is a very simple but not very accurate procedure, in common with the majority of nephelometric methods. In addition, the selectivity is not very high since all the ions which precipitate Ag^+ in acid solutions will interfere.

Limit of sensitivity: 2 ppm.

OPERATING PROCEDURE

For the best results, the solution should contain 200 μg of chloride in 20 ml. This is acidified with 10 ml of 0.1 *N* nitric acid, and mixed with 10 ml of 0.1% silver nitrate. The mixture is held at 40° for thirty minutes. Blanks should be prepared at the same time.

The addition of surface-active materials has been recommended[1].

cf. A. B. LAMB, P. W. CARLETON AND W. B. MELDRUM, *J. Am. Chem. Soc.*, 42 (1920) 251.

Absorption curve of chlorides

M. A. DESESA AND L. B. ROGERS, *Anal. Chim. Acta*, 6 (1952) 534.

(1) P. BLANC, P. BERTRAND AND LIANDIER, *Chim. Anal.*, 38 (1956) 156.

Hypochlorites

A large number of redox indicator dyes (e.g. indigo carmine) are selectively oxidized in an alkaline solution by hypochlorites.

Chlorates

Standard solutions. These can be made from potassium chlorate of guaranteed purity.

$$ClO_3^-/KClO_3 = 0.681$$

Colorimetry

In sufficiently concentrated hydrochloric acid, chlorine is formed. The methods indicated for chlorine may then be used.

With o-Tolidine

D. Williams and G. S. Haines, *Ind. Eng. Chem., Anal. Ed.*, 17 (1945) 538.

With Benzidine

E. A. Burns, *Anal. Chem.*, 32 (1960) 1800.

Perchlorates

Standard solutions. Perchloric acid of guaranteed purity may be used. The solution should be diluted about ten times *(N)* before an acidimetric titration is performed. After standardization, it may be diluted further to a suitable concentration.

$KClO_4$ can also be used:

$$ClO_4/KClO_4 = 0.718$$

Colorimetry

With methylene blue. With ClO_4^-, methylene blue gives a sparingly soluble violet compound which can be extracted by chloroform.

Interfering ions. BF_4^-, MnO_4^-, ReO_4^-, TcO_4^-, and IO_4^- have the same properties.

Reagents

Methylene blue, 1.6% in water.

Chloroform.

Standard solution of $NaClO_4$ – $NaClO_4 \cdot 2H_2O$ is weighed out, dissolved, and diluted to obtain standard solutions.

OPERATING PROCEDURE

The solution is neutralized to pH 5–7. 1 ml of reagent is added, followed by water to about 30 ml. Extraction is performed with 20 ml portions of chloroform (5 times). The extract is made up to a known volume, 100 ml, and a colorimetric measurement is performed at 655 mμ.

cf. D. F. BOLTZ, *Colorimetric Determination of Non-Metals*, Interscience, 1958.

Chromium

$Cr = 52.0$

(1) Separation of trace amounts

(i) Extraction of Cr(VI). Cr(VI) can be extracted from an HCl medium by hexone (4-methylpentan-2-one). Only large amounts of Fe(III) will interfere.

cf. H. A. BRYAN AND J. A. DEAN, *Anal. Chem.*, 29 (1957) 1289.
P. D. BLUNDY, *Analyst*, 83 (1951) 555.

With trioctylphosphine oxide.
C. K. MANN AND J. C. WHITE, *Anal. Chem.*, 30 (1958) 989.

(ii) Extraction of perchromic acid. Perchromic acid can be separated from large amounts of V(V) by extraction with ethyl acetate.

cf. R. K. BROOKSHIER AND H. FREUND, *Anal. Chem.*, 23 (1951) 1110.

(iii) Extraction of the oxinates. Cr(VI) is not extracted, and may thus be separated from V(V) by extracting vanadium oxinate with chloroform at pH 4.

cf. E. B. SANDELL, *Ind. Eng. Chem., Anal. Ed.*, 8 (1936) 336.

(iv) With cupferrates. Cr(III) is separated from V(V) and from Fe(III) by extracting or precipitating the cupferrates of the latter in acid solutions. Cr(III) remains in the aqueous phase.

(v) Standard solutions. $K_2Cr_2O_7$ and K_2CrO_4 of guaranteed purity are available.

$$2Cr/K_2Cr_2O_7 = 0.353$$

A 0.1000 *N* dichromate solution contains 1.733 mg of Cr per millilitre.

(2) Colorimetry

When large amounts of chromium are present, colorimetry of the yellow chromate CrO_4^{2-} can be carried out at pH > 9. It is also possible to determine dichromates colorimetrically in an acid medium; the results then depend on the acidity of the solution. Such colorimetric determinations can be very accurate.

Trace amounts may be estimated by a more sensitive method involving a reaction between the dichromate and diphenylcarbazide.

(i) Oxidation to chromate and dichromate. The Cr(III) which is usually present must be oxidized to Cr(VI).

The following methods are suitable for the estimation of trace amounts:

(a) Attack by fused sodium carbonate plus sodium peroxide. The green manganate formed is destroyed by dilution followed by boiling.

(b) Aerial oxidation during the carbonate fusion (if nitrate were added, the platinum crucible would be attacked and an interfering coloration would be obtained).

(c) Oxidation can be carried out in an acid medium with persulphate in the presence of Ag^+ ions, but a haze of AgCl is formed with traces of chlorides from the reagents, and the colorimetry is upset.

Boiling perchloric acid is not suitable for trace amounts; it is however advantageous, since it eliminates Cl^- and SO_4^{2-}, leaving the ferric solutions colourless.

Oxidation by periodate in a strongly acidic medium is a slow process; it should be carried out at pH > 2. Under these conditions, it is the best method.

Bismuthate may also be used.

Oxidation may also be performed with boiling permanganate. The excess is then destroyed by N_3^-; the same applies to Ce(IV).

NOTE. When oxidation is carried out with persulphate or bismuthate, a little hydrogen peroxide is produced, the last traces of which are difficult to destroy by boiling. In such a case, the diphenylcarbazide colorimetry yields erroneous results.

cf. B. E. SALTZMAN, *Anal. Chem.*, 24 (1952) 1016.
P. D. BLUNDY, *Analyst*, 83 (1958) 555.

(ii) Colorimetry of the chromate. The pH is adjusted to > 9, for example, by the addition of borax (borate buffer).

1 to 10 ppm can be estimated in this way. A precise colorimetric determination may be performed. At 456 mμ, the molar absorption coefficient is $\varepsilon \sim 300$; at 375 mμ it is $\varepsilon \sim 5000$.

Interfering ions. Interference is obtained from a large number of ions which precipitate at this pH. Colloidal $Fe(OH)_3$ may interfere. Mo(VI) and W(VI) do not interfere. V(V) interferes little at 375 mμ. MnO_4^- interferes, but may be reduced selectively in an acid medium with a solution of oxalic acid, added drop by drop. The presence of a little Cu(II) makes the solution blue. U(VI) and Ce(IV) also colour the solution.

(iii) Colorimetry of the dichromate. The colour depends on the acidity, which should therefore be fixed. Between 2 to 40 ppm can be estimated to within 0.6%. The absorption maximum is in the ultraviolet at 349 mμ, $\varepsilon \sim 1500$;

Cr(III) does not absorb at this wavelength, and the blank test is therefore simplified. It is possible to work within the visible spectrum at the shortest wavelength permitted by the apparatus.

Interfering ions. Fe(III) gives a coloration if chloride or sulphate ions are present. H_3PO_4 interferes, since unstable complexes are formed with Cr(VI).

Many other coloured ions can interfere: Cu(II), Co(II), Ni(II), U(VI), Ce(IV), and V(V).

If trace amounts are being determined colorimetrically, the result of the blank test obtained after reduction of the dichromate solution can be subtracted provided the parasitic colorations are not too excessive.

When manganese is present in the form of MnO_4^-, it may be reduced selectively by the dropwise addition of a solution of oxalic acid.

In addition, Cr and Mn may easily be estimated by working at two wavelengths.

OPERATING PROCEDURE (S. Lacroix)

25 ml of solution containing less than 5 mg of chromium (III) are mixed with 400 mg of potassium periodate and the solution is boiled. Caustic soda is added until the pH is between 2 and 3 (indicator paper). If Fe(III) is present, ferric periodate may precipitate. The solution is boiled gently for 5 minutes, and then 5 ml of 9 *N* (1/4) sulphuric acid and 1 ml of concentrated (45 *N*) phosphoric acid are added. The precipitate is dissolved by boiling, if necessary. The volume is made up to 100 ml after cooling. The usual blank test is performed on the solution as it was before oxidation. It is in this case necessary to take the presence of the slightly coloured Cr(III) into account.

At 349 mμ

M. J. CARDONE AND J. COMPTON, *Anal. Chem.*, 24 (1952) 1903.

In steels

F. E. EBORALL, *Metallurgia*, 35 (1946) 104.
H. COX, *Metallurgia*, 37 (1948) 270.
A. A. R. WOOD, *Analyst*, 78 (1953) 54.

Cr + Mn at two wavelengths

R. W. SILVERTHORN AND J. A. CURTIS, *Metals and Alloys*, 15 (1942) 245.
S. LACROIX AND M. LABALADE, *Anal. Chim. Acta*, 3 (1949) 262.
J. J. LINGANE AND J. W. COLLAT, *Anal. Chem.*, 22 (1950) 166.

In aluminium and bauxites

S. LACROIX AND M. LABALADE, *Anal. Chim. Acta*, 3 (1949) 262.

Absorption curves and estimation of Cr(VI) + Cr(III)

D. T. ENGLIS AND L. A. WOLLERMAN, *Anal. Chem.*, 24 (1952) 1983.

(iii) With s-diphenylcarbazide (1.5-diphenylcarbohydrazide). Under acid conditions, $Cr_2O_7^{2-}$ ions give a violet coloration with this reagent. The coloration can be extracted with butanol.

Colorimetry is carried out at 540 mμ. The molar absorption coefficient ε is 30,000. The coloration is stable.

Interfering ions. Oxidizing agents destroy the compound formed. Consequently, if persulphate has been used to oxidize Cr(III) to Cr(VI), its excess must be destroyed. Oxidation can be carried out advantageously with MnO_4^-.

Mo(VI) gives a slight coloration, (at 540 mμ, ε ∼ 500), which disappears on the addition of oxalate ions. Fe(III) interferes only slightly. V(V) gives a red-brown coloration which disappears after 15 minutes. Nevertheless, if the latter is present in a large amount, it should preferably be separated.

Hg(I), Hg(II), Ag(I), Cu(II), Pb(II), and Au(III) interfere to some extent, and Co(II), and Ni(II), a little more.

Reagents

Potassium permanganate: 3 g in 1 l of water (0.02 *M*).

Sodium azide: 5 g in 100 ml of water.

Sulphuric acid 5 *N*: 10 ml of concentrated sulphuric acid are poured into 60 ml of water.

Diphenylcarbazide: 0.5 g in 100 ml of acetone containing 0.5 ml of 5 *N* sulphuric acid.

Standard solution of chromium. 0.2263 g of potassium dichromate of guaranteed purity are dissolved in 1 l of water. This solution contains 80 μg of chromium per ml. 5 ml of this solution is mixed with 25 ml of water, 1 to 2 ml of 5 *N* sulphuric acid, and a few milligrams of sodium sulphite. The solution is boiled gently to expel the excess of SO_2, and the volume made up to 200 ml. This solution contains 2 μg of chromium per ml.

Operating procedure

1 ml of 5 *N* sulphuric acid, followed by 0.5 ml of permanganate are added to 10 ml of solution, this portion containing 2 to 15 μg of chromium. The mixture is heated for 20 minutes on a water bath. The excess of permanganate is destroyed by adding azide to the hot solution, one drop every 5 to 10 seconds. Azide should be added until all the brown colour has disappeared, and any excess should be avoided. The solution is immediately cooled and transferred to a 25 ml measuring flask; 1 ml of diphenylcarbazide is added, and the mixture held for 1–2 minutes. Colorimetry is then performed at 540 mμ.

COMPARE

B. E. Saltzman, *Anal. Chem.*, 24 (1952) 1016.
L. Erdey and J. Inczedy, *Acta Chim. Hung.*, 4 (1954) 289.
P. F. Urone, *Anal. Chem.*, 27 (1955) 1354.
T. L. Allen, *Anal. Chem.*, 30 (1958) 447.

In rocks

E. B. Sandell, *Ind. Eng. Chem., Anal. Ed.*, 8 (1936) 336.
C. F. J. van der Walt and A. J. van der Merwe, *Analyst*, 63 (1938) 809.

In titanium alloys

G. Norwitz and M. Codell, *Anal. Chim. Acta*, 9 (1953) 546.

In soils

A. M. M. Davidson and R. L. Mitchell, *J. Soc. Chem. Ind.*, 59 (1940) 232.

In steels (down to 0.05%)

W. Koch, *Techn. Mitt. Krupp, Forschungsber.*, 2 (1938) 37.

In biochemistry

H. J. Cahnmann and R. Bisen, *Anal. Chem.*, 24 (1952) 1341.
B. E. Saltzman, *Anal. Chem.*, 24 (1952) 1016.
C. H. Grogan, H. J. Cahnann and E. Lethco, *Anal. Chem.*, 27 (1955) 983.

Extraction of perchromic acid with ethyl acetate (Fe(III), Mn(II), V(V), and Mo(VI) do not interfere)

A. Glasner and M. Steinberg, *Anal. Chem.*, 27 (1955) 2008.

Extraction of Cr(VI) by hexone followed by reaction with diphenylcarbazide (ε = 27,000 at 540 mμ)

J. A. Dean and M. L. Beverly, *Anal. Chem.*, 30 (1958) 977.

Cobalt

Co = 58.9

(1) Separation of trace amounts

(i) Precipitation and extraction with nitrosonaphthol. Cobalt forms compounds with 1-nitroso-2-naphthol and with 2-nitroso-1-naphthol, which are sparingly soluble in water and may be extracted by numerous solvents, such as chloroform (see Colorimetry).

cf. E. COGAN, *Anal. Chem.*, 32 (1960) 973.

(ii) Extraction of the thiocyanate. See Colorimetry.

(iii) Extraction of the thiocyanate tetraphenylarsonium complex with chloroform. See Colorimetry.

(iv) Extraction of dithizonate. Co(II) can be extracted, most easily at pH 7–9, with carbon tetrachloride. The Co(II) passes back into the aqueous phase under acid conditions, and may thus be separated from Cu(II), Ag(I), and Hg(II). Ni(II) follows Co(II).

cf. G. H. ELLIS AND J. F. THOMPSON, *Ind. Eng. Chem., Anal. Ed.* 17 (1945) 254.

(v) Extraction of the diethyldithiocarbamate. The compounds formed with Co(II) and Cu(II) can be extracted with carbon tetrachloride from solutions of pH 6.5; these ions may thus be separated from many other elements. The addition of citrate prevents the precipitation of hydroxides.

REAGENTS

Ammonium citrate, crystalline, in 40% solution.
Diethyldithiocarbamate, 0.1% in water.
Bromothymol blue, 0.05% in alcohol.

OPERATING PROCEDURE

1 ml of citrate and 2 drops of indicator are added to the solution. The pH is adjusted to about 6.5, and 5 ml of reagent are added, followed by carbon tetrachloride in portions of 10–20 ml; the extraction is performed after shaking for 10 minutes. More reagent is then added, and the operation is repeated.

(vi) Using anion-exchangers. In 9 *M* HCl, Co(II) can be fixed in the form of chloride ions on a suitable resin – for example, Dowex 1–X8. Cu(II), Zn(II), and Fe(III), are also collected and may thus be separated from Ni(II), Mn(II), etc. On eluting with 4 *N* HCl, only Co(II) is eluted from the column.

cf. R. E. THIERS, J. F. WILLIAMS AND J. H. YOE, *Anal. Chem.*, 27 (1955) 1725.
W. J. WILLIAMS, *Talanta*, 1 (1958) 88.
D. H. WILKINS, *Anal. Chim. Acta*, 20 (1959) 271.

(vii) Internal electrolysis. Cobalt may be separated in this way from large amounts of Zn(II).

cf. J. C. FIFE, *Analyst*, 66 (1941) 192.

(viii) Standard solutions. These can be made from cobalt sulphate heated at 400–500° to constant weight; anhydrous $CoSO_4$ is obtained.

Alternatively, the chloride $CoCl_2 \cdot 6H_2O$ and the nitrate $Co(NO_3)_2 \cdot 6H_2O$, of guaranteed purity, may be used.

$$Co/CoSO_4 = 0.380$$
$$Co/CoSO_4 \cdot 7H_2O = 0.2095$$
$$Co/CoCl_2 \cdot 6H_2O = 0.2475$$
$$Co/Co(NO_3)_2 \cdot 6H_2O = 0.2024$$

(2) Colorimetry

The method involving thiocyanate is simple but not very sensitive. In the presence of tetraphenylarsonium, a complex which can be extracted by chloroform is obtained and subsequent colorimetry can easily be made specific. Colorimetry with nitroso R salt is sensitive but preliminary separations are necessary. The method involving the extraction of the compound with 2-nitroso–1-naphthol is both sensitive and selective.

(i) With thiocyanate. In a sufficiently concentrated thiocyanate solution, Co(II) gives a blue complex which can be extracted with ketones, alcohols and ethers. In aqueous solution, acetone, which diminishes the dielectric constant, also favours the formation of this complex.

Sensitivity. The method can be made 20 times more sensitive by extracting with an equal volume of solvent. $\varepsilon = \sim 1000$ at 620 mμ and 6000 at 310 mμ in iso-amyl alcohol; $\varepsilon = 20{,}000$ in cyclohexane at 620 mμ.

The colour fades with time.

Colorimetry with extraction. This method should be used if large amounts of Ni(II) and Fe(III) are present.

Interfering ions. A certain number of ions give colorations which can be extracted with a mixture of amyl alcohol and ether:

Fe(III) gives rise to red complexes which can be destroyed by ammonium difluoride or pyrophosphate, or by reduction with ascorbic acid. Alternatively, when Fe(III) is present, spectrophotometry can be carried out at two wavelengths – 425 and 525 mμ.

V gives an extractable blue coloration; tartrate in an acetic buffer is then added.

Cu(II) can be complexed by thiosulphate or by thiourea.

Bi(III) gives an orange coloration.

REAGENTS

(a) Sodium thiosulphate, crystalline, 125 g; *solid trisodium phosphate,* 31 g; water, 1000 ml.

(b) Ammonium thiocyanate in 60% aqueous solution.

10 ml of (b) are added to 8 ml of (a).

OPERATING PROCEDURE

The mixture of reagents is added to 5 ml of solution of about pH 1, containing 10 to 100 ppm of Co(II). 7.5 ml of ether and 2.5 ml of amyl alcohol are added, and the mixture shaken vigorously. After separating, colorimetry is performed at 620 mμ or preferably in the ultraviolet range at 312 mμ.

Colorimetry without extraction. Interfering ions. Besides the ions indicated above, the following also interfere:

Ni(II) gives a not very appreciable green coloration which disappears when pyrophosphate is added. Large amounts of Ni(II) interfere.

Cr(III), U(VI), and Mo(VI) give colorations. Hg(I), Ag(I), and W(VI) precipitate. Cr(VI) and Ce(IV) also interfere. Hg(II) consumes the thiocyanate.

The following do not interfere even when present in amounts exceeding that of the Co(II) by a factor of 100 : Al(III), Sb(III), As(III and V), Be(II), Bi(III), Cd(II), Mn(II), P(V), Pb(II), Th(IV), Zn(II), and Zr(IV).

Acidity of the solution strongly influences the coloration obtained. The acid contents of the standard solutions and the solutions to be estimated should therefore be the same.

REAGENTS

Ammonium thiocyanate in 50% solution.

Sodium pyrophosphate, 10% in water.

OPERATING PROCEDURE

The initial solution should contain between 200 μg to 2 mg of cobalt. The acidity (hydrochloric acid) should be adjusted to 1 *N*. 0.1 ml of thiocyanate is added, the red colour due to Fe(III) appears. Pyrophosphate is then added, drop by drop, until the colour vanishes; the volume required is noted, and an excess of half as much again added, followed by 10 ml of thiocyanate. The total volume is made up to about 45 ml, and 50 ml of acetone are added. The volume is again made up to 100 ml and colorimetry is performed at 600 mμ.

COMPARE

R. S. Young and A. J. Hall, *Ind. Eng. Chem., Anal. Ed.*, 18 (1946) 264.
N. Uri, *Analyst*, 72 (1947) 478.
H. M. Putsché and W. F. Malooly, *Anal. Chem.*, 19 (1947) 236.
S. Ikeda, *Japan Analyst*, 2 (1953) 218.

In rocks
E. B. Sandell and R. W. Perlich, *Ind. Eng. Chem., Anal. Ed.*, 11 (1939) 309.

In metallurgical products
S. Hirano and M. Suzuki, *Japan. Analyst*, 2 (1953) 316.
J. Kinnunen, B. Merikanto and B. Wennerstrand, *Chemist-Analyst*, 43 (1954) 21.

Extraction by hexone
R. A. Sharp and G. Wilkinson, *J. Am. Chem. Soc.*, 77 (1955) 6519.

Extraction by butyl phosphate
L. M. Milnick, *Dissert. Abstr.*, 14 (1954) 760.

Simultaneous estimation of Fe(III), Co(II), and Cu(II) at three wavelengths
R. E. Kitson, *Anal. Chem.*, 22 (1950) 664.

Extraction of Co(II)-thiocyanate-tetraphenylarsonium complex by chloroform (ε = 1500 at 620 mμ)
L. P. Pepkowitz and J. L. Marley, *Anal. Chem.*, 27 (1955) 1330.
R. Lundquist, G. E. Markle and D. F. Boltz, *Anal. Chem.*, 27 (1955) 1731.

Extraction by tributylammonium thiocyanate (Ni(II) or Cr(VI) interferes)
M. Ziegler, O. Glemser and E. Preisler, *Angew. Chem.*, 68 (1956) 436.

Extraction with acetylacetone
J. O. Hibbits, A. F. Rosenberg and R. T. Williams, *Talanta*, 5 (1960) 250.

(ii) With nitroso R salt (sodium 1-nitroso-2-naphthol-3, 6-disulphonate).

A red coloration is obtained. The fact that the reagent itself is highly coloured must be taken into account, both in the calibration and in the blank test. The measurement can be made at 565–570 mμ, since the reagent possesses low absorption in this region of the spectrum.

It is also possible to work at pH 8.

Sensitivity. This is the most sensitive method; ε = 15,000 at 520 mμ.

Interfering ions. The method is not very specific and it is often necessary to carry out a preliminary separation.

The medium should contain citrate in order to avoid the precipitation of hydroxides.

NH_4^+ interferes, since at the pH considered, the cobalt can be complexed by NH_3.

The following interfere only slightly: Fe(III), Mn(II), Zn(II), Cd(II), Pb(II), and Sn(IV). Large amounts of Fe(III) can interfere, owing to the yellow colour of its citrate complexes.

Cu(II) and Ni(II) require an excess of reagent, but no interference takes place even when the amounts of these ions exceed that of Co by a factor of 1000 and 50 respectively.

Oxidizing and reducing agents (including Fe(II)) interfere. Many causes of interference can be suppressed in a fluoride medium[1].

REAGENTS

Nitroso R salt, 0.2% in water (stored in the absence of light).

Citric acid, 4.2 g in 100 ml.

Buffer solution: boric acid, 6.2 g; disodium phosphate, 35.6 g; 1.00 *N* caustic soda, 500 ml. The solution is made up to 1000 ml.

Standard solution of cobalt containing 0.0100% of Co.

OPERATING PROCEDURE

The solution containing 1 to 10 μg of cobalt in 5 ml, must be substantially neutral. If this is not the case, the acid present should be expelled by evaporating to dryness. 10 ml of citric acid and 1.2 ml of buffer solution are added. The pH should be in the neighbourhood of 8. The solution is shaken with 0.50 ml of reagent, accurately measured, boiled for 1 minute and mixed with 10 ml of concentrated nitric acid; boiling is continued for a further 1 minute. The solution is then cooled in a dim light and made up to 10 ml. Colorimetry is performed at 565–570 mμ.

COMPARE

T. C. J. OVENSTON AND C. A. PARKER, *Anal. Chim. Acta*, 4 (1950) 142.

H. W. BERKHOUT AND G. H. JONGEN, *Chem. Weekblad*, 49 (1953) 506.

(1) W. H. SHIPMEN, S. C. FOTI AND W. SIMON, *Anal. Chem.*, 27 (1955) 1240.

See also in Ti and Zr

D. F. WOOD AND R. T. CLARK, *Talanta*, 2 (1959) 1.

In soils

A. M. M. DAVIDSON AND R. L. MITCHELL, *J. Soc. Chem. Ind.*, 59 (1940) 232.

In biochemistry

N. S. BAYLISS AND E. W. PICKERING, *Ind. Eng. Chem., Anal. Ed.*, 18 (1946) 446.

B. E. SALTZMAN, *Anal. Chem.*, 27 (1955) 284.

In the presence of Fe(III)

F. W. HAYWOOD AND A. A. R. WOOD, *J. Soc. Chem. Ind.*, 62 (1943) 37.

J. N. PASCUAL, W. H. SHIPMAN AND W. SIMON, *Anal. Chem.*, 25 (1954) 1830.

In zinc

T. KATSURA, *Japan Analyst*, 4 (1955) 574.

In minerals

B. D. GUERIN, *Analyst*, 81 (1956) 409.

In Ni

J. KINNUNEN AND B. WENNERSTRAND, *Chemist-Analyst*, 42 (1953) 33.

In metallurgical products

R. S. YOUNG, E. T. PINKNEY AND R. DICK, *Ind. Eng. Chem., Anal. Ed.*, 18 (1946) 474.

In the presence of Cu(II)

A. J. HALL AND R. S. YOUNG, *Anal. Chem.*, 22 (1950) 497.

In Al

W. STROSS AND G. STROSS, *Metallurgia*, 45 (1952) 315.

(v) With 1-nitroso-2-naphthol. The red complex formed with cobalt can be extracted by various solvents. Cobalt can thus be separated from many ions. The excess of reagent, which is coloured, is also extracted. The operation is carried out in an acetic buffer.

Sensitivity. $\varepsilon = 26{,}500$ at 317 mμ in chloroform.

Interfering ions. Mn(II) is not extracted. Cu(II) and Ni(II), which are extracted, pass back into aqeous phase in the presence of 0.005 N hydrochloric acid. Fe(III) can be complexed with F^-. The precipitation of hydroxides can be prevented by the addition of citrate.

REAGENTS

Sodium fluoride in 5% solution.

Acetic buffer: 12 ml of acetic acid + 28 g of crystalline sodium acetate in 100 ml.

1-Nitroso-2-naphthol solution: 100 mg are dissolved in 20 ml of 0.1 N (1/100) caustic soda and, after cooling, the solution is made up to 200 ml and filtered.

OPERATING PROCEDURE

2 ml of fluoride, 5 ml of acetic buffer (if necessary) and 1 ml of reagent are added to 10 ml of the neutralized solution. The mixture is left to stand for 10 to 15 minutes, and extracted with chloroform.

The chloroform solution is washed several times with 0.005 N hydrochloric acid, and is then made up to 25 ml; colorimetry is performed at 425 mμ.

COMPARE

E. Boyland, *Analyst*, 71 (1946) 230.
H. Baron, *Z. Anal. Chem.*, 140 (1953) 173.
R. Lundquist, G. E. Markle and D. F. Boltz, *Anal. Chem.*, 27 (1955) 1731.

In soils

H. Almond, *Anal. Chem.*, 25 (1953) 166.

In steels

A. Claassen and A. Daamen, *Anal. Chim. Acta*, 12 (1955) 547.

(vi) Other methods

Extraction of the chloride. Cobalt chloride may be separated from nickel by extraction with tributyl phosphate.

cf. V. T. Athavale, S. V. Gulavane and M. M. Tillu, *Anal. Chim. Acta*, 23 (1960) 487.

2-Nitroso-1-naphthol. Iso-amyl acetate. Cu, Fe, Mn, Ni, Pd, Sn do not interfere.
Sensitivity. $\varepsilon = 15{,}000$ to $25{,}000$ at 530 mμ.

cf. L. J. Clark, *Anal. Chem.*, 30 (1958) 1153.

Diethyldithiocarbamate in ethyl acetate. After extraction, the organic phase is shaken with $HgCl_2$, which displaces Bi(III), Fe, Ni(II), Mn(II), and Cu(II), but not Co(II)[1].
The molar extinction coefficient, $\varepsilon = 14{,}000$ at 367 mμ in CCl_4.
It is also possible to estimate Ni(II) and Cu(II) simultaneously[2].

COMPARE

(1) R. Pribil, J. Jenik and M. Kobrova, *Coll. Trav. Chim. Tchec.*, 19 (1954) 470.
(2) J. M. Chilton, *Anal. Chem.*, 25 (1953) 1274.

Absorption of $CoSO_4$ [$NiSO_4$, $CuSO_4$, $Cr_2(SO_4)_3$].

J. Gagnon, *Chemist-Analyst*, 43 (1954) 15.

Copper

Cu = 63.5

(1) Separation of trace amounts

Copper may be separated by the familiar method of precipitating copper sulphide, using Pb(II) as the entraining agent.

cf. A. E. BALLARD AND C. D. W. THORNTON, *Ind. Eng. Chem., Anal. Ed.*, 13 (1941) 893.

(i) Extraction of the compound with dithizone. See Colorimetry.

(ii) Extraction of the compound with diethyldithiocarbamate. See Colorimetry.

(iii) Extraction of the compound with salicylaldoxime. Chloroform or amyl acetate are used. The complex formed may also be determined directly by flame photometry. Fe(III) is complexed by citrates.

cf. J. A. DEAN AND J. H. LADY, *Anal. Chem.*, 28 (1956) 1887.

(iv) Electrolysis. Copper, which may easily be reduced to the metallic state, is deposited from a medium containing sulphuric and nitric acids and may in this way be separated from Ni(II), Zn(II), Pb(II), As(III), As(V), and small amounts of Cd(II) and Fe(III).

The advantage of NO_3^- is that it is reduced at the cathode in preference to H^+ ions and thus prevents the evolution of hydrogen. If the amount of nitric acid is however excessive, the medium becomes too strongly oxidizing and the deposition of copper is incomplete. HNO_2, which interferes, can be expelled by boiling or destroyed by the addition of a little urea. Large amounts of Fe(III) interfere since this ion is cathodically reduced to Fe(II), which is subsequently reoxidized at the anode. This interferes with the reduction of Cu(II). Fe(III) can be complexed by EDTA, and As(V) and Sb(V) by F^-.

The following ions still interfere: As(III), Sb(III), Sn(IV), Mo(VI), Au(III), Pt(IV), Ag(I), Hg(II), Bi(III), Se(IV), Te(IV), SCN^-, and Cl^-.

COMPARE

In steels

W. S. LEVINE AND H. SEAMAN, *Ind. Eng. Chem., Anal. Ed.*, 16 (1944) 80.
L. SILVERMAN, W. GOODMAN AND D. WALTER, *Ind. Eng. Chem., Anal. Ed.*, 14 (1942) 236.

In aluminium alloys

S. WEINBERG, *Ind. Eng. Chem., Anal. Ed.*, 17 (1945) 197.

Semimicro- and microelectrolysis

W. M. MAC NEVIN AND R. A. BOURNIQUE, *Ind. Eng. Chem., Anal. Ed.*, 12 (1940) 431.
L. H. BRADFORD AND P. L. KIRK, *Ind. Eng. Chem., Anal. Ed.*, 13 (1941) 64.

Cu-Ni

A. J. LLACER, J. A. SOZZI AND A. BENEDETTI-PICHLER, *Ind. Eng. Chem., Anal. Ed.*, 13 (1941) 507.

In cast iron

W. M. MACNEVIN AND R. A. BOURNIQUE, *Ind. Eng. Chem., Anal. Ed.*, 15 (1943) 759.

(v) Standard solutions. These can be made from pure copper; 0.5 g is dissolved in 10 ml of 6 *N* (1/2) nitric acid and the solution is made up to 500 ml.

Alternatively, solutions of pure $CuSO_4 \cdot 5H_2O$ or of $CuCl_2 \cdot 2H_2O$ may be prepared.

$$Cu/CuSO_4 \cdot 5H_2O = 0.2540$$
$$Cu/CuCl_2 \cdot 2H_2O = 0.373$$

(2) Colorimetry

The most important methods are the following:

With diethyldithiocarbamate. By extracting the compound formed, minute traces may be estimated. The method can be made highly selective.

With dithizone. The method is highly selective, and may be used to detect smaller quantities than any other procedure.

With diquinolyl. Although less sensitive, this method is reasonably selective.

Differential colorimetry permits the estimation of large amounts of Cu(II) with high accuracy.

(i) With diethyldithiocarbamate. This method is often simple and rapid. With Cu(II) the reagent gives a yellow to brown coloration or a precipitate. The coloration forms instantaneously and is stable for 1 hour.

The coloured compound is soluble in carbon tetrachloride, chloroform, amyl alcohol, *n*-butanol, amyl acetate, etc. Such solutions are stable for 1 hour when completely protected from light. As a matter of principle, the extractions should be carried out in the dark if high accuracy is required. Chloroform solutions are the most stable. Sensitivity: $\varepsilon = 13{,}000$ at 436 mμ in carbon tetrachloride.

Interfering ions. Most ions which interfere can be complexed by citrate and ethylenediaminetetraacetic acid (EDTA). The selectivity may be increased still further by extraction.

Under such conditions, 100 mg of the following elements do not interfere: alkali metals, alkaline-earth metals, Al, As(III) and (V), borate, Be(II), Cd(II), Ce(III), Cr, Co, Fe, Ga, In, Mo(VI), Ni, Pb, Sb(V), Se(IV) and (VI), Tl(I), Te (VI), Th, Ti, U, V(V), W(VI), Zr, and Zn.

Hg(II) interferes since the mercuric complex is more stable than that of the copper and extraction of the latter only commences after all the mercury has passed across; however, the mercuric compound does not absorb at 436 mμ. It is therefore possible to extract copper by adding a sufficient excess of the reagent (about 2.5 mg of diethylthiocarbamate per milligram of mercury is required).

Silver consumes the reagent and shows a slight absorption at 436 mμ; 1 mg of silver corresponds to approximately 1 μg of copper. The silver can be separated by precipitating AgCl.

1 mg of gold is equivalent to 50 μg of copper; 5 mg of platinum ≡ 2 μg of copper; 1 mg of osmium ≡ 25 μg of copper; and 1 mg of palladium ≡ 10 μg of copper. Sb(III) interferes with the extraction of the copper; Sb(V) does not interfere. The former may therefore be oxidized. Similarly, Te(IV) interferes, but can be oxidized to Te(VI) which does not, and Tl(III) interferes but may be reduced to Tl(I). Bismuth interferes; the chloroform solution can be washed with 5 *N* hydrochloric acid, when the bismuth passes into the aqueous phase, leaving the copper in the solvent.

Reagents

Ammonium citrate: 150 ml of water and 200 g of citric acid are added to 210 ml of concentrated ammonia. The solution is made slightly ammoniacal and is purified by the addition of diethyldithiocarbamate followed by extraction with chloroform.

EDTA solution: 10 g of the disodium salt are dissolved in 100 ml of water.

Diethyldithiocarbamate: 100 mg of reagent are dissolved in 100 ml of water. This solution remains stable for 2 to 3 days when stored in a brown bottle.

Cresol red: 100 mg are dissolved in 100 ml of alcohol.

Chloroform.

Operating procedure

5 ml of citrate, 10 ml of EDTA, and then 2 drops of cresol red are added to 20 ml of solution containing 20 to 100 μg of copper. The mixture is neutralized with ammonia until a red coloration just appears (pH about 8.5), diluted to about 70 ml, and treated with 5 ml of diethyldithiocarbamate. The complex is extracted into 10 ml of chloroform by vigorous agitation for 2 minutes. The organic phase is transferred into a 25 ml measuring flask, and is extracted once more with 5 ml of chloroform. The volume is made up to 25 ml with the solvent, and this solution is filtered through a dry filter to remove droplets of water. The solution is collected in the colorimetric cell and the measurement is performed at 436 mμ.

COMPARE

In steels

A. CLAASSEN AND L. BASTINGS, *Z. Anal. Chem.*, 153 (1956) 30.

In metals

H. POHL, *Anal. Chim. Acta*, 12 (1955) 54.

In lead

K. LOUNAMAA, *Z. Anal. Chem.*, 150 (1956) 7.

In nickel

K. RIEDEL, *Z. Anal. Chem.*, 159 (1957) 25.
A. JEWSBURY, *Analyst*, 78 (1953) 363.

In copper-containing metals

H. J. CLULEY, *Analyst*, 79 (1954) 561.
C. DOZINEL, *Modern Methods of Analysis of Copper and its Alloys*, Elsevier, Amsterdam 1963.

In aluminium

M. JEAN, *Anal. Chim. Acta*, 11 (1954) 79.

In water

E. N. JENKINS, *Analyst*, 79 (1954) 209.

In plants

K. L. CHENG AND R. H. BRAY, *Anal. Chem.*, 25 (1953) 655.
W. A. FOSTER, *Analyst*, 78 (1953) 614.

(ii) With dithizone. At about pH 0.5 and above, copper dithizonate, which is violet-red, can be extracted into chloroform or carbon tetrachloride.

The violet compound is only extracted very slowly from strongly acidic solutions.

At 430 mμ, the molar extinction coefficient, ε, is 22,000. At 580 mμ $\varepsilon =$ 25,000, and at 620 mμ $\varepsilon = 35{,}000$.

Interfering ions. Pd(II), and Au(III) are extracted. Ag(I), Hg(II) and Bi(III), which also tend to come across, can be complexed with 0.1 *M* iodide.

If large amounts of zinc are present, it is extracted partially; if the organic phase is washed with 0.1 *N* HCl, copper may be determined in the presence of zinc, even when the concentration of the latter exceeds that of copper by a factor of 10^5.

The oxidizing action of Fe(III) may be suppressed by the addition of hydroxylamine hydrochloride. Sn(II) should be oxidized to Sn(IV).

At a pH lower than 2, 10 mg of nickel and cobalt are not extracted.

REAGENTS

Dithizone solution containing 6.5 mg dithizone per litre of carbon tetrachloride (see p. 109).

Ammonia, 2 *N* (1/7).

Standard solution of copper: 1 g of copper of guaranteed purity is dissolved in the minimum amount of 1/2 nitric acid, and the volume made up to 1 litre.

OPERATING PROCEDURE

The pH of 20 ml of solution containing 5 to 10 μg of copper is adjusted to 2 to 3; 20 ml of dithizone solution are added, and the mixture shaken vigorously for 3 minutes. The colorimetric determination is performed at 510 or 620 mμ.

COMPARE

G. IWANTSCHEFF, *Das Dithizon*, Verlag Chemie, Weinheim, 1958.
S. L. MORRISON AND H. L. PAIGE, *Ind. Eng. Chem., Anal. Ed.*, 18 (1946) 211.
P. M. HEERTJES, *Chem. Weekblad*, 42 (1946) 91.

In rocks
E. B. SANDELL, *Ind. Eng. Chem., Anal. Ed.*, 9 (1937) 464.

In soils
R. S. HOLMES, *Soil Sci.*, 59 (1945) 77.

In zinc
H. FISCHER AND G. LEOPOLDI, *Metall u. Erz*, 35 (1938) 86.

In mineral oils
A. G. ASSAF AND W. C. HOLLIBAUCH, *Ind. Eng. Chem., Anal. Ed.*, 12 (1940) 695.

In titanium
H. W. PENDER, *Anal. Chem.*, 30 (1958) 1915.

In biochemistry
D. M. HUBBARD AND E. C. SPETTEL, *Anal. Chem.*, 25 (1953) 1245.

(iii) With 2.2'-diquinolyl. The coloured complex formed may be extracted with amyl alcohol.

Sensitivity. $\varepsilon = 5{,}500$ at 540 mμ in iso-amyl alcohol.

Interfering ions. The following do not interfere in a ratio of 1000/1: Al, As, Ba, Ca, Cd, Co, Fe(III) and (II), Mg, Mo, Mn, Ni, Sb, Sn, Ti, V, W and Zn.

REAGENTS

Hydroxylamine hydrochloride: 10% in water.

Tartaric acid: 10% in water.

Diquinolyl solution: 20 mg of the reagent are dissolved in 100 ml of amyl alcohol.

Standard solution of copper: 1 g of guaranteed purity copper is dissolved in the minimum amount of 1/2 nitric acid. The solution is diluted to 1 litre. A solution containing 5 μg of copper per ml is then prepared by dilution.

OPERATING PROCEDURE

5 ml of hydroxylamine hydrochloride and 5 ml of tartaric acid are added to 20 ml of a solution containing 20 to 80 μg of copper. The pH is adjusted to between 5 and 6 with dilute ammonia. Accurately measured 10 ml of 2,2'-diquinolyl solution are added, and the mixture is shaken for 1 to 2 minutes. After settling, the organic phase is filtered if necessary, and the colorimetric determination is performed at 545 mμ.

COMPARE

R. J. GUEST, *Anal. Chem.*, 25 (1953) 1484.
J. HOSTE, J. EECKHOUT AND J. GILLIS, *Anal. Chim. Acta*, 9 (1953) 263.

In sea water, rocks, biochemistry

J. P. RILEY AND P. SINHASENI, *Analyst*, 83 (1958) 299.

(iv) Other methods

With 1.5-diphenylcarbohydrazide ($\varepsilon = 160{,}000$)

R. W. TURKINGTON AND F. N. TRACY, *Anal. Chem.*, 30 (1958) 1699.

Precision colorimetry of Cu^{2+}

R. BASTIAN, *Anal. Chem.*, 21 (1949) 972.

Fluorine

F = 19.00

(1) Separation of trace amounts

(i) Pyrohydrolysis. Superheated steam, either in the presence or absence of oxygen, is passed over the solid fluorine-containing substance at temperatures between 400 and 1000° C, according to the circumstances. The reaction is performed, either in a platinum, a nickel[1], or in a fused silica tube[2,3], in the presence of a fluxing agent if necessary[2]. HF is liberated by the reaction

$$MF_2 + H_2O \rightarrow 2HF + MO$$

This method may be applied to compounds of Al(III), Bi(III), Ce(III), Th(IV), U(IV), U(VI), V, Mg(II), Zn(II), Zr(IV), Na(I), and Ca(II).

The reaction may proceed very rapidly (*e.g.* to completion in 20 minutes).

COMPARE

J. C. WARF, W. D. CLINE AND R. D. TEVEBAUGH, *Anal. Chem.*, 26 (1954) 342.
(1) C. D. SUSANO, J. C. WHITE AND J. E. LEE JR., *Anal. Chem.*, 27 (1955) 453.
F. W. DYKES, G. L. BOOMAN, M. C. ELLIOTT AND J. E. REIN, *U.S. Atomic Energy Comm.* IDO – 14405, 1957.

In U, UF_4, UF_6

J. O. HIBBITS, *Anal. Chem.*, 29 (1957) 1760.
(2) G. M. GILLIES, N. J. KEEN, B. A. J. LISTER AND D. REES, *Atomic Energy Research Establishment* C/M 225, Oct. 26, 1954.

In AlF_3

L. V. HAFF, C. P. BUTLER AND J. D. BISSO, *Anal. Chem.*, 30 (1958) 984.

In the rare earths

C. V. BANKS, K. E. BURKE AND J. W. O'LAUGHLIN, *Anal. Chim. Acta*, 19 (1958) 239.
(3) R. H. POWELL AND O. MENIS, *Anal. Chem.*, 30 (1958) 1546.

In SiO_2 and Al_2O_3

L. W. GAMBLE, W. E. PRICE AND W. H. JONES, *Anal. Chem.*, 32 (1960) 189.

(ii) Distillation of hexafluorosilicic acid. Although this is the most frequently used method, it often presents difficulties. The substance to be analyzed is heated to a suitable temperature in perchloric, sulphuric, or phosphoric acid and in the presence of silica. A steam distillation is carried out and the azeotropic H_2SiF_6/water mixture is collected in dilute caustic soda.

NOTES (1) If perchloric acid is used, the temperature should be about 135° C. The apparatus should preferably be fitted with ground glass joints to safeguard against explosions, which frequently occur in the presence of organic substances (cork or rubber) if the temperature is accidentally allowed to rise.

After distillation, fluorine can be estimated either by the lead chlorofluoride method (gravimetric or volumetric), directly with thorium nitrate, or by a suitable colorimetric method.

(2) If sulphuric acid is used, the operation can be accelerated by heating to about 165°. This is particularly useful with compounds difficult to decompose. Small amounts of sulphuric acid are carried over.

N.B. The distillate may, if necessary, be concentrated, working in a neutral solution, pH 6–8; a platinum vessel should be used if the amount of fluorine involved is small. The solution should never be evaporated to dryness.

Interfering ions. With apatite, fluorite, and acid-soluble solids, no difficulties are encountered if the interfering ions indicated below are absent.

The presence of organic substances prevents treatment with perchloric acid because of the inherent danger of explosion. Such substances may be destroyed by moderate calcination in an alkaline medium.

If the solid is not soluble in acids, it should first be attacked either by fused sodium carbonate, possibly in the presence of silica, or by fused caustic soda.

The gelatinous silica obtained when dealing with silicates, particularly after attack by the dry method, hinders the separation. When the ratio of SiO_2 to F by weight exceeds 50, it becomes very difficult to distil all the fluorine.

When phosphates are present, a little phosphoric acid passes over, which subsequently interferes when thorium nitrate is the reagent used for the estimation. A second distillation is the generally necessary.

Nitric acid distils over and does not interfere.

Hydrochloric acid would interfere, but may be kept back by means of silver sulphate.

Borates give rise to the formation of HBF_4 which distils over; the fluorine present can however be estimated quite satisfactorily with thorium.

When metals which complex F^- are present, the procedure described below is adopted.

COMPARE

Distillation of trace amounts

J. SAMACHSON, N. SLOVIK AND A. E. SOBEL, *Anal. Chem.*, 29 (1957) 1888.

In blood

L. SINGER AND W. D. ARMSTRONG, *Anal. Chem.*, 31 (1959) 105.

(iii) Extraction of tetraphenylantimonium fluoride. The solvent is CCl_4. Complexing ions such as Al^{3+}, Fe^{3+}, etc. interfere.

cf. K. D. MOFFETT, J. R. SIMMLER AND H. A. POTRATZ, *Anal. Chem.*, 28 (1956) 1356.

(iv) With ion-exchangers. F^- may, under basic conditions and in the presence of EDTA, be separated by anion exchangers from phosphates and small amounts of Al(III), Fe(III), Mg(II) and Cd(II). F^- is eluted by 0.05 *M* ammonium chloride at pH 8.8.

COMPARE

A. C. D. NEWMAN, *Anal. Chim. Acta*, 19 (1958) 471.
H. M. NIELSEN, *Anal. Chem.*, 30 (1958) 1009.
I. ZIPKIN, W. D. ARMSTRONG AND L. SINGER, *Anal. Chem.*, 29 (1957) 310.
W. FUNASAKA, M. KAWANE, T. KOJIMA AND Y. MATSUDA, *Japan Analyst*, 4 (1955) 514

Separation from large amounts of Al(III). F^- can be separated by the elution of Al(III) in 0.2 *N* caustic soda.

The separation may even be carried out at an F/Al ratio of 0.01%.

Preparation of the column. About 30 g of Amberlite IRA 400, size graded to retain only the 0.3 mm fraction, are weighed out and left overnight in distilled water. A column with a cross-section of 1 cm^2 is then packed with the resin to a height of 54 cm. 100 ml of 2 *N* caustic soda are passed through, and the column is washed with 100 ml of distilled water. 100 ml of 0.2 *N* caustic soda is then passed through, at a rate of 2 to 2.5 ml/minute.

OPERATING PROCEDURE

50 ml of the solution (containing 100 mg of Al(III) and at least 10 μg of fluoride) are passed through the column at a rate of 2 to 2.5 ml/minute. The aluminium(III) is then eluted, at the same rate of flow, using a 0.2 *N* solution of caustic soda; elution is continued until the volume recovered is between 280 and 300 ml. The fluoride is then eluted with a *N* solution of caustic soda. For trace amounts, 100 ml of eluate are sufficient, while large amounts of fluoride (greater than or equal to 10 mg) require 250 ml.

cf. J. COURSIER AND J. SAULNIER, *Anal. Chim. Acta*, 14 (1956) 62.

(v) Standard solutions. If the results obtained by colorimetry are to be accurate, it is necessary to start with hydrofluoric acid of guaranteed purity, which is titrated against caustic soda. For trace amounts it is however sufficient to weigh out NaF of guaranteed purity, previously dried to constant weight.

$$F/NaF = 0.452.$$

(2) Colorimetry

The available colorimetric methods are all indirect: a coloured compound is formed by the action of an appropriate reagent on a cation which can be complexed by F^-. Addition of fluoride ions to the solution causes the partial destruction of the coloured compound and consequently an attenuation of the colour. The diminution in the optical density of the solution permits the amount of fluorine added to be determined.

Interfering ions. Interference is obtained from ions which give sufficiently stable complexes with F^-. Thus Si(IV) does not usually interfere, but Al(III), Zr(IV), B(III), etc. give rise to errors. If the latter ions are present, colorimetry should be performed after the distillation of H_2SiF_6.

A recently described method capable of very high accuracy employs the ferric sulphosalicylate complex. For estimating trace amounts, certain precautions can be omitted, or alternatively certain other complexes possessing a more sensitive coloration may be used.

(i) Ferric sulphosalicylate complex. Addition of F^- diminishes the intensity of the initial violet colour, owing to the formation of ferrifluoride complexes. The results are governed by pH which should be fixed at a suitable level—2.9—by means of a chloroacetic buffer mixture.

Accuracy and sensitivity. Under normal conditions, 50 to 80 ppm of F^- can be estimated, the limiting concentration being 0.4 ppm. The accuracy may reach 0.1% with a suitable apparatus. The estimated is carried out at the absorption maximum at 500 mμ.

REAGENTS

It is extremely important the coloration and pH of the reagent used should remain invariant, since the reagent is used both to establish the calibration curve and to carry out the estimations.

Ferric nitrate: 3 g of crystalline ferric nitrate in 1 litre of 0.034 *N* perchloric acid.

Buffer solution: 18.9 g of monochloroacetic acid and 100 ml of standardized normal caustic soda are made up to 1000 ml. The pH of this solution should be between 2.85 and 2.90.

Sulphosalicylic acid: 9.5 g in 1 litre.

Reagent: 100 ml of the ferric nitrate solution, 40 ml of sulphosalicylic acid, and 6.85 ml of normal caustic soda are made up to 250 ml with the buffer solution in a volumetric flask. The solution is left for 5 hours before use, and is stable for 10 days. Checks of the constancy of the optical density of the solution should be performed at regular intervals.

Standard solution of fluoride: this is made from hydrofluoric acid of guaranteed purity. After diluting about 20 times, 25 ml are placed in a corrosion-resistant vessel and are titrated against N caustic soda with phthalein.

The volume is then adjusted to give a standard solution containing about 80 mg of fluorine per litre. This solution may be kept in glass containers.

OPERATING PROCEDURE

25 ml of the initial solution should contain 1.2 to 2 mg of fluorine. The solution is neutralized with 0.1 N caustic soda until phthalein changes to pink and is then fully decolorized with 0.1 N perchloric acid. This solution is poured into a mixture of 50 ml of buffer solution and 10 ml of reagent. The volume is made up to 100 ml, and the colorimetric estimation is then performed.

NOTES (1) Beer's law is not obeyed, so that a calibration curve should be plotted, using titrated solutions of sodium fluoride.
(2) Hexafluorosilicic acid solutions obtained during the separation of fluorine by distillation may also be estimated in this way. Since the compounds which distil over have a composition slightly different from H_2SiF_6, it is necessary to carry out a calibration with distillates obtained from known amounts of fluoride.

cf. S. LACROIX AND LABALADE, *Anal. Chim. Acta*, 4 (1950) 68.

(ii) With the Zr(IV) – Eriochrome cyanine R reagent. Zirconium gives a red coloration with Eriochrome cyanine R. Since however fluorides give very stable complexes with zirconium, the intensity of the coloration is diminished by the addition of F^-.

The method is very sensitive (limit 0.02 ppm).

Interfering ions. Less than 10 ppm of iron do not interfere.

Al(III) interferes, but this effect may be reduced by making the solution alkaline with caustic soda and then adding the acidic solutions of zirconium and Eriochrome Cyanine. In an alkaline medium, the fluoride complexes of aluminium are destroyed: on acidification the free fluoride ions are preferentially claimed by the zirconium.

Phosphates interfere above 2.5 ppm.

Sulphates react similarly to fluorides, but with a much lower sensitivity.

REAGENTS

Reagent A: 1.8 g of Eriochrome Cyanine R dissolved in 1 l of water.

Reagent B: 0.265 g of zirconium oxychloride ($ZrOCl_2 \cdot 8H_2O$) dissolved in 50 ml of water. 700 ml of concentrated hydrochloric acid are added, and the volume is made up to 1 l.

Diluted (8 N) hydrochloric acid.
Reference solution: 10 ml of reagent A, followed by 10 ml of dilute acid are added to a 100 ml flask; the volume is made up to 50 ml.
Standard solution of fluoride containing 5 μg of F^- per ml.

OPERATING PROCEDURE

5 ml of reagent A, followed by 5 ml of reagent B are added to 25 ml of solution containing 10 to 50 μg of fluoride. The volume is made up to 50 ml and colorimetry is performed at 527 mμ, the zero of the apparatus being adjusted with to the reference solution.

COMPARE

S. MEGREGIAN, *Anal. Chem.*, 26 (1954) 1161.
L. L. THATCHER, *Anal. Chem.*, 29 (1957) 1709.
B. J. MACNULTY, G. I. HUNTER AND D. G. BARRETT, *Anal. Chim. Acta*, 14 (1956) 368.

In air

D. F. ADAMS AND R. K. KOPPE, *Anal. Chem.*, 31 (1959) 1249.

(iii) Other methods. Alizarin-zirconium reagent. The limiting concentration is 0.1 ppm. The method can be used up to 1 mg of F^-.

COMPARE

W. L. LAMAR, *Ind. Eng. Chem., Anal. Ed.*, 17 (1945) 148.
H. E. BUMSTED AND J. C. WELLS, *Anal. Chem.*, 24 (1952) 1595.
W. M. SHAW, *Anal. Chem.*, 26 (1954) 1212.

Haematoxylin-Al(III) reagent

G. J. HUNTER, B. J. MACNULTY AND E. A. TERRY, *Anal. Chim. Acta*, 8 (1953) 351.
B. J. MACNULTY AND G. J. HUNTER, *Anal. Chim. Acta*, 9 (1953) 425.

Pentahydroxyflavone-Al(III) complex (0.2 ppm)

H. H. WILLARD AND C. A. HORTON, *Anal. Chem.*, 24 (1952) 862.

Fe(III)-'Ferron' complex (0.2 ppm)

P. URECH, *Helv. Chim. Acta*, 25 (1942) 1115.

In air

D. F. ADAMS, *Anal. Chem.*, 32 (1960) 1312.

Ti(IV)-H_2O_2 complex

D. MONNIER, R. VAUCHER AND P. WENGER, *Helv. Chim. Acta*, 31 (1948) 929; 33 (1950) 1.

Gallium

Ga = 69.7

(1) Separation of trace amounts

(i) Established methods of separation In the case of trace amounts of Ga(III), the following methods may be used:
Precipitation by ammonia. Entraining agent: Al(III).
Precipitation of the sulphide. Ga_2S_3 precipitates at pH 5. Entraining agent: As(III).

(ii) Extraction of the chloride. The solvent is *iso*-propyl or ethyl ether. $HGaCl_4$ can be extracted from 7 *N* HCl. Some loss of Ga(III) occurs however, when the ethyl ether is evaporated off.

Gallium may in this way be separated from numerous elements including Al(III); Fe(III), Tl(III), Mo(VI), and traces of In(III), V(V), Zn(II), etc. are however also extracted.

Fe(III) may be reduced before the extraction, as also may Tl(III) by Ti(III). Mo(VI) can be precipitated by a little Pb(II).

REAGENTS

Titanous chloride: $TiCl_3$ at 200 g per litre.
Hydrochloric acid, 7 *N* (7/12).
Iso-propyl ether.

OPERATING PROCEDURE

2 ml of titanous chloride are added to 10 ml of solution made 7 *N* with respect to hydrochloric acid. After 5 minutes, 10 ml of ether are added and the mixture is shaken to 30 seconds. A second extraction with 5 ml of solvent is then carried out, the extracts are combined and washed twice with 1 ml of 7 *N* hydrochloric acid and 0.5 ml of $TiCl_3$.

COMPARE

H. ONISHI AND E. B. SANDELL, *Anal. Chim. Acta,* 13 (1955) 159.

In rocks

F. CULKIN AND J. P. RILEY, *Analyst,* 83 (1958) 208.

Various solvents

G. W. C. MILNER, A. J. WOOD AND J. L. WOODHEAD, *Analyst,* 79 (1954) 272.

(iii) Extraction of the iodide.

cf. H. M. IRVING AND F. C. ROSSOTTI, *Analyst*, 77 (1952) 801.

(iv) Extraction of the thiocyanate. The solvent is a mixture of ether and tetrahydrofuran.

cf. H. SPECKER AND E. BANKMANN, *Z. Anal. Chem.*, 149 (1956) 97.

(v) Extraction of the mixed Cl^--Rhodamine B complex with benzene. See Colorimetry.

(vi) Separation by cupferron. Gallium may be precipitated or extracted from a 5 *N* sulphuric acid solution, and may thus be separated from large amounts of Zn(II), Al(III), and In(III). Sn(IV), Ti(IV), Zr(IV), V(V), and Fe(III) accompany the Ga(III). One of these substances may, if necessary, be used as an entraining agent.

cf. J. A. SCHERRER, *J. Res. Natl. Bur. Std.*, 15 (1935) 585.
E. GASTINGER, *Z. Anal. Chem.*, 14 (1953) 244.

(vii) Extraction of the oxinate. See Colorimetry.

(viii) Separation by ion-exchangers. With a hydrochloric acid solution, separation from Al(III), In(III), etc. may be effected. Ga(III) is collected in the form of anions.

COMPARE

K. A. KRAUS, F. NELSON AND G. W. SMITH, *J. Phys. Chem.*, 58 (1954) 11.
R. KLEMENT AND H. SANDMANN, *Z. Anal. Chem.*, 145 (1955) 325.

In an SCN^- medium, from Fe(III)
J. KORKISCH AND F. HECHT, *Mikrochim. Acta*, 1230 (1956).

(ix) Standard solutions. 1 g of the pure metal is dissolved in 20 ml of 6 *N* (1/2) hydrochloric acid and the solution is made up to 1 litre.

(2) Colorimetry

(i) With Rhodamine B. In 6 *N* hydrochloric acid and in the presence of Rhodamine B, Ga(III) gives a violet-red compound which can be extracted with benzene.

Sensitivity: $\varepsilon = 60{,}000$.

Interfering ions. Sb(V), Au(III), Tl(III), Fe(III), etc. give analogous reactions (see Antimony). Sb(III), Tl(I), Al(III), In(III), Zn(II), etc., do not interfere. Though Fe(III) interferes, it can be separated from Ga(III) by extracting Ga(III) chloride in the presence of Ti(III), which reduces the Fe(III) to Fe(II).

REAGENTS

Hydroxylamine hydrochloride, 10% in water.
Rhodamine B: 0.5 g in 100 ml of 6 *M* (1/2) hydrochloric acid.

OPERATING PROCEDURE

5 ml of solution made 6 *M* (1/2) with respect to hydrochloric acid are placed in a separating funnel and treated with 0.4 ml of reagent and 1.0 ml of hydrochloride solution. After mixing, 10 ml of benzene are added, the mixture is shaken for 1 minute, and the liquid is the allowed to settle. The aqueous phase is removed with a little benzene. The benzene phase is filtered through glass wool and the first portions are rejected.

Colorimetry is performed at 565 mμ, and the results are compared with a calibration curve.

COMPARE

H. ONISHI AND E. B. SANDELL, *Anal. Chim. Acta,* 13 (1955) 159.

In rocks

F. CULKIN AND J. P. RILEY, *Analyst,* 83 (1958) 208.

Malachite green

H. JANKOVSKY, *Talanta,* 2 (1959) 29.

(ii) By means of the oxinate. Gallium oxinate can be extracted with chloroform at pH > 2. The method is very unselective. Extraction can be carried out in the presence of cyanide. The colorimetric estimation is carried out at 392 mμ, ε being 6,500.

COMPARE

S. LACROIX, *Anal. Chim. Acta,* 1 (1947) 260.
T. MOELLER AND A. J. COHEN, *Anal. Chem.,* 22 (1950) 686.
C. L. LUKE AND M. E. CAMPBELL, *Anal. Chem.,* 28 (1956) 1340.

(3) Fluorimetry

Of the two methods available, the one involving 'Chrome blue' is very selective but not accurate. This method is suitable except when Al(III) is present. The oxine method is more accurate but not selective; it is nevertheless suitable for estimations in the presence of Al(III).

(i) Oxine. Gallium oxinate dissolves in chloroform at pH > 2.0 to give a solution with a yellow-green fluorescence. In chloroform solution, the oxine itself exhibits only negligible fluorescence.

The method is suitable for the estimation of trace amounts. Thus 0.3 μg of Ga(III) per millilitre of chloroform can be determined to within 10%, by comparison with standard solutions. The limiting concentration is 0.025 μg per millilitre of the solvent.

A number of compounds exhibit the same reaction, and preliminary separations are therefore necessary. Al(III) is not extracted at pH 2.0.

In rocks

E. B. SANDELL, *Ind. Eng. Chem., Anal. Ed.*, 13 (1941) 844, and *Anal. Chem.*, 19 (1947) 63.

In bauxites

S. LACROIX, *Anal. Chim. Acta*, 2 (1948) 167.

(ii) 'Chrome blue'. 2,2'-Hydroxy-4-sulphonaphthalene-azo-naphthalene (Colour Index 202) reacts with gallium to give an orange fluorescence soluble in alcohols.

Only Al(III) gives a similar reaction.

COMPARE

A. WEISSLER AND C. E. WHITE, *Ind. Eng. Chem., Anal. Ed.*, 18 (1946) 530.
G. CHARLOT, *Anal. Chim. Acta*, 1 (1947) 236.
I. LADENBAUER, J. KORKISCH AND F. HECHT, *Mikrochim. Acta*, 1076 (1955).

Germanium

Ge = 72.6

(1) Separation of trace amounts

(i) Precipitation of the hydroxide. The hydroxide of germanium may be entrained by those of Fe(III), Al(III), etc.

(ii) Precipitation of the sulphide GeS_2. GeS_2 is sparingly soluble in 6 *N* (1/2) hydrochloric acid, but forms complexes which are stable in an acetic buffer medium.

(iii) Precipitation with tannin in oxalic acid. Fe, Al, V, Zr, and Th remain in solution.

cf. H. HOLNESS, *Anal. Chim. Acta*, 2 (1948) 254.

(iv) Distillation of $GeCl_4$. $GeCl_4$ boils at 86°. The distillation is carried out in an apparatus incorporating a distillation column in order to avoid mechanical entrainment.

4 mg of Ge(IV) can be distilled.

Separation from fluorine. 15 ml of concentrated sulphuric acid and a little glass wool are added to the solution. The temperature is raised to 160° and maintained at this value during distillation, while water is added drop by drop; any fluorine present then passes over in the form of H_2SiF_6. When the distillation of fluorine is complete, as indicated by means of a suitable reagent (*e.g.* alizarin-zirconium or salicylate-Fe(III), see Fluorine), the conditions are adjusted to bring over the $GeCl_4$.

For this purpose, 20 ml of 20% sodium chloride solution are added to the solution drop by drop, followed by 25 ml of water, the distillation temperature being maintained at 160°. The distillate is collected in 10 ml of 5 *N* (1/2) caustic soda contained in a plastic of a paraffin-coated metallic receiver. (Glass would be attacked by the caustic soda and the silica thus introduced into the solution would interfere during subsequent colorimetry of the germanium).

Interfering ions. Some $AsCl_3$ distills over as well. This can be prevented either by operating in a current of chlorine[1] or, more simply, by adding permanganate and hydrochloric acid[2], alternatively, copper arsenide may be precipitated[3].

The distillation can be carried out in a current of carbon dioxide[4].

Precipitated silica, even if it is made insoluble, retains Ge(IV).

cf. S. T. PAYNE, *Analyst*, 77 (1952) 278.
(1) J. A. SCHERRER. *J. Res. Natl. Bur. Std.*, 16 (1936) 253.
(2) H. LUNDIN, *Trans. Electrochem. Soc.*, 63 (1933) 149.
(3) W. C. AITKENHEAD AND A. R. MIDDLETON, *Ind. Eng. Chem., Anal. Ed.*, 1 (1938) 633.
(4) W. GEILMANN AND K. BRUNGER, *Z. Anorg. Allgem. Chem.*, 196 (1931) 312.

Separation from P(V)

C. L. LUKE AND M. E. CAMPBELL, *Anal. Chem.*, 25 (1953) 1588.

(v) Extraction of the chloride. Ge(IV) can be extracted with carbon tetrachloride from $\geqslant 9\ N$ HCl. In effect, only As(III) accompanies the Ge(IV), and since As(V) is not extracted, Ge can be separated from As within the limits[1] 10^6 to 10^{-6}. Care must be taken against the loss of $GeCl_4$ by volatilization.

Traces of any other elements extracted may be removed by washing with 9 *N* HCl.

REAGENTS

Hydrochloric acid, 9 *N*.
Carbon tetrachloride.

OPERATING PROCEDURE

15 ml of CCl_4 are added to 25 ml of 9 *N* HCl solution. The mixture is shaken vigorously for 2 minutes. A second extraction is carried out with 15 ml of solvent.

The extracts are combined and washed with 5 ml of 9 *N* hydrochloric acid.

COMPARE

W. A. SCHNEIDER, JR., AND E. B. SANDELL, *Mikrochim. Acta*, (1954) 263.
(1) W. FISCHER, W. HARRE, W. FREESE AND K. G. HACKSTEIN, *Angew. Chem.*, 66 (1954) 165.

In minerals

E. H. STRICKLAND, *Analyst*, 80 (1955) 548.
G. SAUVENIER AND G. DUYCKAERTS, *Anal. Chim. Acta*, 16 (1957) 592.

With hexone

P. SENISE AND L. SANT AGOSTINO, *Mikrochim. Acta*, (1956) 1446 (see Colorimetry).

(vi) Anion-exchangers in concentrated HCl solution. Ge(IV) is bound to the resin.

Separation from As(V)

F. NELSON AND K. A. KRAUS. *J. Am. Chem. Soc.*, 77 (1955) 4508.

Mixed exchanger

T. R. CABBELL, A. A. ORR AND J. R. AYES, *Anal. Chem.*, 32 (1960) 1602.

(vii) Standard solutions. The pure metal is dissolved in a mixture of $H_2SO_4 + HNO_3 + HF$, and the solution is heated until abundant white fumes of sulphuric acid are produced. After dilution, germanium may be estimated in the form of GeO_2.

Alternatively, if GeO_2 is available, this is digested with fused carbonates and taken up in sulphuric acid.

$$Ge/GeO_2 = 0.694.$$

(2) Colorimetry

Of the methods available, two are analogous to those which are used for silica and the phosphates.
(1) The formation of yellow germanomolybdate.
(2) When the germanomolybdate is reduced, a blue coloration is obtained. The first method is more accurate, but not as sensitive to small traces as the second.
(3) Alternatively, phenylfluorone may be used.

(i) Germanomolybdate. Ge(IV) combines with molybdates ions to give yellow $Ge(Mo_2O_7)_6^{8-}$, which is analogous to the silicomolybdate complex.

The colour depends on the pH, concentrations of the reagents and on the temperature. The medium may consist of 0.1–0.2 *N* sulphuric acid.

Sensitivity. $\varepsilon \sim 2000$ at 440 mμ.

Interfering ions. Since the germanomolybdate complex is not very stable, ions which complex Ge(IV) and Mo(VI) diminish the coloration. Many such ions exist, and even SO_4^{2-} and Cl^- exert an effect. Thus, since after the distillation of $GeCl_4$, Cl^- is present in excess, the same amount should be added to the standards.

Si(IV), As(V), and P(V) give analogous coloration, and should be removed. In fact, it is in most cases necessary to carry out a previous separation by distilling or extracting $GeCl_4$.

REAGENTS

Sulphuric acid, (1/40).

Ammonium molybdate, 5% solution.

OPERATING PROCEDURE

7.5 ml of *N* (1/40) sulphuric acid are added to the neutralized solution, the volume is made up to 50 ml, and 2.5 ml of molybdate are added. The colorimetric estimation is performed at 440 mμ after 5 minutes. The result of a blank test carried out with the reagents themselves, which may contain silica, (silica gives only a very slight coloration at this pH), should be subtracted.

cf. R. E. KITSON AND M. G. MELLON, *Ind. Eng. Chem., Anal. Ed.*, 16 (1944) 128.

(ii) Reduction of the germanomolybdate. The germanomolybdate complex may be reduced (with the formation of a blue coloration) while the excess of molybdic reagent remains unchanged. For this purpose, the medium should be made sufficiently acid to avoid the reduction of the latter as far as possible; if however the acidity is too high, reduction of the germanomolybdate would proceed too slowly. The germanomolybdate is produced under the conditions described above 0.1–0.2 *N* sulphuric acid—and is then reduced by Fe(II) in 1.5–2.0 *N* sulphuric acid.

The absorption maximum of the molybdenum blue formed by the reduction of Mo(VI) is at about 625 mμ and that of the blue coloration obtained by the reduction of the germanomolybdate complex at about 820–830 mμ. It is therefore advantageous to carry out the colorimetry at the latter wavelength. The coloration varies with time.

Sensitivity. $\varepsilon \sim$ 10,000 at 830 mμ.

Interfering ions. Approximately the same ions interfere as in the preceding case. The presence of Cl^- slows down the reaction and in such cases, the colorimetry should be performed after 30 minutes.

For the estimation of 3 ppm of Ge(IV) at 830 mμ, the following do not interfere to within 2%: 500 ppm of Al(III), Cr(III), Cd(II), Co(II), Cu(II), Ni(II), Mn(II), Zn(II), As(V), As(III), Ag(I), Cl^-, citrate, NO_3^-, $C_2O_4^{2-}$, and W(VI); 25 ppm of P(V); 50 ppm of F^-.

Si(IV), Ba(II), Bi(III), Fe(III), Pb(II), and V(V) interfere.

REAGENTS

These are the same as in the preceding case, together with a 0.4% solution of Mohr's salt in 3.2 *N* (9%) sulphuric acid.

OPERATING PROCEDURE

The solution is neutralized (indicator paper) and sulphuric acid is then added to make it about 0.2 *N*, 2 ml of molybdate are added, immediately followed by 25 ml of Mohr's salt. The volume is made up to 50 ml, and the mixture is kept for 20–30 minutes. Colorimetry is then performed at 825 mμ. (S. LACROIX unpublished work).

cf. E. R. SHAW AND J. F. CORWIN, *Anal. Chem.*, 30 (1958) 1314.

(iii) With phenylfluorone. With phenylfluorone, Ge(IV) gives a sparingly soluble red compound.

Si(IV) and As(III) do not interfere.

This method of colorimetry can be specific provided that Ge(IV) has been separated in a hydrochloric acid medium.

Chlorine (oxidizing agents + chloride) will interfere.
Sensitivity. $\varepsilon \sim 10{,}000$ at 510 mμ.

REAGENTS

Phenylfluorone (2,3,7-trihydroxy-9-phenyl-6-fluorone). 0.05 g are dissolved in 50 ml of methanol and 1 ml of hydrochloric acid, and the volume is made up to 500 ml with methyl alcohol.

Gum arabic. 0.5 g are dissolved in 50 ml of hot water. This solution should be prepared afresh every day.

A solution buffered at pH 4.5. 450 g of sodium acetate are dissolved in about 350 ml of water. 240 ml of acetic acid are added, and the volume is then made up to 1 litre.

OPERATING PROCEDURE

To 10 ml of solution, preferably containing 5 to 10 μg of germanium, are added 1.5 ml of 20 *N* (1/2) sulphuric acid, 10 ml of buffer solution, and 1 ml of gum arabic, stirring after each addition. A final addition of 10 ml of the reagent is made. A precipitate of sodium acetate forms. The solution is allowed to stand for 5 minutes, made up to 50 ml with *N* (1/10) hydrochloric acid, and the measurement is carried out immediately at 510 mμ.

cf. C. L. LUKE AND M. E. CAMPBELL, *Anal. Chem.*, 28 (1956) 1273.
H. I. CLULEY, *Analyst*, 76 (1951) 523.
E. H. KUNSTMANN AND E. F. E. MULLER, *Analyst*, 84 (1959) 324.

(iv) Extraction with phenylfluorone. The complex formed is stable in benzyl alcohol. The excess of reagent dissolves in the solvent. Since a number of ions interfere, Ge(IV) should first be isolated. As(III) and Ti(VI) do not interfere.

REAGENTS

Phenylfluorone. 30 mg of reagent are dissolved in a mixture of 85 ml of ethanol and 5 ml of sulphuric acid (1/7), with slight warming. The volume is made up to 100 ml with alcohol.

OPERATING PROCEDURE

5 ml of phenylfluorone are added to 25 ml of solution, which should be 0.5 *N* with respect to hydrochloric acid and should contain 1.5 to 15 μg of germanium. After standing for at least 5 minutes, the complex is extracted first with 10 ml and then twice with 5 ml of benzyl alcohol, each time shaking for 3 minutes. The organic phases are collected in a 25 ml measuring flask and made up to

volume with ethanol. The measurement is performed at 505 mμ against a blank sample prepared under the same conditions.

cf. A. HILLEBRANT AND J. HOSTE, *Anal. Chim. Acta*, 18 (1958) 569.

With oxidized hematoxylin

H. NEWCOMBE, W. A. E. MCBRYDE, J. BARTLETT AND F. E. BEAMISH, *Anal. Chem.*, 23 (1951) 1023.

Gold

Au = 197.2

(1) Separation of trace amounts

(i) The dry method. This procedure allows traces of gold to be separated from large samples (*cf.* the references on p. 346), and is the most widely used method.

(ii) Precipitation in the form of metallic gold. Gold can be precipitated by reduction with Sn(II), Zn, or Mg, using Hg, Hg_2Cl_2, Te as the entraining agents. Consecutive calcination leaves only Au.

REAGENTS

TeO_2, 0.1% in N (1/10) hydrochloric acid.
$SnCl_2$, aqueous, 10% in N (1/10) hydrochloric acid.

OPERATING PROCEDURE

A 25–100 ml portion of the solution, containing 1 to 25 μg of gold, is made 1 to 2 N in hydrochloric acid. 0.2 ml of Te(IV) and 2 ml of stannous chloride are added, the mixture is boiled for 5 to 10 minutes, and the resulting brown precipitate produced is separated by filtration through sintered glass and washed with $N/2$ HCl. It is then dissolved in 1 ml of aqua regia, the residue is washed with water, and the solution is evaporated to dryness on a water bath. The residue is taken up in a little dilute hydrochloric acid.

cf. W. B. POLLARD, *Analyst*, 62 (1937) 597.

Gold may be separated from Pt(IV) and from Pd(II) by reduction with hydroquinone or with metol.

(iii) Precipitation by ammonium sulphide. 1 μg of gold may be recovered from 1 litre of solution in this manner, using Pb(II) as the entraining agent.

cf. C. FINK AND G. L. PUTNAM, *Ind. Eng. Chem., Anal. Ed.*, 14 (1942) 468.
G. MILAZZO, *Anal. Chim. Acta*, 3 (1949) 126.

(iv) Extraction of $HAuCl_4$ and $HAuBr_4$. Ether, *iso*-propyl ether, or ethyl acetate may be used as solvents.

The platinum group metals are not extracted, except for Os(VIII). Fe(III) is extracted to a small extent, but can be complexed with H_3PO_4.

REAGENTS

Hydrobromic acid, concentrated, 7 *N*.

Hydrobromic acid, 4 *N*.

Iso-propyl ether free from alcohols.

OPERATING PROCEDURE

A portion of the solution containing 100 μg of gold is adjusted to a volume of 10–12 ml, and 5 ml of conc. hydrobromic acid are added. The solution is extracted twice with 15 ml of *iso*-propyl ether. The extracts are washed with 5 ml of 4 *N* hydrobromic acid and the gold is transferred back to the aqueous solution by shaking first with 20 ml and then twice with 10 ml portions of water. A further 1 ml of concentrated hydrobromic acid is added to the solution and the latter is heated on a water bath to expel the ether.

Colorimetry can then be carried out at 380 mμ.

cf. W. A. E. McBryde and J. H. Yoe, *Anal. Chem.*, 20 (1948) 1094.
J. H. Yoe and L. G. Overholser, *J. Am. Chem. Soc.*, 61 (1939) 2058.

(v) Extraction in the presence of rhodamine B. See Colorimetry.

(vi) Extraction in the presence of rhodanine. The solvent is *iso*-amyl acetate.

cf. T. M. Cotton and A. A. Woolf, *Anal. Chim. Acta*, 22 (1960) 192.

(vii) Standard solutions. About 1 g of gold is dissolved in 71 ml of concentrated hydrochloric acid and 3 ml of concentrated nitric acid. The solution is diluted as necessary.

(2) Colorimetry

(i) With rhodamine B. In the form of $AuCl_4^-$, Au(III) gives rise to a violet complex with the cation of rhodamine B. This compound can be extracted with benzene and *iso*-propyl ether. The extraction is governed by the concentration of hydrochloric acid and other chlorides.

Interfering ions. Sb(V), Tl(III), W(VI), Hg(II) (> 250 μg), Fe(III) (> 100 μg) and Sn(IV) (> 10 μg) also give extractable colorations. When these metals are present, the gold may be separated by precipitation with hydroxylamine hydrochloride, using tellurium as an entraining agent.[1]

REAGENTS

Hydrochloric acid, 6 M. 250 ml of water are added to 250 ml of concentrated hydrochloric acid.

Ammonium chloride, saturated: 150 g of ammonium chloride are dissolved in 500 ml of water.

Rhodamine B, 0.04%: 200 mg of rhodamine B are dissolved in 500 ml of water.

Iso-propyl ether.

OPERATING PROCEDURE

2.5 ml of 6 *M* hydrochloric acid and 5.0 ml of ammonium chloride are added to 5 ml of solution containing 10 to 20 μg of gold: the volume is made up to 15 ml. 5 ml of rhodamine B are added, followed by 10 ml (accurately measured) of *iso*-propyl ether. The mixture is vigorously shaken 100 times, allowed to settle, and the colorimetric determination is then performed at 565 mμ.

cf. B. J. MACNULTY AND L. B. WOOLLARD, *Anal. Chim. Acta,* 13 (1955) 154.
(1) H. ONISHI, *Mikrochim. Acta,* (1959) 9.

With methyl violet (ε ∼ 115,000 at 600 mμ: trichloroethylene).

L. DUCRET AND H. MAUREL, *Anal. Chim. Acta,* 21 (1959) 74.

(ii) p-Dimethylaminobenzylidenerhodanine. The reagent produces a violet-red coloration in the presence of gold salts in neutral or slightly acid solutions. The compound formed is sparingly solute in the latter but dissolves in a mixture of chloroform and benzene.

Sensitivity. The molar extinction coefficient, ε ∼ 20,000.

Interfering ions. Interference is obtained with Ag(I) (20 ppm of Ag = 1 ppm of gold), Pd(II), Pt(II), Hg(II), and Cu(I). Hg can be separated by calcination, and Ag(I) can be precipitated in the form of chloride. Some fluoride should be added when Fe(III) is present. Many neutral salts interfere by flocculating the colloid, which is otherwise stable.

REAGENT

p-Dimethylaminobenzylidenerhodanine, 0.03% in alcohol, 1 ml. 13 ml of benzene and 356 ml of chloroform should be added.

OPERATING PROCEDURE

The solution is, if necessary, freed from the excess of acid by evaporation. It should contain 0.2 to 2 μg of gold per 5 ml. 3 drops of concentrated nitric acid are added, followed by 1 ml of reagent. The mixture is shaken vigorously, allowed to stand for 20 minutes, and then compared against standards.

cf. E. B. SANDELL, *Anal. Chem.,* 2 (1948) 253.

In biochemistry

S. NATELSON AND J. L. ZUCKERMAN, *Anal. Chem.,* 23 (1951) 653.

(3) Other methods

Colorimetry of bromide at 254 mμ

W. A. E. McBryde and J. H. Yoe, *Anal. Chem.*, 20 (1948) 1094.

Colorimetry of chloride at 312 mμ

F. Vydra and J. Celikovsky, *Chem. Listy*, 51 (1957) 768.

Oxidation of o-tolidine ($\varepsilon \sim 50{,}000$)

H. Schreiner, H. Brantner and F. Hecht, *Mikrochim. Acta*, 36/37 (1951) 1056.

By means of dithizone

L. Erdey and G. Rady, *Z. Anal. Chem.*, 135 (1952) 1.

In minerals

M. Shima, *Japan Analyst*, 2 (1953) 96.

Hydrogen

H = 1.008

Colorimetric methods of estimation are rarely used. Dyes such as methylene blue may however be reduced, in the presence of a catalyst such as finely divided palladium.

cf. M. PESEZ, *Bull. Soc. Chim.*, (1946) 692.
L. SILVERMAN AND W. BRADSHAW, *Anal. Chim., Acta*, 15 (1956) 31.

Indium

In = 114.8

(1) Separation of trace amounts

Of the more common separations, it is possible to precipitate the sulphide at pH 4.5. Tartaric acid may be added if it is necessary to separate the indium from Al(III), Ti(IV), etc.

In dilute sulphuric acid, In_2S_3 is coprecipitated with other sulphides.

(i) Extraction of the iodide complex. InI_3 can be extracted with ether from 0.5–2.5 *N* hydriodic acid, and may in this way be separated from Al(III), Fe(II), Be(II), Ga(III), etc.: F^-, CN^-, phosphates, and citrates do not interfere[1].

The iodide may also be extracted with cyclohexanone[2]: the extraction takes place from a medium 0.2 *N* in H_2SO_4 and 0.25 *M* in KI. This method permits a separation of In from Ga(III), Al(III), Fe(III), etc.

REAGENTS

Cyclohexanone.

Sulphuric acid, 10 N: 100 ml of concentrated sulphuric acid are added to 260 ml of water.

Potassium iodide.

OPERATING PROCEDURE

1 ml of 10 *N* sulphuric acid and 2 g of potassium iodide are added to 50 ml of solution. The iodide is extracted with 50 ml of cyclohexanone, shaking for 3 minutes.

(1) H. M. IRVING, F. J. C. ROSSOTI AND J. G. DRYSDALE, *Nature*, 169 (1952) 619; *Analyst*, 77 (1952) 801.

(2) H. HARTKAMP AND H. SPECKER, *Angew. Chem.*, 68 (1956) 678; *Talanta*, 2 (1959) 67. E. JACKWERTH AND H. SPECKER, *Z. Anal. Chem.*, 177 (1960) 327.

Extraction of the bromide with isopropyl ether (Zn(II) is not extracted)

L. KOSTA AND J. HOSTE, *Mikrochim. Acta*, (1956) 790.

T. A. COLLINS, JR. AND J. H. KANZELMAYER, *Anal. Chem.*, 33 (1961) 245.

(ii) Extraction of the oxinate with chloroform: see Colorimetry.

(iii) Extraction of the dithizonate: see Colorimetry.

(iv) Separation by means of anion-exchangers. Starting with a hydrochloric acid solution, In(III) may be separated from Cd(II), Al(III), Ga(III), Fe(III), Mn(II), As(III). Sn(IV) follows In(III).

cf. K. A. Kraus, F. Nelson and G. W. Smith, *J. Phys. Chem.*, 58 (1954) 11.
D. Jentzsch, I. Frotscher, G. Schwerdtfeger and G. Sarfert, *Z. Anal. Chem.*, 144 (1955) 8.
R. Klement and H. Sandmann, *Z. Anal. Chem.*, 145 (1955) 325.

(v) Microelectrolysis. Indium is deposited from an ammoniacal oxalate medium at 70–80° C.

cf. G. L. Royer, *Ind. Eng. Chem., Anal. Ed.*, 12 (1940) 439.

(vi) Standard solution. 1 g of the pure metal is dissolved in 4 to 5 *N* (1/8) sulphuric acid, and the solution is then diluted as required.

cf. T. Moeller, *J. Am. Chem. Soc.*, 62 (1940) 2444.

(2) Colorimetry

(i) Oxine. With 8-hydroxyquinoline, In(III) gives a compound which dissolves in chloroform to give a yellow solution. This reaction is unfortunately common to many other ions, so that careful preliminary separations are necessary.

The oxinate is extracted at pH 3.5. Under these conditions, small amounts of Pb(II), Zn(II), Sn(IV), Hg(II), Mn(II), and Cr(III) do not interfere. A number of other compounds are extracted, including those of Al(III) and Ga(III). The method has not yet been successfully applied to the separations of indium from large amounts of Zn(II).

Sensitivity. $\varepsilon = 6700$ at 395 mμ.

Reagents

Buffer solution: A solution of 1 g of potassium hydrogen phthalate in 100 ml of water, adjusted to pH 3.5 with the aid of a pH-meter.

Oxine solution: 1 g of 8-hydroxyquinoline is dissolved in 200 ml of chloroform.

Operating procedure

5 ml of buffer followed by 20 ml of the solution of oxine in chloroform are added to 20 ml of the test solution; the latter should contain 20 to 200 μg of indium and should have a pH of about 3.5. The mixture is shaken vigorously for 30 seconds, allowed to settle and the organic phase is then transferred into a 50 ml measuring flask. The volume is made up with chloroform.

Colorimetry is performed at 400 mμ.

cf. T. MOELLER, *Ind. Eng. Chem., Anal. Ed.*, 15 (1943) 270.
S. LACROIX, *Anal. Chim. Acta*, 1 (1947) 260.
C. L. LUKE AND M. E. CAMPBELL, *Anal. Chem.*, 28 (1956) 1340.
G. K. SCHWEITZER AND G. R. COE, *Anal. Chim., Acta*, 24 (1961) 311.

By fluorescence
E. B. SANDELL, *Ind. Eng. Chem., Anal. Ed.*, 13 (1941) 844.
R. BOCK AND K. G. HACKSTEIN, *Z. Anal. Chem.*, 138 (1953) 337.

5,7-Dibromooxine
J. E. JOHNSON, M. C. LAVINE AND A. J. ROSENBERG, *Anal. Chem.*, 30 (1958) 2055.

(ii) Dithizone. The pH at which indium dithizonate is extracted depends on the solvent used.
(1) With carbon tetrachloride, indium is extracted between pH 5 and 6. The metal is not extracted from ammoniacal solutions, and is therefore left in the aqueous phase when Zn(II), Pb(II), Cd(II), and Bi(III), for example, are extracted at pH 9.
(2) With chloroform the extraction takes place between pH 8 and 9.5. Zn(II), Bi(III), and Cu(II) can be separated by previous extraction at a pH of about 4.5–5.

Alternatively, Zn(II), Cd(II), Cu(II), etc., may be complexed with cyanide[1]. Moderate amounts of cyanide do not interfere with the extraction of indium dithizonate at pH $\sim$ 9. It is therefore possible to determine 10 μg of indium in the presence of 10 mg of zinc.

Large amounts of citrate, tartrate and phosphate interfere.

Pb(II) and Sn(II) accompany In(III), but not Sn(IV), Bi(III), and Tl(I and III). Rhodium and iridium interfere.

REAGENTS

Dithizone: 10 mg of dithizone are dissolved in 1000 ml of chloroform.

Ammonia, 2 N: 14 ml of concentrated ammonia solution are added to 86 ml of water.

Potassium cyanide, 10% in water.

OPERATING PROCEDURE

1 ml of the cyanide solution is added to 15 ml of solution containing 10 to 50 μg of indium. The pH is adjusted to $\sim$ 9 by means of ammonia and indium is extracted with 5 ml portions of dithizone. The extracts are made up to 25 ml with chloroform and the colorimetric estimation is performed at 510 mμ.

cf. K. E. KLEINER AND L. V. MARKOVA, *Zh. Analit. Khim.*, 8 (1953) 279.
(1) C. L. LUKE AND M. E. CAMPBELL, *Anal. Chem.*, 28 (1956) 1340.

Iodine and iodides

I = 126.9

(*i*) *Standard solutions.* Iodine or potassium iodide of guaranteed purity can be weighed out. Normally a standardized 0.1 N iodine solution is available. Alternatively, a given amount of iodine may be liberated by the action of a standardized strong acid on an excess of an iodate-iodide mixture.

$$I/KI = 0.765$$

(*ii*) *Colorimetry of iodine.* Iodine can be separated from the aqueous solution either by boiling or by extraction with carbon tetrachloride.

The element is most often estimated by microvolumetric methods, except in the case of very small traces.

The sensitivity can be increased by oxidizing the iodine to the iodate. The excess of oxidizing agent is then destroyed, and the iodate reduced by an excess of I^- in an acid medium, *i.e.*

$$\tfrac{1}{2}I_2 \rightarrow 1IO_3^- \rightarrow 3I_2$$

cf. W. Crouch, Jr., *Anal. Chem.*, 33 (1962) 1698.

(*iii*) I_3^- *in aqueous solution.* The method is very sensitive at 350 mμ: $\varepsilon = 25{,}000$.

cf. J. J. Custer and S. Natelson, *Anal. Chem.*, 21 (1949) 1005.
C. Herbo and J. Sigalla, *Anal. Chim. Acta*, 17 (1957) 199.

(*iv*) *Using a starch paste.* The blue coloration which $I_2 + I^-$ give with starch paste permits the estimation of 0.5 to 5 μg of iodine to be accomplished. The coloration produced by reaction with α-naphthoflavone is more sensitive.

Reagents

Starch paste: a slurry made of 0.5 g of starch and 2.5 g of water is added gradually to 200 ml of water. The mixture is boiled for 15 minutes with stirring, and then cooled and treated with 0.25 g of salicylic acid which acts as an antiseptic and stabilizes the reagent.

cf. W. G. Gross, L. K. Wood and J. S. McHargue, *Anal. Chem.*, 20 (1948) 900.
F. G. Houston, *Anal. Chem.*, 22 (1950) 493.
J. F. Lambert, *Anal. Chem.*, 23 (1951) 1247, 1251.

(*v*) *After extraction.* When iodine is extracted with chloroform, a purple solution is obtained. A still more sensitive yellow coloration is produced by mixing the free iodine solution with an alcoholic solution of iodide, with an absorption maximum at 365 mμ.

The extraction of iodine with chloroform is governed by the presence, in the aqueous solution, of halides which complex I_2. The calibration should therefore be performed under identical conditions.

REAGENT

Potassium iodide, 1 g/l in 95% alcohol.

OPERATING PROCEDURE

The solution is extracted twice with 5 ml of chloroform, shaking for 1 minute. The extracts are transferred to a 25 ml measuring flask, and 2 ml of the iodide solution are added. The volume is made up to 25 ml with alcohol.

cf. T. C. J. OVENSTON AND W. T. REES, *Anal. Chim. Acta,* 5 (1951) 123.

(vi) Redox indicator. In the presence of Hg(II), which removes I^-, iodine can oxidize *o*-tolidine. The limiting concentration is 0.1 ppm.

cf. J. K. JOHANNESSON, *Anal. Chem.,* 28 (1956) 1475.

(vii) Colorimetry of the iodides. (1) Iodides are oxidized to iodine by numerous oxidizing agents: iodate, Fe(III), Br_2, HNO_2. The iodine may then be estimated. (2) The reader is referred to the methods for estimating chlorides: ferrichloride complexes, coloration of ferric thiocyanate, Hg(II)-diphenylcarbazide, etc.

(viii) Estimation of traces by their catalytic effect. Iodine and its derivatives catalyze the oxidation of As(III) by ceric salts or permanganate in acid solutions. The rates of the reactions, which are normally very slow, depend on the amount of iodine present. Very minute traces of iodide or iodine can thus be detected.

cf. 0.001 to 0.5 μg in 5 ml.
H. F. W. KIRKPATRICK, *Analyst,* 78 (1953) 348.
B. ROGINA AND M. DUBRAVCIC, *Analyst,* 78 (1953) 594.
F. LACHIVER, *Ann. Chim.,* 10 (1955) 92.
R. A. BARKLEY AND T. G. THOMPSON, *Anal. Chem.,* 32 (1960) 154.

In sea water
M. DUBRAVCIC, *Analyst,* 80 (1955) 146.
B. ZAK (D. F. BOLZ, ed.), *Colorimetric Determination of Non-Metals,* Interscience, 1958.

Iridium

Ir = 192.2

(1) Separations

See Metals of the Platinum Group.

(2) Colorimetry

(i) With stannous chloride. $IrCl_6^{2-}$ is reduced in HBr, giving an intense yellow coloration, $\varepsilon = 50{,}000$. The method is not accurate; Pt, Pd and Rh interfere.

cf. S. S. BERMAN AND W. A. E. MCBRYDE, *Analyst*, 81 (1956) 566.

(ii) By means of redox indicators. Ir(IV) oxidizes numerous dyes. Pt, Pd, and Rh interfere.

With dianisidine[1], $\varepsilon = 12{,}000$.

cf. (1) S. S. BERMAN, F. E. BEAMISH AND W. A. E. MCBRYDE, *Anal. Chim. Acta*, 15 (1956) 363.

(iii) Oxidation with Ce(IV). Rh does not interfere.

A. D. MAYNES AND W. A. E. MCBRYDE, *Analyst*, 79 (1954) 230.

Review of methods

F. E. BEAMISH AND W. A. E. MCBRYDE, *Anal. Chim. Acta*, 9 (1953) 349; 18 (1958) 551.

Iron

Fe = 55.85

(1) Separation of trace amounts

Of the more familiar methods of separation, the following may be used:

Precipitation of the sulphide. Ammonium sulphide is used, with Cd(II) as the entraining agent. The sulphide may also be precipitated from a tartaric acid solution.

Precipitation of the hydroxide. Entraining agent: $Mn(OH)_4$. If necessary, the volume of the solution may be reduced by evaporation, to remove the excess of acid. 5 ml of 2 *N* (1/6) hydrochloric acid are added, and the solution diluted to 100 ml. 5 drops of 1% potassium permanganate are added, followed by a slight excess of ammonia, and finally 1 ml of alcohol. The mixture is heated until $Mn(OH)_4$ is precipitated; the latter is filtered, washed and dissolved in a little 2 *N* (1/6) hydrochloric acid.

Many other hydroxides can be used as entraining agents.

(i) Cupferron; precipitation or extraction.

Extraction

R. E. PETERSON, *Anal. Chem.*, 24 (1952) 1850.

(ii) Oxine; precipitation or extraction.

K. MOTOJIMA AND H. HASHITANI, *Japan Analyst*, 7 (1958) 33.

(iii) Extraction of the thiocyanate; see Colorimetry.

(iv) Extraction of ferrous o-phenanthroline perchlorate. Iron can be separated from V, Cr, Mn, Ni, Zn, etc. by this method, using nitrobenzene as the solvent.

cf. D. W. MARGERUM AND C. V. BANKS, *Anal. Chem.*, 26 (1954) 200.

(v) By ion exchangers. By cation exchangers[1]: P(V), Mo(VI), etc. are separated. By anion exchangers[2]: in HCl solution, Fe(III) is determined (see p. 133).

(1) R. PECSOK AND R. PARKHURST, *Anal. Chem.*, 27 (1955) 1920.

R. KLEMENT, *Z. Anal. Chem.*, 145 (1955) 9.
(2) C. K. MANN AND C. L. SWANSON, *Anal. Chem.*, 33 (1961) 459.
L. L. LEWIS, M. J. NARDOZZI AND L. M. MELNICK, *Anal. Chem.*, 33 (1961) 1351.

(vi) Standard solutions. (1) 1000 mg of pure powdered iron are dissolved in 10 ml of 4 *N* (1/10) sulphuric acid. 3 ml of concentrated nitric acid are added if it is necessary to oxidize ferrous iron. The volume is made up to 1 litre.
(2) Alternatively, Mohr's salt $FeSO_4 \cdot (NH_4)_2SO_4 \cdot 6H_2O$ or ferric alum $Fe_2(SO_4)_3 \cdot (NH_4)_2SO_4 \cdot 24H_2O$ of guaranteed purity may be used.

$FeCl_2 \cdot 4H_2O$ and $FeSO_4 \cdot 7H_2O$ are also available.

$$Fe/FeSO_4 \cdot (NH_4)_2SO_4 \cdot 6H_2O = 0.1423$$

$$Fe/Fe(SO_4)_3 \cdot (NH_4)_2SO_4 \cdot 24H_2O = 0.1158$$

(2) Colorimetry

A number of reagents give satisfactory results, but only the two most important methods will be described. In particular cases, it may be advantageous for reasons of selectivity, to use one of the other reagents mentioned in the bibliography on page 276.

The thiocyanate method is simple and allows trace amounts to be estimated; the accuracy is however rather low unless special precautions are observed, and the method is not specific.

The *o*-phenanthroline method is specific and in addition permits the estimation of minute traces. The coloration is stable.

(i) Thiocyanate. Fe^{3+} ions give rise to red thiocyanate complexes in acid solutions. This is the most frequently used method, although it possesses numerous disadvantages. In fact, the ferrithiocyanate complexes are unstable and the colour is therefore governed by concentrations of the reagents, ionic strength of the solution, and the presence of numerous ions, even those exhibiting little complexing power, such as Cl^- and SO_4^{2-}. The colour is also a function of pH. Intensity of the coloration is increased by diminishing the dielectric constant of the medium by additions of acetone or dioxan.

Fe^{3+} is gradually reduced by SCN^-; the reduction is more rapid in the presence of light, particularly if the colorimetric cell is illuminated by an intense beam (*e.g.* colorimeters with filters). This reaction may be prevented by adding an oxidizing agent, such as persulphate or hydrogen peroxide.

The complex can be stabilized with respect to light by a mixture of acetone and methyl ethyl ketone[1].

$Fe(SCN)_3$ can be extracted with alcohols or ethers. This compound is also red.

Accuracy and sensitivity. $\varepsilon \sim 7000$ at 480 mμ, or, in the presence of acetone,

$\varepsilon \sim 15{,}000$. The results are only reproducible if the conditions of the operation are maintained strictly invariant.

The influence of temperature is considerable.

Interfering ions. Certain ions have little interfering effect. Thus, in concentrations equal to 250 times that of the Fe(III), the error is less than 2% for acetates, As(V), Br^-, citrates, Cl^-, HCN, HCO_2H, NO_3^-, H_3PO_4, silica, SO_4^{2-}, tartrates, and Al(III).

Pyro- and metaphosphates should be present in concentrations lower than 5 ppm; fluorides < 30 ppm (400 ppm in the presence of acetone), and oxalates < 30 ppm in the presence of acetone.

The following do not interfere in the presence of acetone: B(III) < 150 ppm, As(III) < 500 ppm, Sn(IV) < 50 ppm, Hg (II) < 50 ppm; V(V) does not interfere up to 25 mg, since its blue colour does not absorb radiation of 480 mμ.

Cu(II), Co(II), Bi(III), Ti(IV), Ru, Os, and Mo(VI) give rise to coloured complexes with SCN^-; W(VI) precipitates in the form of hydroxide. Cu(I) and Ag(I) precipitate in the form of thiocyanates. Hg(II) consumes SCN^-.

The acidity should be fixed. In general, a concentration of thiocyanate of 0.3 *M* is used. If an accuracy of 1% is required, this concentration should be established to within $\pm$ 1%.

Reagents

15% potassium thiocyanate solution. The purity must be checked if it is desired to estimate very minute traces. For this purpose, a small amount of reagent is treated with an equal volume of acetone. If a red coloration appears, a little Al(III) is added to act as an entraining agent and precipitated by ammonia. The solution is neutralized after filtering.

Hydrochloric or nitric acid. The purity should be checked. In the case of nitric acid, air should be bubbled through the acid to eliminate nitrous vapours.

Operating procedure

5 ml of concentrated hydrochloric acid and then 12 ml of thiocyanate are added to 20 ml of solution containing 50 to 300 μg of ferric iron. The volume is made up to 50 ml and colorimetry is performed at 480 mμ.

cf. Of the many available references, *e.g.*
J. T. Woods and M. G. Mellon, *Ind. Eng. Chem., Anal. Ed.*, 13 (1941) 551.
C. A. Peters and C. L. French, *Ind. Eng. Chem., Anal. Ed.*, 13 (1941) 604.
H. Cox, *Analyst*, 69 (1944) 235.
(1) P. Baily, *Anal. Chem.*, 29 (1957) 1534.

In titanium
J. M. Thompson, *Anal. Chem.*, 25 (1953) 1231.

Extraction of Fe(III) thiocyanate. By using butyl phosphate Fe(III) can be separated, in particular from Al(III), Ni(II), Sb(III) and Sn(IV).

M. AVEN AND H. FREISER, *Anal. Chim. Acta*, 6 (1952) 412.
L. MELNICK, H. FREISER AND H. F. BEEGHLY, *Anal. Chem.*, 25 (1953) 856.
L. M. MELNICK, *Dissert. Abstr.*, 14 (1954) 760.

With ether + tetrahydrofuran

H. SPECKER AND H. HARTKAMP, *Z. Anal. Chem.*, 140 (1953) 353.

In steels

H. SPECKER, M. KUCHTNER AND H. HARTKAMP, *Z. Anal. Chem.*, 142 (1954) 166.

Extraction of iron tributylammonium thiocyanate with amyl acetate. The method is very sensitive: $\varepsilon \sim 20{,}000$ at 480 mμ. The compound formed is very stable.

M. ZIEGLER, O. GLEMSER AND N. PETRI, *Z. Anal. Chem.*, 154 (1957) 81.

(ii) 1,10-Phenanthroline. Fe^{2+} reacts with 1,10-phenanthroline to give a red complex ion containing 3 molecules of 1,10-phenanthroline per ion of Fe^{2+}.

The coloration remains stable for over six months and is independent of the pH between 2 and 9. The complex formed is very stable and the coloration is only slightly affected by an excess of reagent and by the presence of certain anions. If however, accurate results are to be obtained, a certain number of factors (all involving slight errors) must be taken into account: order of addition of the reagents, time, temperature, pH, interfering ions (see below).

Sensitivity. At 490 mμ, the molar absorption coefficient $\varepsilon = 10{,}600$; at 505 mμ, $\varepsilon = 11{,}000$.

Fe(III) must first be reduced to Fe(II) with hydroxylamine hydrochloride or with hydroquinone.

Interfering ions. A number of ions interfere, either by the formation of precipitates or by giving a coloration. If the results of colorimetry may be approximate (to within 1 to 2%), the effect of these ions may be neglected up to a certain concentration, or minimized by working at a suitable pH. Details of the permissible conditions are given below:

During the estimation of 2 ppm of iron, it is possible to tolerate: 50 ppm of Cd(II), 10 of Zn(II), and 1 of Hg(II); an excess of reagent must be present in all three cases, in order to redissolve the precipitate. There should not be more than: 10 ppm of Hg(I) (the pH should lie between 3–9), 500 ppm of Be(II) (pH 3.0 to 5.5) or Mo(VI) (pH 5.5), 5 ppm of W(VI), 10 ppm of Cu(II) (pH 2.5–4.0), 2 ppm of Ni(II), 10 ppm of Co(II) (pH 3–5), 20 ppm of Sn(II) (pH 2–3), 50 ppm of Sn(IV) (pH 2.5), 30 ppm of Sb(III), 20 ppm of Cr(III), 500 ppm of oxalate and tartrate (pH > 6 or < 3), 50 ppm of $P_2O_7^{4-}$ (pH > 6), 10 ppm of CN^-, 20 ppm of PO_4^{3-} (pH 2–9), 500 ppm of F^- (pH > 4). Bi(III) interferes unless EDTA is present.

Many other hydroxides and phosphates may precipitate. This may often be avoided by adding citrate to the solution.

ClO_4^- precipitates the compound.

Up to 500 ppm of the following ions do not interfere: $CH_3CO_2^-$, Cl^-, ClO_3^-, Br^-, I^-, NO_3^-, SO_4^{2-}, SO_3^{2-}, SCN^-, citrate, As(V), As(III), Al(III), Pb(II), Mn(II), and the alkaline earth and alkali metals.

REAGENTS

Hydroxylamine hydrochloride, 10% in water.

Sodium acetate, 25% (2 *M*) solution.

1,10-Phenanthroline hydrochloride, 0.5% in water.

OPERATING PROCEDURE

The slightly acidic solution is placed in a 25 ml volumetric flask. The amount of acetate necessary to bring the pH to about 3.5 (bromophenol blue) is determined in a separate sample. The pH is then adjusted accordingly. 1 ml of hydroxylamine hydrochloride is added to reduce Fe(III), followed by 1 ml of reagent. After standing for 1 hour, the solution is made up to 25 ml and colorimetry is carried out at 480–520 mμ.

NOTE. It is also possible to estimate Fe(II) in the presence of Fe(III) by omitting the hydroxylamine reduction.

COMPARE

Spectrophotometric study

W. B. FORTUNE AND M. G. MELLON, *Ind. Eng. Chem., Anal. Ed.,* 10 (1938) 60.

S. L. BANDEMER AND P. J. SCHAIBLE, *Ind. Eng. Chem., Anal. Ed.,* 16 (1944) 317.

E. ASMUS, *Z. Anal. Chem.,* 122 (1941) 81.

D. W. MARGERUM AND C. V. BANKS, *Anal. Chem.,* 26 (1954) 200.

Estimation of large amounts to within 0.1%

J. P. MEHLIG AND H. R. HULETT, *Ind. Eng. Chem., Anal. Ed.,* 14 (1942) 859.

In biochemistry

F. C. HUMMEL AND H. H. WILLARD, *Ind. Eng. Chem., Anal. Ed.,* 10 (1938) 13.

D. L. DRABKIN, *J. Biol. Chem.,* 140 (1941) 387.

H. COWLING AND E. J. BENNE, *J. Assoc. Official Agr. Chem.,* 25 (1942) 555.

In silicates

H. R. SHELL, *Anal. Chem.,* 22 (1950) 326.

In glass

A. GOTTLIEB, *Mikrochim. Acta,* 35 (1950) 320.

In foodstuffs

C. HOFFMAN, T. R. SCHWEITZER AND G. DALBY, *Ind. Eng. Chem., Anal. Ed.*, 12 (1940) 454.

In the presence of Bi

D. G. HOLMES, *Analyst*, 82 (1957) 528.

In tantalum

J. HASTINGS, T. A. MCCLARITY AND E. J. BRODERICK, *Anal. Chem.*, 26 (1954) 379.

(3) Other methods

(i) 2,2'-Dipyridyl. $\varepsilon \sim 8000$ at 522 mμ. The reactions and properties somewhat resemble those of 1,10-phenanthroline.

cf. L. GERBER, R. I. CLAASSEN AND C. S. BORUFF, *Ind. Eng. Chem., Anal. Ed.*, 14 (1942) 364
M. L. MOSS AND M. G. MELLON, *Ind. Eng. Chem., Anal. Ed.*, 14 (1942) 862.

Precision colorimetry

J. P. MEHLIG AND M. J. SHEPHERD, JR., *Chemist-Analyst*, 36 (1947) 52.

In biochemistry

H. BOREL, *Biochem. Z.*, 314 (1943) 359.
A. J. WOIWOD, *Biochem. J.*, 41 (1947) 39.

In beverages

P. P. GRAY AND I. M. STONE, *Ind. Eng. Chem., Anal. Ed.*, 10 (1938) 415.
H. L. ROBERTS, C. L. BEARDSLEY AND L. V. TAYLOR, JR., *Ind. Eng. Chem., Anal. Ed.*, 12 (1940) 365.

(ii) Sulphosalicylic acid. $\varepsilon \sim 6000$ at 430 mμ.

cf. L. C. E. KNIPHORST, *Chem. Weekblad*, 42 (1946) 311.

Ferron. (7-Iodo-8-hydroxyquinoline-5-sulphonic acid). $\varepsilon \sim 4000$ at 610 mμ.

cf. J. H. YOE AND R. T. HALL, *J. Am. Chem. Soc.*, 59 (1937) 872.
P. HANISET, *Ing. Chim. Belg.*, 33 (1950) 51.

In wines

J. ROUBERT, *Chim. Anal.*, 38 (1956) 134.

(iii) Extraction of the compound with 'ferron' and tributylammonium. Phosphates do not interfere.

cf. M. ZIEGLER, O. GLEMSER AND N. PETRI, *Mikrochim. Acta*, (1957) 215.

(iv) Precision colorimetry.

In $HClO_4$ at 260 mμ

R. Bastian, R. Weberling and F. Palilla, *Anal. Chem.*, 28 (1956) 459.

M. Ishibashi, T. Shigematsu, Y. Yamamoto, M. Tabushi and T. Kitagawa, *Bull. Chem Soc. Japan*, 29 (1956) 57.

In a Cl^- medium. – in Zr

L. Silverman and K. Trego, *Anal. Chim. Acta*, 19 (1958) 299.

In a sulphate medium (ε ∼ 2000 at 320 mμ)

R. Bastian, R. Weberling and F. Palilla, *Anal. Chem.*, 25 (1953) 284.

(v) Comparison and review of methods.

J. T. Woods and M. G. Mellon, *Ind. Eng. Chem., Anal. Ed.*, 13 (1941) 551.

Lead

Pb = 207.2

(1) Separation of trace amounts

Of the more common methods, the following can be used:

(i) Precipitation of the sulphide with Cu(II) as the entraining agent.

(ii) Precipitation of the sulphate with Sr(II) as the entraining agent. The precipitate is then redigested in fused carbonates.

cf. I. T. ROSENQVIST, *Am. J. Sci.*, 240 (1942) 359.

(iii) Precipitation of the phosphate, with Ca(II) as the entraining agent.

cf. A. E. BALLARD AND C. D. W. THORNTON, *Ind. Eng. Chem., Anal. Ed.*, 13 (1941) 893.

(iv) Precipitation of the hydroxide, with Fe(II) as the entraining agent.

cf. A. E. BALLARD AND C. D. W. THORNTON, *Ind. Eng. Chem., Anal. Ed.*, 13 (1941) 893.

(v) Electrolysis in the form of PbO_2. In this method the lead is deposited at the anode.

cf. G. NORWITZ, *Z. Anal. Chem.*, 132 (1951) 165.

In copper alloys
H. C. J. SAINT, *Analyst*, 83 (1958) 88.

(vi) Extraction of lead iodide. Methyl *iso*-propyl ketone (or methyl *iso*-butyl ketone) are used as the solvent.

Zn(II), Cd(II), Cu(II), As(III), Sb(III), Sn(IV) and Bi(III) are partially extracted. A previous extraction of the thiocyanates enables the separation of Zn(II), Cd(II), Sn(IV) and Bi(III) to be performed.

REAGENTS

Hydrochloric acid, 0.5 N: 5 ml of concentrated hydrochloric acid are added to 95 ml of water.

Methyl iso-butyl ketone: the solvent is shaken with 0.5 *N* hydrochloric acid.

Potassium iodide.

OPERATING PROCEDURE

1 ml of concentrated hydrochloric acid and 3 g of solid potassium iodide are added to 20 ml of solution. When solution is complete, 10 ml of solvent are added and the mixture is shaken for 1 minute. A second extraction is performed.

The lead(II) can be made to pass back into an aqueous phase consisting either of caustic soda at 5 g per litre or of a solution of sodium citrate, ammonia, and potassium cyanide. The aqueous solution obtained may then be estimated colorimetrically with dithizone.

cf. P. W. WEST AND J. K. CARLTON, *Anal. Chim. Acta*, 6 (1952) 406.

In biochemistry

W. M. McCORD AND J. W. ZEMP, *Anal. Chem.*, 27 (1955) 1171.

In rocks

A. D. MAYNES AND W. A. E. McBRYDE, *Anal. Chem.*, 29 (1957) 1259.

(vii) Extraction of the diethyldithiocarbamate. Carbon tetrachloride is used at pH 7.0.

cf. H. C. LOCKWOOD, *Analyst*, 79 (1954) 143.

Estimation in alloys

J. KINNUNEN AND B. WENNERSTRAND, *Chemist-Analyst*, 43 (1954) 65.

(viii) With ion-exchangers. Pb(II) in *N* HCl is collected by an anionic resin [Amberlite IRA 400] and may thus be separated, in particular from Cu(II) and Fe(III).

In foodstuffs

E. I. JOHNSON AND R. D. A. POLHILL, *Analyst*, 82 (1957) 238.

(ix) Standard solutions. These can be made by dissolving the pure metal in dilute nitric acid.

Pure lead nitrate dried at 110° may also be used.

The dilute solutions should be acidified with 0.1 *N* (1/100) nitric acid. They are not stable during storage.

$$Pb/Pb(NO_3)_2 = 0.626.$$

(2) Colorimetry

(i) Dithizone. The method involving dithizone, which is the only important one, permits the estimation of very small traces of lead. It can be quite accurate

($\pm$ 1%) and is consequently a suitable method for the rapid estimation of even quite large amounts. Its disadvantage is however that preliminary separations may often be required.

With Pb(II), dithizone produces a red complex which may be extracted with chloroform from solutions at pH 7.5 to 11.5, and with carbon tetrachloride at pH 7 to 10. Pb(II) passes back into an aqueous phase containing 0.02 N HCl.

Sensitivity. $\varepsilon = 60{,}000$ to $70{,}000$ at 520 mμ in CCl_4.

Interfering ions. The extraction is carried out in the presence of cyanide, (which complexes the elements of the zinc group), and of citrate which prevents precipitation of the hydroxides of the ammonia group. The precipitation of alkaline earth phosphates can be prevented by the addition of hexametaphosphate[1]. Under these conditions, Sn(II), Tl(I), and Bi(III) interfere; the last-mentioned can however be separated by dithizone, at pH 3.0, and subsequently passes quantitatively into the chloroform layer. Sn(II) can be oxidized to Sn(IV). Lead can be separated from 1 mg of Tl(I) by the extraction of its dithizonate at pH 6.0–6.4. Any oxidizing interfering ions may be reduced with hydroxylamine.

Mn(II) catalyzes the oxidation of dithizone by air.

Colloidal Zr(IV) may retain Pb(II).

Alternatively, Pb(II) may be isolated beforehand, by extracting lead iodide or the diethyldithiocarbamate compound. In this way 0.5% of lead can be estimated in the presence of the Zn, Cd, Hg, Ag, As, Al, Sn, and other metals.

Two techniques for the estimation of lead with dithizone will be described; a rapid titration for large amounts (50–100 μg), and a more sensitive method necessitating careful control of the reagents, whereby 0.5 μg may be estimated.

Titration. The red lead dithizonate is extracted with 0.5 ml portions of dithizone until the presence of an excess of the reagent is indicated by the appearance of a green coloration.

Reagents

Dithizone solution: 25 mg per litre of chloroform. The reagent need not be purified. The solution should be standardized shortly before use, with a known solution of Pb(II) (10 ppm) by the procedure given below.

Buffer solution: 6 ml of concentrated ammonia, 5 g of ammonium chloride and 100 ml of water.

Sodium citrate, 10% solution.

Hydroxylamine hydrochloride, 10% solution.

Potassium cyanide, 10% solution.

Operating procedure

10 ml of solution, preferably containing 50 to 100 μg of lead, are neutralized with ammonia to the change-point of thymol blue (indicator paper). 5 ml of

citrate, 2 ml of cyanide, 1 ml of hydroxylamine hydrochloride and 10 ml of buffer solution are added, and the complex is extracted with 0.5 ml portions of dithizone. After each extraction, the solution is rinsed with 0.5 ml of chloroform. The extraction is stopped as soon as the colour changes from red to mauve. The result is then straddled to within 0.5 ml of dithizone, or to about $\pm$ 2.5 μg of lead.

A blank test should be carried out with the reagents, which must be purified if the amount of lead which they contain is significant.

Colorimetry. This method permits the estimation of 0.5 to 100 μg of lead.

Principle. A first extraction is made up at pH 9–9.5 in the presence of citrate and cyanide. Pb(II) and any Bi(III) present pass into the chloroform layer. The solvent is shaken with a buffered solution at pH 3.4 when only Pb(II) passes into the aqueous phase. This solution is brought to about pH 10–11 and the Pb(II) is extracted.

REAGENTS

Thymol blue, 0.1% in alcohol.

Ammonium or sodium citrate, 50 g per litre.

Potassium cyanide, 50 g per litre.

Ammonia, diluted 1/2.

Cyanide-ammonia solution: 100 ml of dilute ammonia, 20 ml of sodium sulphite at 10 g per litre, and 20 ml of potassium cyanide at 50 g per litre are mixed together.

pH 3.4 buffer solution: 1 g of potassium hydrogen phthalate is dissolved in 200 ml of water. The pH is adjusted to 3.4 (pH-meter) with hydrochloric acid, and the volume is made up to 1 litre.

Dithizone solution: 25 mg of dithizone are dissolved in 100 ml of chloroform; this solution should be purified.

Solution diluted to 25 mg per litre: the preceding solution is diluted by a factor of 10 with chloroform.

OPERATING PROCEDURE

2 to 3 drops of thymol blue and 10 ml of citrate are added to 25 ml of solution containing 10 to 50 μg of Pb(II) and the pH is adjusted to 9.0–9.5 with concentrated ammonia. 10 ml of cyanide are added and the complex is extracted with 20 ml portions of dilute dithizone, shaking for 1 minute during each extraction until the dithizone remains green. The extracts are combined and shaken for 1 minute with 25 ml of the buffer solution (pH 3.4). The lead passes into the aqueous phase. To this are added 75 ml of ammoniacal cyanide solution and 25 ml of dilute dithizone; the mixture is shaken for 1 minute, and colorimetry is then performed at 510 mμ. A blank test is performed with the same reagents.

COMPARE

A. D. MAYNES AND W. A. E. MCBRYDE, *Anal. Chem.*, 29 (1957) 1259.
(1) E. I. JOHNSON AND R. D. A. POLHILL, *Analyst*, 80 (1955) 364.

Separation from Bi(III) and Tl(I)

P. A. CLIFFORD, *J. Assoc. Offic. Agr. Chem.*, 26 (1943) 26.

Separation from Bi(III)

K. BAMBACH AND R. E. BURKEY, *Ind. Eng. Chem., Anal. Ed.*, 14 (1942) 904.

In rocks

E. B. SANDELL, *Ind. Eng. Chem., Anal. Ed.*, 9 (1937) 464.

In fruit

O. B. WINTER, H. M. ROBINSON, F. W. LAMB AND E. J. MILLER, *Ind. Eng. Chem., Anal. Ed.*, 7 (1935) 265.

In metals

G. W. C. MILNER AND J. TOWNEND, *Anal. Chim. Acta*, 5 (1951) 584.
K. OTA AND S. MORI, *Japan Analyst*, 5 (1956) 442.

In steels

L. G. BRICKER AND K. L. PROCTOR, *Ind. Eng. Chem., Anal. Ed.*, 17 (1945) 511.
M. JEAN, *Précis d'analyse chimique des aciers et des fontes*, Dunod, Paris, 1949.

In nickel

S. YOKOSUKA, *Japan Analyst*, 4 (1955) 99.

In monazite

R. A. POWELL AND C. A. KINSER, *Anal. Chem.*, 30 (1958) 1139.

In biological liquids

J. CHOLAK, D. M. HUBBARD AND R. E. BURKEY, *Anal. Chem.*, 20 (1948) 671.
H. M. IRVING AND E. J. BUTLER, *Analyst*, 78 (1953) 571.
S. L. TOMPSETT, *Analyst*, 81 (1956) 330.
W. M. MCCORD AND J. W. ZEMP, *Anal. Chem.*, 27 (1955) 1171.

In foodstuffs

H. C. LOCKWOOD, *Analyst*, 79 (1954) 143.
H. V. HART, *Analyst*, 76 (1951) 692.

In pharmaceutical products

K. BAMBACH, *Ind. Eng. Chem., Anal. Ed.*, 12 (1940) 63.

In organic compounds

F. NEUMANN, *Z. Anal. Chem.*, 155 (1957) 340; *Analyst*, 84 (1959) 127.

Review

F. X. Mayer and P. Schweda, *Mikrochim. Acta,* (1956) 485.

Colorimetry in a homogeneous water-acetone phase

D. G. M. Diaper and A. Kulesis, *Canadian J. Chem.*, 32 (1957) 1278.

(ii) Absorptiometry in the ultraviolet in hydrochloric acid in the presence of Bi(III) and Tl(I).

C. Merritt, Jr., H. M. Hershenson and L. B. Rogers, *Anal. Chem.*, 25 (1953) 572.

Lithium

Li = 6.94

Normally, flame-spectrophotometry is preferred to colorimetric estimations. In fact, the latter often necessitate laborious preliminary separations.

(1) Separations

Of those available, the following are worthy of mention.

(i) Volatilization of LiCl.

cf. M. H. Fletcher, *Anal. Chem.*, 21 (1949) 173.

(ii) The use of ion-exchangers. With suitable ion-exchangers, it is possible to separate 0.5 μg of Li(I) from as much as 500 mg of Ca(II)[1].

(1) H. Hering, *Anal. Chim. Acta*, 6 (1952) 340.
R. C. Sweet, W. Rieman and J. Beukenkamp, *Anal. Chem.*, 24 (1952) 952.

Li/Mg
D. H. Wilkins, *Anal. Chim. Acta*, 2 (1959) 116.

Review of methods of separation of Li(I) and Na(I)
B. Grüttner, *Z. Anal. Chem.*, 133 (1951) 36.

(iii) Electrolysis at a mercury cathode. A number of ions may be separated.

(2) Colorimetry

(i) 'Thoron'. With 'thoron', lithium gives an orange coloration in alkaline solutions. The reaction is more sensitive ($\varepsilon = 6000$ at 486 mμ) in the presence of acetone.

Interfering ions. All ions which precipitate in alkaline media will interfere. Less than 100 μg of Ca(II) and Mg(II) may be present. Amounts of sodium lower than 500 μg and 100 μg or more of Rb(I) and Cs(I) do not interfere.

Reagents

Acetone.

Potassium hydroxide, 200 g per litre.

'Thoron' [*2-(2-hydroxy-3,6-disulpho-1-naphthylazo)-benzenearsonic acid*]: 200 mg are dissolved in 100 ml of water.

OPERATING PROCEDURE

0.2 ml of potassium hydroxide, 7 ml of acetone, and 1 ml of 'thoron' are added to 1 ml of neutral solution containing from 1 to 10 μg of lithium. After 30 minutes, the colorimetric estimation is carried out at 485 mμ against a reference solution containing all the reagents.

cf. P. F. THOMASON, *Anal. Chem.*, 28 (1956) 1527.

(ii) Fluorimetry of the oxinate. In an alkaline medium in 95% alcohol.

cf. C. E. WHITE, M. H. FLETCHER AND J. PARKS, *Anal. Chem.*, 23 (1951) 478.

Magnesium

Mg = 24.32

(1) Separation of trace amounts

(i) Electrolysis at a mercury cathode. A large number of interfering ions can be separated whilst Mg(II) remains in solution.

J. A. MAXWELL AND R. P. GRAHAM, *Chem. Revs.*, 46 (1950) 471.

(ii) Extraction with oxine. A number of elements can be separated by extraction of the oxinates at pH 5.0.

In rocks

J. P. RILEY, *Anal. Chim. Acta*, 19 (1958) 413; 21 (1959) 317.

In compounds of the alkali metals

O. A. KENYON AND G. OPLINGER, *Anal. Chem.*, 27 (1955) 1125.

In zinc

H. POHL, *Metall*, 10 (1956) 709.

Alternatively, magnesium oxinate can be extracted from an alkaline solution (see Colorimetry).

(iii) Extraction of the acetylacetonates. The compounds formed by Fe(III), Cu(II), V(V), U(VI), Al(III), and Mn(II) are extracted by CCl_4.

cf. E. ABRAHAMCZIC, *Mikrochim. Acta*, 33 (1947) 208; *Angew. Chem.*, 61 (1949) 96.

(iv) Ion-exchangers. In citric acid solutions Mg(II) can be separated from Ba(II), Sr(II), and Ca(II).

cf. F. NELSON AND R. A. KRAUS, *J. Am. Chem. Soc.*, 77 (1955) 801.

(v) Standard solutions. The pure metal may be dissolved in 0.1 *N* (1/100) hydrochloric acid.

Alternatively, standard solutions may be prepared from guaranteed purity $MgSO_4 \cdot 7H_2O$, $Mg(NO_3)_2 \cdot 6H_2O$, or $MgCl_2 \cdot 6H_2O$.

$$Mg/MgSO_4 \cdot 7H_2O = 0.0987.$$
$$Mg/Mg(NO_3)_2 \cdot 6H_2O = 0.0949.$$
$$Mg/MgCl_2 \cdot 6H_2O = 0.1196.$$

(2) Colorimetry

A number of dyes give coloured adsorption compounds with magnesium hydroxide. The accuracy of all such methods is low.

Magnesium oxinate can be extracted with chloroform and then estimated colorimetrically. Eriochrome Black T gives a stable coloured complex, but Ca(II) undergoes the same reaction. Such techniques should be preceded by preliminary separations, which are sometimes not very sharp. Flame spectrophotometry is frequently to be preferred for the estimation of trace amounts.

(i) Thiazole Yellow. Thiazole yellow and 'titan yellow' dissolve in water giving a yellow coloration. In the presence of colloidal magnesium hydroxide they are adsorbed, with a pink coloration.

The method possesses the disadvantages associated with all adsorption techniques: the coloration is affected by many variables such as time, temperature, concentration, operating procedure, pH, etc. The results may also vary with the origin of the dye.

The colloid may be stabilized by the addition of starch, hydroxylamine, polyvinyl alcohol, or sodium phosphate[1], and is then stable for 2 days in the absence of light.

Sensitivity. Molar extinction coefficient: $\varepsilon = 1500$ at 535 mμ.

The method is not, however, accurate ($\pm$ 2 to 5%).

Interfering ions. Interference is obtained from many ions and preliminary separations are often essential. P(V) interferes above 100 ppm, Ca(II) below 500 ppm enhances the colour, but can be complexed with mannitol. Many ions (Cu(II), Ni(II) and Al(III), since $Mg(OH)_2$ adsorbs AlO_2^-, also Sn(IV), Ag(I), Hg(I) and (II), Cd(II), Co(II), Pb(II), Si(IV), Li(I), Fe(III), Zn(II), and La(III)) precipitate, sometimes giving colorations by adsorbing the dye. Ti(IV) can be complexed with H_2O_2 to give a colourless solution at pH $\geqslant 12$. Mn(II) oxidizes in air but this may be prevented by adding hydroxylamine hydrochloride which in addition stabilizes the coloration due to Mg(II). Sb(III), As(III), and As(V) prevent the appearance of the coloration to some extent. NH_4^+ present in large amounts, ($>$ 500 ppm), interferes because of its buffering effect on the solution. Proteins interfere. $C_2O_4^{2-}$ has no appreciable effect.

REAGENTS

Caustic soda, 10 *N* (400 g per litre).

Hydroxylamine hydrochloride, 5% in water.

Polyvinyl alcohol, 2% in water.

Thiazole yellow, 0.5% in 50% alcohol, kept in a brown bottle.

Thiazole yellow, 0.01%: 2 ml of the preceding solution are mixed with 5 ml of 0.5% polyvinyl alcohol and made up to 100 ml with water.

OPERATING PROCEDURE

5 ml of hydroxylamine hydrochloride, 4 ml of polyvinyl alcohol, 5 ml of 0.01% thiazole yellow, and 3.5 ml of 10 *N* caustic soda are added to 30 ml of neutral solution containing 30–200 μg of magnesium. The solution is allowed to stand for 15 minutes at 25 $\pm$ 0.5°, and colorimetry is carried out at 540 mμ within the next half hour.

cf. O. A. KENYON AND G. OPLINGER, *Anal. Chem.*, 27 (1955) 1125.
T. A. MITCHELL, *Analyst*, 79 (1954) 280.
A. BUSSMANN, *Z. Anal. Chem.*, 148 (1956) 413.
(1) A. MEHLICH, *J. Assoc. Offic. Agr. Chemists*, 39 (1956) 518.

In aluminium

M. BOURSON AND S. FAYETTE, *Chim. Anal.*, 31 (1949) 33.

In titanium

KOZO MOMOKI, *Japan Analyst*, 4 (1955) 581.
H. J. G. CHALLIS AND D. F. WOOD, *Analyst*, 79 (1954) 762.

In plants and soils

A. J. STERGES AND W. H. MCINTIRE, *Anal. Chem.*, 22 (1950) 351.
H. Y. YOUNG AND R. F. GILL, *Anal. Chem.*, 23 (1951) 751.

In the presence of the alkali metals

O. A. KENYON AND G. OPLINGER, *Anal .Chem.*, 27 (1955) 1125.

(ii) Extraction of the oxinate. Hydrated magnesium oxinate is soluble in 5% butyl cellosolve, from which it may be extracted by chloroform[1]. Mg^{2+} can also be extracted at pH 11 $\pm$ 0.5 by a chloroform solution of oxine in the presence of *n*-butylamine.

Sensitivity. ε = 5,600 at 380 mμ.

Interfering ions. Fluorides and EDTA prevent the extraction. Oxalates, cyanides, sulphates, phosphates, tartrates and citrates do not interfere.

Concentrations of the oxine and of the *n*-butylamine are such that the alkali metals and the alkaline earth metals and Cr, Mo, W, As, Sb, B, Se, Te, and Be do not interfere.

Sn(IV) prevents the extraction of Mg^{2+} when present in amounts exceeding 5 mg.

Moderate amounts of Ti(IV), V(V), and U(VI) can be complexed with H_2O_2; Zn(II), Cd(II), Ni(II), Co(II), and Fe(III) can be complexed with cyanide.

15 mg of Al(III) does not interfere in the presence of triethylamine.

The following interfere and should be separated or extracted with oxine before the addition of *n*-butylamine: In(III), Ga(III), Tl(III), Sn(II), Pb(II), Zr(IV), Th(IV), Bi(III), Nb(V), Ta(V), Mn(II), and the rare earth metals.

The operating procedure described permits Mg^{2+} to be determined in the presence of small residues of interfering ions.

REAGENTS

Oxine solution, 0.1% in chloroform.
Potassium sodium tartrate, 20% in water.
Ammonia solution, M (1/12).
n-Butylamine.
Potassium cyanide.
Hydrogen peroxide, 30% (110-volume).

OPERATING PROCEDURE

To 30 ml of solution containing 20 to 200 μg of magnesium, are added 5 ml of tartrate and, (if V(V), Ti(IV), or U(VI) is present), 2 ml of hydrogen peroxide. 1 *M* ammonia solution is added to give a pH of about 9 and the compound is extracted with 20 ml of the oxine solution; the extraction is repeated until the chloroform phase becomes colourless.

0.5 to 1 g of potassium cyanide and 1 ml of *n*-butylamine are added to the aqueous solution, and the pH is adjusted to 11.0 ± 0.5 with concentrated ammonia; the mixture is shaken with 50 ml of oxine for 1 minute and colorimetry is performed at 380 mμ.

Alternatively, the solution may be extracted twice with 20 ml of the oxine solution, shaking each time for 30 seconds. The extracts are transferred to a 50 ml measuring flask, 2 ml of methanol are added, and the volume is made up to 50 ml with chloroform.

(1) C. L. LUKE, *Anal. Chem.*, 28 (1956) 1443.
(2) F. UMLAND AND W. HOFFMANN, *Anal. Chim. Acta*, 17 (1957) 234.

(iii) With Eriochrome Black T. The reagent gives a red complex with Mg^{2+} at about pH 10. The coloration is not stable.

Sensitivity. Molar extinction coefficient: $\varepsilon \sim 20{,}000$ at 520 mμ.

Interfering ions. A number of other ions including Ca^{2+} also give coloured complexes.

REAGENTS

Eriochrome Black T, 0.1% in methanol. The reagent is dissolved in the hot, and filtered.
Buffer solution of pH 10.2: 0.75% ammonium chloride in 1 *M* ammonia.

OPERATING PROCEDURE

25 ml of buffer solution are added to the neutral solution containing less than

100 μg of Mg(II). The volume is made up to 90 ml, and the solution is stirred. 10.0 ml of the reagent are added with further stirring, and the volume is made up to 100 ml. The measurement is performed at 520 mμ in comparison with a blank.

COMPARE

A. E. Harvey, Jr., J. M. Komarmy and G. M. Wyatt, *Anal. Chem.*, 25 (1953) 498.
W. Dirschell and H. Breuer, *Mikrochim. Acta*, 49 (1953) 322.
J. K. R. Gasser, *Analyst*, 80 (1955) 482.
H. Pohl, *Z. Anal. Chem.*, 155 (1957) 263.

Ca(II) and Mg(II)

A. Young, T. R. Sweet and B. B. Baker, *Anal. Chem.*, 27 (1955) 356.

In biochemistry

R. Levine and J. R. Cummings, *J. Biol. Chem.*, 221 (1956) 735.

In Al

G. Selzer and M. Ariel, *Anal. Chim. Acta*, 14 (1958) 496.

In zinc alloys

H. Pohl, *Metall*, 10 (1956) 709.

Manganese

Mn = 54.9

(1) Separation of trace amounts

Preliminary separations are often unnecessary.

Of the more common separations, the following may be used:

(i) Precipitation by caustic soda, using Fe(III) as the entraining agent.

(ii) Precipitation with sodium peroxide, using Fe(III) as the entraining agent. Manganese may in this way be separated from Cl^-, Cr, V, Mo, and W.

(iii) By ion exchangers. In 10 *M* HCl, Mn(II) can be determined by anion exchanger. Similarly Cr(III), Al(III), Tl(I), Ni(II), and rare earths can be separated.

Elution in 8 *M* HCl separates Mn(II).

K. A. Kraus and G. E. Moore, *J. Am. Chem. Soc.*, 75 (1953) 1460.
D. H. Wilkins and L. E. Hibbs, Jr., *Anal. Chim. Acta*, 20 (1959) 427.

(iv) Standard solutions. 1 g of 99.9% metal may be dissolved in a dilute acid. Guaranteed purity $MnCl_2 \cdot 4H_2O$ may be weighed out.

Normally, a standardized 0.1 *N* permanganate solution is used. If this is diluted or acidified, it should be stabilized by the addition of a little solid potassium periodate.

A permanganate solution may be converted into Mn(II) by acidifying a known volume of the former with 9 *N* (1/4) sulphuric acid and reducing it by the dropwise addition of oxalic acid, in the hot, until the colour just disappears. The volume of the solution is then standardized.

1 ml of 0.1000 *N* MnO_4^- = 1.098 mg of Mn.

(2) Colorimetry

The colorimetry of MnO_4^- is a method which is both accurate, specific and sensitive.

(i) Colorimetry of MnO_4^- after oxidation by periodate. Mn(II) can be oxidized to MnO_4^- with periodate. The addition of phosphoric acid prevents the precipitation of manganese oxide and manganese periodate, and possibly of ferric

periodate, decolorizes the ferric solution, and stabilizes the MnO_4^-. A nitric or sulphuric acid solution may be used. Solutions obtained by this method are stable indefinitely.

Oxidation may be performed with persulphate in the presence of Ag^+ ions, but this practice is confined to cases when only minute traces of Mn(II) are present, (less than 0.1 ppm), since the periodate oxidation is very slow. The disadvantages of this method are the lack of stability of the MnO_4^- solutions obtained, formation of oxygen bubbles on the walls of the colorimetric cells, formation of a haze of silver chloride from traces of chlorides present in the reagents (this haze can be prevented by complexing Cl^- by the addition of Hg(II)), and formation of colorations, *e.g.* with Ti(IV).

Colorimetry is performed in the neighbourhood of 530 mμ.

Accuracy and sensitivity. The molar extinction coefficient ε is 2420 at 546 mμ and 2230 at 520 mμ.

The accuracy is limited only by that of the spectrophotometric apparatus, since the errors inherent in the chemical procedure are negligible. Colorimetry can therefore be used as a general method for the determination of manganese, and represents the best method if a precise spectrocolorimeter is available.

Interfering ions. Mo(VI), V(V), and Ti(IV) give colorations. At 530 mμ, $\varepsilon_{Mo} = 0.97$, $\varepsilon_{V} = 269$, $\varepsilon_{Ti} = 97.1$. Cr(VI) interferes owing to its coloration; this may be eliminated by estimating Mn(VII) at 575 mμ.

Fe(III) is decolorized by the addition of phosphoric acid.

The effect of coloured ions may be eliminated by means of a blank test, performed after the destruction of MnO_4^- by a little hot oxalic acid; the difference in the optical densities permits the calculation of MnO_4^-.

Large amounts of Cl^- should be eliminated by heating with H_2SO_4 until white fumes appear.

The following interfere: As(III), citrate and tartrate ions, NO_2^-, $C_2O_4^{2-}$, CN^-, SCN^-, Bi(III), U(VI), and Sn(IV).

The following do not interfere: NH_4^+, ClO_4^-, CH_3CO_2H, As(V), B(III), HF, NO_3^-, H_3PO_4, $H_4P_2O_7$, SiO_2, Al(III), Sb(III), Ba(II), Ca(II), Sr(II), Be(II), Cd(II), Pb(II), Li(I), Mg(II), Ag(I), Th(IV), Zn(II), and Zr(IV); 40 ppm of Ni(II), 200 ppm of Sn(IV), and 20 ppm of Co(II).

Operating procedure

There should be less than 1 mg of Mn(II) per 50 ml of solution. 10 ml of concentrated sulphuric acid or 15–20 ml of concentrated nitric acid are added to such a portion, followed by 5–10 ml of concentrated phosphoric acid; 0.3–0.4 g of potassium periodate are then added, and the solution is heated for five to ten minutes at about 90°. The volume is made up to 100 ml after cooling, and the colorimetric estimation is performed at 522 mμ.

(ii) Estimation of Cr(VI) and Mn(VII) at two wavelengths. $Cr_2O_7^{2-}$ and MnO_4^- possess sufficiently different colours to permit the estimation of one in the presence of the other by working at two wavelengths.

Principle. Mn(II) is first oxidized in an acid medium with periodate; the oxidation is then continued at pH 2.1 to complete the oxidation of the chromium.

A colorimetric examination is first carried out at 520 mμ; the MnO_4^- is then reduced with oxalic acid, another measurement is made at 520 mμ, and the MnO_4^- is obtained from the difference.

A measurement is then performed at 440 mμ on the solution from which the MnO_4^- has been eliminated; the value obtained from the solution before oxidation constitutes the blank, and the chromium may then be obtained by difference.

OPERATING PROCEDURE

25 ml of slightly acid solution containing less than 50 ppm of Mn(II) and less than 180 ppm of Cr(III), are mixed with 4 ml of 9 *N* (1/4) sulphuric acid and 500 mg of potassium periodate. The solution is boiled for 5 minutes, 40 ml of *N* caustic soda are added, and boiling is continued for a further 5 minutes (the pH should lie between 2 and 3; indicator paper). 5 ml of 9 *N* (1/4) sulphuric acid and 1 ml of concentrated (45 *N*) phosphoric acid are added, and the precipitate is dissolved by boiling. After cooling, the volume is made up to 100 ml, and the optical density D_1 is determined at 520 mμ.

A 50 ml portion is boiled for 5 minutes to concentrate it, and is then treated with 0.1 *M* oxalic acid, drop by drop until the coloration of the MnO_4^- disappears; an extra drop is added. After cooling, the volume is made up to 50 ml, and the optical density D_2 is determined. The difference between D_1 and D_2 at 520 mμ gives the content of manganese. The optical density D_3 is determined at 440 mμ on the same solution.

25 ml of the initial solution are mixed with 3 ml of 9 *N* (1/4) sulphuric acid and 1 ml of concentrated (45 *N*) phosphoric acid. D_4 is determined at 440 mμ. The difference between D_3 and D_4 permits the chromium to be obtained from calibration curves which take the initial colour of Cr(III) into account.

COMPARE

J. P. MEHLIG, *Ind. Eng. Chem., Anal. Ed.*, 11 (1939) 274.
G. P. ROWLAND, *Ind. Eng. Chem., Anal. Ed.*, 11 (1939) 442.

In steels

P. ROCQUET, *Rev. Mét.*, 44 (1947) 158.
R. W. SILVERTHRON AND J. A. CURTIS, *Metals and Alloys*, 15 (1942) 245.
M. D. COOPER, *Anal. Chem.*, 25 (1953) 411.

In aluminium

S. LACROIX AND M. LABALADE, *Anal. Chim. Acta*, 3 (1949) 262.

In glasses

F. HECHT AND A. GOTTLIEB, *Mikrochemie*, 35 (1950) 329.

In caustic soda

R. F. MORAN AND A. P. MCCUE, *Ind. Eng. Chem., Anal. Ed.*, 18 (1946) 556.
D. WILLIAMS AND R. V. ANDES, *Ind. Eng. Chem., Anal. Ed.*, 17 (1945) 28.

In biochemistry

J. W. COOK, *Ind. Eng. Chem., Anal. Ed.*, 13 (1941) 48.

(3) Other methods

Oxidation with persulphate

F. NYDAHL, *Anal. Chim. Acta*, 3 (1949) 144.

With bismuthate

J. H. KIYOTA AND T. YAMAMOTO, *Nippon Kagaku Zasshi*, 76 (1955) 1179.

Precision colorimetry (Precision ± 0.1%)

I. G. YOUNG AND C. F. HISKEY, *Anal. Chem.*, 23 (1951) 506.

Cr + Mn

M. D. COOPER, *Anal. Chem.*, 25 (1953) 411.
C. F. HISKEY AND D. FIRESTONE, *Anal. Chem.*, 24 (1952) 342.
R. BASTIAN, R. WEBERLING AND F. PALLILA, *Anal. Chem.*, 22 (1950) 160.

Mercury

Hg = 200.6

(1) Separation of trace amounts

NOTE. When hydrochloric acid solutions of mercuric ions are evaporated, some loss of $HgCl_2$ may occur.

(i) Precipitation of the sulphide. The sulphide may be precipitated by means of hydrogen sulphide under acid conditions, using Cu(II) as an entraining agent. Alternatively, the solution may be passed through a precipitate of CdS, whereby HgS is formed.

cf. A. E. BALLARD AND C. F. H. THORNTON, *Ind. Eng. Chem., Anal. Ed.*, 13 (1941) 893.

(ii) Reduction to mercury. This is carried out with hydrazine[1] or hydroxylamine in ammoniacal solutions and in the presence of tartrate to prevent the precipitation of basic salts. Ag is also precipitated.

In a highly concentrated hydrochloric acid medium, Hg may be precipitated by means of Sn(II)[2]. Alternatively, Fe, Al, Zn, etc. may be used, but many other metals are also precipitated: Cu[3], Sb, Ag, Au, etc.

Hg may also be liberated by calcining HgO at 700° C in the presence of CaO. Even traces can be separated by this method.

(1) J. J. LINGANE AND R. S. KLINE, *Anal. Chim. Acta*, 15 (1956) 410.
(2) J. N. BARTLETT AND W. W. MCNABB, *Anal. Chem.*, 19 (1947) 484.
(3) G. JANGG, *Z. Anal. Chem.*, 183 (1961) 255.

(iii) Distillation of the metal.

cf. W. L. MILLER AND L. E. WACHTER, *Anal. Chem.*, 22 (1950) 1312.
J. LACQ, *Anal. Chim. Acta*, 20 (1959) 195.
H. E. BROOKES AND L. E. SOLOMON, *Analyst*, 84 (1959) 622.

(iv) Extraction of the dithizonate. See Colorimetry.

(v) Extraction of the diethyldithiocarbamate. The compound formed can be extracted with carbon tetrachloride from a complexing medium: carbonate, EDTA, and cyanide.

Only Cu(II), Tl(III), and Bi(III) are also extracted.

fc. E. A. HAKKILA AND G. R. WATERBURY, *Anal. Chem.*, 32 (1960) 1340.

(vi) By electrolysis. Even very slight traces can be separated using gold or copper electrodes.

cf. D. PAVLOVIC AND S. ASPERGER, *Anal. Chem.*, 31 (1959) 939.

(vii) Traces of mercury in the air. A number of methods have been suggested. For example, the gases can be bubbled through a hypobromite solution[1], through permanganate in 2 *N* (1/20) sulphuric acid[2], through an iodine-iodide mixture[3], or brought into contact with active charcoal[4].

cf. *Review* in H. GUÉRIN. *Traité de manipulation et d'analyse des gaz.* Masson et Cie, Paris 1952.

(1) R. F. MILTON AND W. D. DUFFIELD, *Analyst*, 72 (1947) 11.
(2) R. G. DREW AND E. KING, *Analyst*, 82 (1957) 461.
(3) E. C. BARNES, *J. Ind. Hyg. Toxicol.*, 28 (1946) 257.
(4) G. A. SERGEANT, B. E. DIXON AND R. G. LIDZEY, *Analyst*, 82 (1957) 27.

(viii) Standard solution. Hg, $HgCl_2$, and HgO of guaranteed purity are available.

If the metal itself is used, it should be dissolved in the minimum amount of 3 *N* (1/4) nitric acid, and the solution then diluted.

$$Hg/HgCl_2 = 0.739.$$
$$Hg/HgO = 0.926.$$

(2) Colorimetry

NOTE. Traces of mercury are generally found adsorbed in laboratories on filter papers, glassware, etc.

(i) With dithizone. The method involving dithizone permits the estimation of very minute traces, is selective in acid solutions, and is generally considered to be the best method. Mercury dithizonate is yellow-orange and stable between pH 0 and 1.

Interfering ions. Au(III), Pd(II), Pt(II) and Ag(I) give analogous reactions, but the latter can be separated by means of a little Cl^-. Large amounts of Bi(III) would be partly extracted. Ions which complex Hg(II), such as large amounts of Cl^-, SCN^-, etc. interfere.

In the presence of a large excess of Cl^-, the extraction may be performed at pH 2.0. Cu(II) interferes, but the extraction can be stopped when all Hg(II) has been extracted and the red-violet colour of the copper dithizonate begins to appear.

In the presence of EDTA[1] and at pH 4–5, only Hg(II) and Ag(I) are extracted. The latter is retained in the aqueous phase when SCN^- is present.

The same applies when an appreciable quantity of chloride is present; Hg(II) can be extracted at pH 2, while Ag(I) stays behind[2].

Fe(III) should be reduced by the normal methods if present in small amounts by hydroxylamine and separated if present in large amounts.

Sensitivity: $\varepsilon \sim 70{,}000$ at 490 mμ.

Reagents

Dithizone, 0.001% in carbon tetrachloride.

Hg(II) standard solution, 0.1%. The solution is diluted to strength immediately before use.

Acetic buffer: 56 g of sodium acetate and 24 ml of pure acetic acid are made up to 100 ml.

Disodium salt of ethylenediaminetetraacetic acid, 5% in water.

Operating procedure

There should be less than 50 μg of Hg(II) per 50–100 ml of solution. 10 ml of buffer solution are added, followed by sufficient EDTA to complex all ions present in the solution. The extraction is performed with a known volume of the dithizone solution—*e.g.* 20 ml .If Ag(I) is present in the solution, a quantity of SCN^- should also be added.

The colorimetry of the dithizonate is performed at 500 mμ, referred to a calibration curve prepared under the same conditions.

COMPARE

(1) V. Vašák and V. Šedivec, *Coll. Trav. Chim. Tchéc.*, 15 (1950) 1076.

(2) H. Friedeberg, *Anal. Chem.*, 27 (1955) 305.

S. S. Yamamura, *Anal. Chem.*, 32 (1960) 1896.

J. H. Reith and K. W. Gerritsma, *Rec. Trav. Chim.*, 64 (1945) 41.

In fruit

D. C. Abbott and E. I. Johnson, *Analyst*, 82 (1957) 1206.

In urine

J. F. Reith and C. P. van Dijk, *Chem. Weekblad*, 37 (1940) 186.

In metals

H. Fischer and G. Leopoldi, *Metall. Erz*, 38 (1941) 154.

In vegetables

D. C. Abbott and E. I. Johnson, *Analyst*, 82 (1957) 1206.

In biochemistry

V. L. Miller and F. Swanberg, *Anal. Chem.*, 29 (1957) 391.

(EDTA). H. Wanntorp and A. Dyfverman, *Arkiv. Kem.*, 9 (1956) 7.

In air

N. S. Kuzyatina, *Zavodsk. Lab.*, 8 (1939) 174.

(3) Other methods

Complex formation with diethyldithiocarbamate – in CCl_4. $\varepsilon = 33{,}000$ at 278 mμ

E. A. Hakkila and G. R. Waterbury, *Anal. Chem.*, 32 (1960) 1340.

Absorption at 235.6 mμ (Mercury vapour lamp) by mercury vapour. 0.1 μg

A. Berton, *Compt. Rend.*, 221 (1945) 464.

C. W. Zuehlke and A. E. Ballard, *Anal. Chem.*, 22 (1950) 953.

Absorption of the thiocyanate complex in the ultraviolet

G. E. Markle and D. F. Boltz, *Anal. Chem.*, 26 (1954) 447.

Traces of mercury in biological liquids

M. R. Jacobs. *The Analytical Chemistry of Industrial Poisons Hazard and Solvents*, Interscience Publ., New York 1949.

F. R. Barrett, *Analyst*, 81 (1956) 294.

Traces of mercury in the air (Selenium sulphide paper)

F. Stitt and Y. Tomimatsu, *Anal. Chem.*, 23 (1951) 1098.

G. A. Sergeant, B. E. Dixon and R. G. Lidzey, *Analyst*, 82 (1957) 27.

Molybdenum

Mo = 96.0

(1) Separation of trace amounts

(i) Precipitation of the sulphide. The sulphide may be precipitated by hydrogen sulphide under acid conditions. MoS_3 only precipitates when certain precautions are taken: if large amounts of molybdenum are present, an appreciable pressure of hydrogen sulphide should be used, and where only trace amounts are involved, an entraining agent such as Sb(III) is required[1]. In this way, Mo may be separated from V, Ti, Cr, Nb, U, etc.

In a tartaric acid medium, with Cu(II) as entraining agent, Mo may be separated from W(VI)[2]. It may also be precipitated in formate or fluoride solutions[3].

COMPARE

(1) R. B. HENRICKSON AND E. B. SANDELL, *Anal. Chim. Acta*, 7 (1952) 57.
(2) H. G. SHORT, *Analyst*, 76 (1951) 710.
P .G. JEFFERY, *Analyst*, 81 (1956) 104.
(3) G. NORWITZ AND M. CODELL, *Anal. Chem.*, 25 (1953) 1438.

Homogeneous precipitation with thioacetamide
W. N. MCNERNEY AND W. F. WAGNER, *Anal. Chem.*, 29 (1957) 1177.

(ii) Precipitation by means of α-benzoin-oxime. Only Mo(VI), W(VI), Cr(VI), V(V), Nb(V), Pd(II), and Ta(V) precipitate in strongly acid solutions. Cr(VI) and V(V) can be reduced before precipitation. In this way, Mo(VI) and W(VI) can be separated from practically all the other elements, including Re(VII). U(VI) interferes. 70% of trace amounts as small a 1 μg may be separated by this method.

REAGENTS

α-Benzoin-oxime, 2% in alcohol. The solution is stable for several months.
Washing solution: 50 ml of the 2% reagent and 10 ml of concentrated sulphuric acid, in 1 litre of water.
Ferrous sulphate, M/10 in N (1/36) sulphuric acid.
Saturated bromine water.

OPERATING PROCEDURE

About 25 ml of solution, acidified, *e.g.* to 5 *N*, is cooled to below 18° and 5 ml of reagent per 10 mg of molybdenum are added. Bromine water is added, drop by drop, until a yellow coloration is obtained; any molybdenum reduced by the reagent is then reoxidized. A further few millilitres of 2% benzoin-oxime are added. After 10 minutes the mixture is filtered and the precipitate is washed with 150 ml of the cold washing solution.

NOTE. The solution should be filtered within a few minutes of precipitation, since the amount of precipitate present is appreciably reduced after even half an hour.

cf. H. B. KNOWLES, *J. Res. Natl. Bur. Std.*, 9 (1932) 1.
H. YAGODA AND H. A. FALES, *J. Am. Chem. Soc.*, 60 (1938) 640.

(iii) Precipitation of the oxinate in the presence of EDTA. Many substances become strongly complexed, but the complex formed by Mo(VI) is sufficiently weak to allow the oxinate to be precipitated quantitatively in an acetate buffer.

Mo(VI) may in this way be separated from: V(IV)[1], Fe(III), Al(III), Be(II), Zn(II), Ni(II), Co(II), Mn(II), Pb(II), Cd(II), Bi(III), Cu(II), and Hg(II). On the other hand, Ti(IV), W(VI), V(V) and U(VI) also precipitate.

cf. R. PŘIBIL AND M. MALAT, *Coll. Trav. Chim. Tchec.*, 15 (1950) 120.
(1) M. MALINEK, *Chem. Listy*, 48 (1954) 30.

(iv) Extraction from HCl. Mo(VI) can be extracted from a sufficiently concentrated HCl medium (6 *M*) with ether; V(V) and W(VI) are separated completely[1].

In the presence of phosphoric acid, Mo(VI) may also be separated from W(VI)[2].

Molybdenum can also be extracted with hexone from a fluoride complexing medium. Only the following interfere: Fe(III), Sn(IV), W(VI), and Bi(III)[3].

COMPARE

(1) M. CODELL, J. J. MIKULA AND G. NORWITZ, *Anal. Chem.*, 25 (1953) 1442.
I. NELIDOW AND R. M. DIAMOND, *J. Phys. Chem.*, 59 (1955) 710.
L. P. ALIMARIN AND V. N. POLYANSKY, *Zh. Analit. Khim.*, 8 (1953) 266.
(2) F. G. ZHAROVSKII, *Zh. Neorg. Khim.*, 2 (1957) 623.
(3) G. R. WATERBURY AND C. E. BRICKER, *Anal. Chem.*, 29 (1957) 129.

(v) Extraction of the compound with α-benzoin-oxime. The solvent is chloroform. W(VI) is also extracted, but not V(V) or Cr(VI). In the presence of phosphoric acid and Fe(II), Mo(VI) may be separated from W(VI) and V(V).

cf. G. B. JONES, *Anal. Chim. Acta*, 10 (1954) 584.
P. G. JEFFERY, *Analyst*, 81 (1956) 104.
H. J. HOENES AND K. G. STONES, *Talanta*, 4 (1960) 250.

(vi) Extraction of the thiocyanate of Mo(V). See Colorimetry.

(vii) Extraction of the cupferrate with chloroform. This is one of the best methods of separation. Fe(III), Ti(IV), V(V), Sn(IV), Cu(II), and W(VI) are also extracted at pH 1.6.

With iso-amyl alcohol

S. H. Allen and M. B. Hamilton, *Anal. Chim. Acta*, 7 (1952) 483.

(viii) Extraction of the ethyl xanthate with chloroform. See Colorimetry.

(ix) Extraction of the acetylacetonate with chloroform. In 6 N H_2SO_4, this method may be used to separate Mo(VI) from Cu(II), Cr(III), W(VI), etc.

cf. J. P. McKaveney and H. Freiser, *Anal. Chem.*, 29 (1957) 290.

(x) Extraction of the compound with dithiol. See Colorimetry.

(xi) With ion-exchangers. An anion exchanger may be used to collect Mo(VI) from 0.5 M HCl; the metal may thus be separated from 10,000 times as much Fe(III), and can then be eluted with 0.04 M SCN^-. A suitable resin is Dowex 50.

(xii) Standard solutions. Apart from the pure metal, MoO_3 is often available and, with a lower guarantee of purity, $(NH_4)_6Mo_7O_{24} \cdot 4\, H_2O$.

$$7Mo/(NH_4)_6Mo_7O_{24}4H_2O = 0.544$$
$$Mo/MoO_3 = 0.667$$

(2) Colorimetry

(i) Complex thiocyanate of Mo(V). The controlled reduction of Mo(VI) in the presence of thiocyanate ions leads to the formation of a complex orange-red Mo(V) thiocyanate. The reduction should not be too vigorous and the acidity should be fixed at a definite level, since several alternative reductions may occur. Thus for example, Mo(III) may be formed or, in a weakly acidic solution 'molybdenum blue' may appear.

Mild oxidizing agents such as NO_3^-, stabilize the coloration. Organic solvents such as acetone and the glycol esters 'cellosolve', 'butyl cellosolve', and 'carbitol' exert a similar effect.

Controlled reduction with I^- or Sn(II) has been suggested as an alternative[1]. The compound formed can be extracted with ethers, esters, and alcohols *e.g.* cyclohexanol and butyl phosphate[2]. The coloration is stable for 1 hour.

Sensitivity. $\varepsilon \sim 15{,}000$ at 475 mμ (*iso*-amyl alcohol).

Interfering ions. Coloured ions may interfere if the molybdenum is not first isolated by extraction. Large amounts of Cr(III) should be separated in the form of CrO_2Cl_2.

Certain other ions give rise to extractable coloured thiocyanate complexes; moderate quantities of Fe(III) are reduced and do not interfere. Pt(IV) interferes. Co(II) interferes if its content exceeds half that of the Mo(VI). Cu(II) precipitates in the form of cuprous thiocyanate and may be separated in this way. W(VI) should be complexed by citrate or tartrate ions, and Ti(IV) by F^-.

Large amounts of Bi(III), V(V) and P(V) interfere. Re(VII) gives the same reaction. U (VI) interferes.

Reagents

Potassium thiocyanate, 10%.

Stannous chloride: 10 g of $SnCl_2 \cdot 2\ H_2O$ are dissolved in 10 ml of concentrated hydrochloric acid and the solution is made up to 100 ml with water.

Solutions of ferrous iron: 1 g of Mohr's salt is dissolved in 100 ml of 0.2 *N* (1/80) sulphuric acid.

Iso-amyl alcohol.

Standard solution of molybdenum: 750 mg of guaranteed purity MoO_3 are dissolved in a few ml of dilute caustic soda. The solution is made slightly acid with HCl and the volume adjusted to 500 ml. The solution is diluted to the point where 1 ml contains 1 μg of Mo(VI).

Operating procedure

2.0 ml of concentrated hydrochloric acid, 1 ml of ferrous solution, 3.0 ml of thiocyanate, and 3 ml of stannous chloride are added to 15 ml of solution containing from 1 to 50 μg of molybdenum. The solution is diluted to 25 ml and an accurately measured 10 ml portion of *iso*-amyl alcohol is added. After shaking vigorously for 1 minute and allowing to settle, the colorimetry is performed at 475 mμ.

Compare

G. E. Markle and D. F. Boltz, *Anal. Chem.*, 25 (1953) 1261.
J. L. Grigg, *Analyst*, 78 (1953) 470.
(1) L. B. Ginzburg and Y. Y. Lure, *Zavodsk. Lab.*, 14 (1948) 538.
(2) L. M. Melnick, *Dissert. Abstr.*, 14 (1954) 760.

In steels

A. Thomas, *Proc. Am. Soc. Testing Materials*, 44 (1944) 769.
R. I. Mays, *Chemist-Analyst*, 35 (1946) 62.
G. V. L. N. Murty, *Metallurgia*, 35 (1947) 167.

In minerals and rocks

E. B. SANDELL. *Ind. Eng. Chem., Anal. Ed.*, 8 (1936) 336.

F. WARD, *Anal. Chem.*, 23 (1951) 788.

In plants and soils

I. BARSHAD, *Anal. Chem.*, 21 (1949) 1148.

J. L. GRIGG, *Analyst*, 78 (1953) 470.

C. M. JOHNSON AND T. H. ARKLEY, *Anal. Chem.*, 26 (1954) 572.

In tungsten steels

L. J. WRANGELL, E. C. BERNHAM, D. F. KUEMMEL AND O. PERKINS, *Anal. Chem.*, 27 (1955) 1966.

In tungsten

R. P. HOPE, *Anal. Chem.*, 29 (1957) 1053.

In beryllium

J. O. HIBBITS, W. F. DAVIS AND M. R. MENKE, *Talanta*, 4 (1960) 104.

(ii) Dithiol. Dithiol, or 4-methyl-1, 2-dimercaptobenzene (toluene-3,4-dithiol) reacts with Mo(VI) to give a deep green compound which is sparingly soluble in acid solutions. This compound is soluble in many solvents such as esters, hydrocarbons, benzene, carbon tetrachloride, etc.

Sensitivity. $\varepsilon \sim 20{,}000$ at 675 mμ.

Interfering ions. In 9–12 N H_2SO_4, W(VI) is not extracted; it can however be extracted at pH 0.5–2[1]. (For the estimation of Mo(VI) and W(VI), see Tungsten).

REAGENTS

Potassium iodide, 50%.

Sodium thiosulphate, 10%.

Tartaric acid, 50%.

Dithiol, 0.2%: 0.5 g of dithiol are dissolved in 250 ml of 1% caustic soda. The solution is left for 1 hour, shaking from time to time. About 4 ml of thioglycollic acid are added, drop by drop, until a permanent opalescence is formed. The solution should be stored in a refrigerator.

Iso-amyl acetate.

OPERATING PROCEDURE

To 25 ml of solution containing from 2 to 10 μg of molybdenum, are added 10 ml of hydrochloric acid (acidity 4 N) and then 1 ml of iodide. The mixture is stirred, left for 10 minutes, decolorized with thiosulphate and 1 ml of tartaric acid is added. 2 ml of dithiol are then added, the solution is stirred for 30 seconds, left for 10 minutes, and finally shaken with 5 ml of *iso*-amyl acetate for 30 seconds. Colorimetry is performed at 680 mμ.

cf. H. G. SHORT, *Analyst*, 76 (1951) 710.
L. J. CLARK AND J. H. AXLEY, *Anal. Chem.*, 27 (1955) 2000.

Separation of W

(1) P. G. JEFFERY, *Analyst*, 81 (1956) 104; 82 (1957) 558.

In U and Nb

C. O. GRANGER, *Analyst*, 83 (1958) 609.

Nickel

Ni = 58.7

(1) Separation of trace amounts

(i) Extraction of the compound formed with dimethylglyoxime, using chloroform. The medium used should contain citrate and ammonia. Nickel may in this way be separated from Fe(III), Al(III), Co(II), etc. Traces of Co(II) and Cu(II) are also extracted, and may be reclaimed from the chloroform phase by shaking with ammonia. Large amounts of Mn(II) interfere, since manganese catalyzes the oxidation of Ni(II) to nickelic dimethylglyoxime, which is insoluble in $CHCl_3$. This may be prevented by adding hydrazine or hydroxylamine hydrochloride.

In the presence of Cu(II) the extraction can be carried out at about pH 6.5 after the addition of thiosulphate.

When the extract is shaken with hydrochloric acid, Ni(II) passes back into the aqueous phase. The dimethylglyoxime can be separated from the Ni(II) by extraction from acid solutions with ether.

REAGENTS

Dimethylglyoxime, 1% in alcohol.
Sodium citrate, crystalline, 10% in water.
Chloroform.

OPERATING PROCEDURE

The solution should contain at least 5 μg of Ni(II) in 5–10 ml of weakly acid solution. 5 ml of citrate are added, the mixture is made slightly ammoniacal, and is mixed with 2 ml of reagent. The extraction is performed with 2–3 ml of chloroform: the chloroform phase is shaken with 5 ml of 1/50 ammonia, the aqueous layer is separated. The operation is repeated with 1–2 ml of chloroform, subsequently added to the first fraction. The nickel is reclaimed into aqueous solution by shaking the chloroform phase with two 5 ml portions of hydrochloric acid: alternatively, absorptiometry may be performed directly on the chloroform solution at 366 mμ (see Colorimetry).

COMPARE

E. B. SANDELL AND R. W. PERLICH, *Ind. Eng. Chem., Anal. Ed.*, 11 (1939) 309.
R. S. YOUNG, E. H. STRICKLAND AND A. LEIBOWITZ, *Analyst*, 71 (1946) 474.
A. CLAASSEN AND L. BASTINGS, *Rec. Trav. Chim.*, 73 (1954) 783.
W. OELSCHLÄGER, *Z. Anal. Chem.*, 146 (1955) 339.

Separation from Cu

W. Nielsch, *Z. Anal. Chem.*, 150 (1956) 114.

Extraction of the compound formed with cyclohexanedione dioxime (Traces of Ni(II) may be estimated in the presence of Co(II))

D. Monnier and W. Haerdi, *Anal. Chim. Acta*, 20 (1959) 444.

(ii) With ion-exchangers. With a 12 *m* HCl medium and an anionic resin, Co(II), Cu(II), Fe(III), etc., are collected while Ni(II) remains in the solution.

K. A. Kraus and G. E. Moore, *J. Am. Chem. Soc.*, 75 (1953) 1460.
G. E. Moore and K. A. Kraus, *J. Am. Chem. Soc.*, 74 (1952) 843.
A. Liberman, *Analyst*, 80 (1955) 595.
D. H. Wilkins, *Talanta*, 2 (1959) 355.

(iii) Separation by electrolysis. The electrolysis can be performed in an ammine complex medium.

Interfering ions. Cu(II), which interferes, may be separated by electrolysis in a medium containing sulphuric and nitric acids. The solution is in this case heated, until white fumes appear, to expel NO_3^- which retards the electrolysis.

Fe(III), Ce(III), and W(VI) cease to interfere in solutions containing citric and oxalic acids.

Ag(I) interferes. Co(II) deposits at the same time. In the presence of Zn(II), the potential must be controlled. A certain number of ions, such as As(III), should be removed. Cl^- does not interfere. Fe(II), V(V), Mo(VI), and W(VI) and large amounts of Mn(II) interfere. Cr(VI) should be reduced. In the presence of Cr(III), hydrazine or oxalate should be added.

Operating procedure

12 g of ammonium sulphate and 25 ml of concentrated ammonia are added to 300 ml of the neutralized solution. Electrolysis is performed and continued for 10 minutes after the solution has become decolorized.

In the presence of Zn(II)

S. Torrance, *Analyst*, 63 (1938) 488.

(iii) Standard solutions. Pure nickel, and nickel salts—sulphate, chloride, and nitrate of guaranteed purity are available.

Accurate results are obtained by weighing out $NiSO_4$ which has previously been heated at 400–500° to constant weight.

$$Ni/NiSO_4 = 0.379$$
$$Ni/NiCl_2 \cdot 6H_2O = 0.2470$$
$$Ni/Ni(NO_3)_2 \cdot 6H_2O = 0.2018$$
$$Ni/NiSO_4 \cdot 7H_2O = 0.2090.$$

(2) Colorimetry

The most suitable method involves the formation of nickelic dimethylglyoxime, although other methods can be useful in particular cases. It is also possible to estimate large amounts of Ni(II) with very high accuracy by the absorption of Ni^{2+}.

(i) Dimethylglyoxime in an oxidizing medium. In the combined presence of dimethylglyoxime and an oxidizing agent (Br_2, I_2, ClO^-, or $S_2O_8^{2-}$) under alkaline conditions, Ni(II) gives wine-red compounds corresponding to a higher state of oxidation[1].

Ammoniacal solutions are generally employed, although media of higher alkalinity may be preferable for the formation of a specific compound[2].

Sensitivity. The molar extinction coefficient is $\varepsilon \sim 15{,}000$ at 465 mμ.

If too much Ni(II) is present, (more than 1.5 ppm at the moment of colorimetry), a part of it will precipitate in the form of Ni(II) dimethylglyoxime.

Interfering ions. Interference is obtained from a number of ions, which give precipitates: Sn(IV), Si(IV), V(V), Al(III), Sb(III), Be(II), Bi(III), Cr(III), Fe(III), Pb(II), Ca(II), Mg(II), Mn(II), Hg(II), Pt(IV), Th(IV), Sr(II), Ti(IV), U(VI), Zr (IV), etc., alternatively, colorations may appear: Cu(II), Au (III), Co(II), and Cr(VI).

Many of the above can be complexed with tartrate or citrate ions. Fe(III), should not however be present in such media, since a greenish yellow coloration is produced. At 520–540 mμ, 200 ppm of Fe(III) absorb approximately as much as 10 ppm of Ni(II): in consequence, if the determination is to be accurate the Ni(II) should first be separated.

REAGENTS

Dimethylglyoxime, 1% in alcohol.
Saturated bromine water.
Standard solution of Ni(II), 0.001%.
Ammonia, 4 N(1/3).

OPERATING PROCEDURE

The solution containing less than 150 μg of nickel is placed in a 100 ml volumetric flask and, if necessary, neutralized with caustic soda. Bromine water is added until a yellow shade is obtained. After 10 minutes, ammonia is added dropwise to decolorize the bromine. An additional extra 10 ml of ammonia are then added, followed by 10 ml of dimethylglyoxime. The solution is made up to 100 ml and, after at least 15 minutes, the colorimetry is carried out at 465 mμ (or at 520 mμ where there is a very small residue of Fe(III) left after separation).

COMPARE

Compounds formed

(1) M. HOOREMAN, *Anal. Chim. Acta*, 3 (1949) 635.
E. BOOTH AND J. D. H. STRICKLAND, *J. Am. Chem. Soc.*, 75 (1953) 3017.
(2) A. CLAASSEN AND L. BASTINGS, *Rec. Trav. Chim.*, 73 (1954) 783.
K. YAMASAKI AND C. MATSUMOTO, *Nippon Kagaku Zasshi*, 77 (1956) 1111.
I. KUDO, *Nippon Kagaku Zasshi*, 77 (1956) 1792.

In steels

M. D. COOPER, *Anal. Chem.*, 23 (1951) 875.
A. CLAASSENS AND L. BASTINGS, *Rec. Trav. Chim.*, 73 (1954) 783.

In copper alloys

G. HAIM AND B. TARRANT, *Ind. Eng. Chem., Anal. Ed.*, 18 (1946) 51.

In aluminium

Aluminium Research Institute, *Analytical Methods for Aluminium Alloys*, Chicago, 1949.
A.L.A.R., *Modern Methods for the Analysis of Aluminium Alloys*, Chapman and Hall, London, 1949.
M. D. COOPER, *Anal. Chem.*, 23 (1951) 880.

In rocks

E. B. SANDELL AND R. W. PERLICH, *Ind. Eng. Chem., Anal. Ed.*, 11 (1939) 309.
A. N. CHOWDHURY AND B. DASSARMA, *Anal. Chem.*, 32 (1960) 820.

In tungsten

K. L. ROHRER, *Anal. Chem.*, 27 (1955) 1200.

In manganese

W. OELSCHLÄGER, *Z. Anal. Chem.*, 146 (1955) 339 and 346.

In beryllium

T. C. J. OVENSTON AND C. A. PARKER, *Anal. Chim. Acta*, 4 (1950) 147.

In biochemistry

M. KENIGSBERG AND I. STONE, *Anal. Chem.*, 27 (1955) 1339.

(3) Other methods

(i) Nickel (II) dimethylglyoxime in chloroform. The estimation can be carried out at 375 mμ, $\varepsilon \sim 3{,}500$, or at 325 mμ, $\varepsilon \sim 5{,}000$.

In copper

W. NIELSCH, *Z. Anal. Chem.*, 150 (1956) 114.

In steels

W. NIELSCH AND L. GIEFER, *Mikrochim. Acta*, (1956) 522.

(ii) Extraction of the compound with α-furyldioxime. $\varepsilon = 17{,}000$ at 438 mμ in dichlorobenzene.

cf. C. G. TAYLOR, *Analyst*, 81 (1956) 369.
A. R. GAHLER, A. M. MITCHELL AND M. G. MELLON, *Anal. Chem.*, 23 (1951) 500.

(iii) Diethyldithiocarbamate. With Ni(II), a yellow-green complex is formed, which may be extracted with *iso*-amyl alcohol.

The spectrophotometry is sensitive, $\varepsilon = 37{,}000$ at 325 mμ, and 60,000 at 385 mμ.

cf. O. R. ALEXANDER, E. M. GODAR AND N. J. LINDE, *Ind. Eng. Chem., Anal. Ed.*, 18 (1946) 206.

In blood
M. L. CLUETT AND J. H. YOE, *Anal. Chem.*, 29 (1957) 1265.

(iv) Precision colorimetry of Ni(II).

M. C. STEELE AND L. G. ENGLAND, *Anal. Chim. Acta*, 20 (1959) 555.
J. KINNUNEN, *Metall u. Erz*, 41 (1944) 158.
R. BASTIAN, *Anal. Chem.*, 23 (1951) 580.

(v) Complex with EDTA. Al(III), Be(II), Ca(II), Cd(II), Cr(III), Fe(III), Hg(II), Mn(II), and Pb(II), do not interfere.

cf. L. D. BRAKE, W. M. MCNABB AND J. F. HAZEL, *Anal. Chim. Acta*, 10 (1958) 39.

Niobium and tantalum

Nb = 91.9   Ta = 180.9

(1) Separation of trace amounts

The hydroxides of Nb(V) and Ta(V) are generally distributed between the precipitates formed and the solutions remaining in the latter in the colloidal state. In general, these elements are present in solution in the form of oxalate, tartrate, fluoride, hydrogen peroxide, and other complexes.

Niobium and tantalum can be precipitated from acid solutions by means of cupferron or arsonic acid. Fe(III) is not precipitated by the latter reagent. Alternatively, the metals may be precipitated from tartrate or oxalate media by means of cupferron. Zr(IV), Ti(IV), Sn(IV), and Fe(III) also precipitate, and one of them may serve as an entraining agent.

The oxinates of many elements can be extracted from tartrate solutions with chloroform; Nb(V) and Ta(V) are left behind in the aqueous phase.

In cassiterite, SnO_2 can be reduced to the metallic state by means of a current of hydrogen at 800° C (the classical method of estimating tin). The residue is taken up in a little hydrochloric acid whereby tin and iron pass into solution while Nb(V) and Ta(V) remain insoluble.

In general, Nb(V), Ta(V), and Ti(IV) are recovered together.

(i) Precipitation with cupferron. In a very weakly acidic solution containing tartrate or oxalate, Nb(V) and Ta(V) can be precipitated with cupferron; Zr(IV), Ti(IV), Fe(III), and Sn(IV) are also precipitated.

The compound formed with Nb(V) is extractable.

cf. T. R. Cunningham, *Ind. Eng. Chem., Anal. Ed.*, 10 (1938) 233.

(ii) Precipitation with arsonic acids. In the absence of complexes and in a strongly acidic solution, small amounts of Nb(V) and Ta(V) can be precipitated by arsonic acids; Zr(IV), Sn(IV) and Ti(IV) are coprecipitated, but Fe(III) remains in solution.

Using phenylarsonic acid

H. Fucke and J. Daubländer, *Tech. Mitt. Krupp. Forschungsber.*, 2 (1939) 174.

(iii) Separations by extraction. A recently developed method of separation involves the extraction of fluoride complexes. By varying the concentration of the hydrofluoric acid and by adding various other acids to the solution, Ta(V)

may be separated from Nb(V)[1,3] or Nb(V) and Ta(V) from Ti(IV), Zr(IV), Sn(IV), Mo(VI), U(VI), W(VI), Fe(III), etc. The principal solvents used are di-*iso*-propyl ketone[1], tributyl phosphate[2], and in particular, methyl *iso*-butyl ketone[3].

Nb(V) may also be extracted from hydrochloric acid with di-*iso*-propyl ketone[4].

Zr(IV) can be separated from Nb(V) by extraction with a mixture of mono- and dibutyl phosphates[5].

Long-chain tertiary amines allow a separation of Nb(V) from Ta(V); for example, tribenzylamine in methylene chloride[6], and methyldioctylamine in xylene[7] may be employed.

(1) P. C. Stevenson and H. G. Hicks, *Anal. Chem.*, 25 (1953) 1517.
(2) D. F. C. Morris and D. Scargill, *Anal. Chim. Acta*, 14 (1956) 57.
(3) G. W. C. Milner, G. A. Barnett and A. A. Smales, *Analyst*, 80 (1955) 380.
G. R. Waterbury and C. E. Bricker, *Anal. Chem.*, 29 (1957) 1474; 30 (1958) 1007.
C. E. Crouthamel, B. E. Hjelte and C. E. Johnson, *Anal. Chem.*, 27 (1955) 507.
M. L. Theodore, *Anal. Chem.*, 30 (1958) 465.
(4) H. G. Hicks and R. S. Gilbert, *Anal. Chem.*, 26 (1954) 1205.
(5) E. M. Scadden and N. E. Ballon, *Anal. Chem.*, 25 (1953) 1602.
(6) J. Y. Ellenburg, G. W. Leddicotte and F. L. Moore, *Anal. Chem.*, 26 (1954) 1045.
(7) G. W. Leddicotte and F. L. Moore, *J. Am. Chem. Soc.*, 74 (1952) 1618.

From a fluoride medium with hexone. Nb(V) and Ta(V) can be extracted from a solution 3 *M* with respect to sulphuric and 10 *M* with respect to hydrofluoric acid, by means of methyl *iso*-butyl ketone (hexone). The method is practically specific. Elements such as Fe(III), Ti(IV), U(VI), Mo(VI), W(VI), Zr(IV), and Sn(IV), etc., do not interfere; interference is however, obtained from chlorides, bromides, and iodides, since they promote the extraction of Fe(III), Mo(VI) and in particular Sn(IV). In such cases, all operations should be performed in polythene vessels.

Reagents

Sulphuric acid/hydrofluoric acid mixture: 83 ml of concentrated sulphuric acid are added to 210 ml of water. After cooling, 210 ml of concentrated hydrofluoric acid are added.

Washing solution: 250 ml of the above mixture of acids are shaken with 50 ml of methyl *iso*-butyl ketone.

Solvent: 250 ml of methyl *iso*-butyl ketone are shaken with 50 ml of the sulphuric acid/hydrofluoric acid mixture.

Hydrogen peroxide, 10 volumes.

Operating procedure

The oxides are dissolved in a few ml of hydrofluoric acid. The volume of the solution is reduced, and 20 ml of the sulphuric/hydrofluoric acid mixture is added.

Alternatively, the oxides may be digested in fused pyrosulphate, dissolving the product in 5 ml portions of the acid mixture with gentle heat.

A 20 ml portion of the solution is shaken with 10 ml of methyl *iso*-butyl ketone for 1 minute. The organic phase is removed and the extraction repeated twice more.

The ketone extracts are combined and washed five time with 5 ml aliquots of the washing solution, shaking for 15 seconds each time.

Nb(V) and Ta(V) are extracted from the organic phase by shaking for 30 seconds with 10 ml of hydrogen peroxide. The extraction should be repeated five times.

(iv) Separation by chromatography. Partition chromatography may be performed either on cellulose[1], or by paper chromatography[2].

Chromatography on cellulose permits the separation of Nb(V) and Ta(V) from Fe(III), Sn(IV), Mo(VI), Ti(IV), and Zr(IV); W(VI) follows Ta(V) and Nb(V) to some extent.

(1) F. H. BURSTALL, P. SWAIN, A. F. WILLIAMS AND G. A. WOOD, *J. Chem. Soc.*, (1952) 1497.
A. F. WILLIAMS, *J. Chem. Soc.*, (1952) 3155.
R. A. MERCER AND R. A. WELLS, *Analyst*, 79 (1954) 339 and 351.
(2) *Prospecting*. E. C. HUNT, A. A. NORTH AND R. A. WELLS, *Analyst*, 80 (1955) 179.
E. BRUNINX, J. EECKOUT AND J. GILLIS, *Mikrochim. Acta*, (1956) 689.

(2) Separations of Nb(V) from Ta(V)

(i) Precipitation with N-benzoyl-N-phenylhydroxylamine. In a medium containing sulphuric and hydrofluoric acids, Nb(V) can be separated from Ta(V), Ti(IV), and Zr(IV), which remain in solution.

In a tartrate medium, Nb(V) precipitates at pH 3.5–6.5. Ta(V) can then be precipitated at pH < 1.5.

In a medium containing tartrate and EDTA, the above separation can be carried out while the other elements remain in solution, except for Ti(IV) (which may be complexed with H_2O_2), Zr(IV), and Mo(VI).

cf. R. W. MOSHIER AND J. E. SCHWARBERG, *Anal. Chem.*, 29 (1957) 947.
A. K. MAJUMDAR AND A. K. MUKHERJEE, *Anal. Chim. Acta*, 19 (1958) 23 and 21 (1959) 245.
F. J. LANGMYRH AND T. HONGSLO, *Anal. Chim. Acta*, 22 (1960) 301.

(ii) Precipitation with oxine. In a tartrate medium, Nb(V) oxinate precipitates at pH 6.5. Ta(V), Sb(III), and Sn(IV) do not precipitate, while Ti(IV) comes out partially.

cf. G. K. Belekar and V. T. Athavale, *Analyst*, 82 (1957) 630.

(iii) Extraction of Nb(V) oxinate. Niobium oxinate can be extracted from a tartrate or citrate medium at about pH = 6–9 by various solvents, and may thus be separated from Ta(V) and W(VI).

cf. I. P. Alimarin and J. M. Gibalo, *Chem. Abstr.*, 17 (1957) 342a.

(iv) Extraction from a fluoride medium. Ta(V) can be separated from Nb(V) by extraction from a medium 3 *M* with respect to hydrochloric acid and 0.4 *M* with respect to hydrofluoric acid using di-*iso*-propyl ketone as the solvent (methyl *iso*-butyl ketone may also be used). Although Nb(V) is extracted to only a negligible extent, if it is present in large amounts, a double extraction of the Ta(V) is necessary.

A certain number of other elements are extracted: Sb(V), Fe(III), Mo(VI), etc.; Nb(V) and Ta(V) should therefore be separated from them beforehand.

REAGENTS

Hydrochloric/hydrofluoric acid mixture: 62 ml of concentrated hydrochloric acid and 5 ml of concentrated hydrofluoric acid are added to 180 ml of water.

Washing solution: 80 ml of concentrated sulphuric acid are added to 165 ml of water, the solution is cooled and treated with 5 ml of concentrated hydrofluoric acid. This mixture is adjusted to equilibrium with the solvent used by shaking with a 25 ml portion of the latter.

Di-iso-propyl ketone: 250 ml of the solvent are shaken with 25 ml of the hydrochloric acid/hydrofluoric acid mixture.

OPERATING PROCEDURE

15 ml of each of the oxides of niobium and tantalum are dissolved in a few ml of concentrated hydrofluoric acid. The solution is evaporated to very small volume and the residue is dissolved in 20 ml of the HCl/HF mixture. The mixture is transferred into a polythene vessel, and shaken with 10 ml of solvent for 2 minutes. The organic phase is run off, and a second extraction is performed. The ketone extracts are combined and washed three times by shaking for 15 seconds with 5 ml of the washing solution.

Tantalum can be re-extracted by shaking the organic phase three times with 10 ml of 10 volume hydrogen peroxide or with 6 *M* hydrochloric acid containing H_3BO_3.

(v) Extraction from a Cl^- medium. In a tartrate solution 7.5 *N* with respect to HCl, Nb(V) is extracted by hexone, and may thus be separated from Ta(V). Colorimetry can then be carried out directly with fluorone.

cf. P. Senise and L. Sant'Agostino, *Anal. Chim. Acta*, 22 (1960) 296.

(vi) Chromatography. With ion-exchangers. Nb(V) can be separated from Ta(V) in a mixture of hydrochloric and hydrofluoric acids.

cf. M. J. Cabell and I. Milner, *Anal. Chim. Acta*, 13 (1955) 258.
C. U. Wetlesen, *Anal. Chim. Acta*, 22 (1960) 189.
J. L. Hague, E. D. Brown and H. A. Bright, *J. Res. Natl. Bur. Std.*, 53 (1954) 261.

On alumina
N. Tikhomiroff, *Compt. Rend. Paris*, 236 (1953) 1263.

Review of separations of Nb(V) from Ta(V)
R. H. Atkinson, J. Steigman and C. F. Hiskey, *Anal. Chem.*, 24 (1952) 477.
H. Shäfer, *Angew. Chem.*, 71 (1959) 153.

Standard solutions. Either Nb_2O_5 free from Ta_2O_5, or Ta_2O_5 free from Nb_2O_5 is digested with fused pyrosulphate and dissolved in an oxalate or tartrate solution.

Operating procedure

100 to 200 mg of the oxide is digested with 6 to 8 g of fused potassium pyrosulphate. The residue may, for example, be taken up with 100 ml of 4% ammonium oxalate and 2 ml of concentrated sulphuric acid. The solution is warmed until completely clear, and is then made up to a known volume.

$$2Nb/Nb_2O_5 = 0.699$$
$$2Ta/Ta_2O_5 = 0.819$$

(2) Colorimetry

Niobium. Of the various possible methods of determining niobium colorimetrically, only one is accurate, namely that involving hydrogen peroxide, and in which the absorption is determined in the ultraviolet. A correction must be made for the absorption due to titanium. Sensitivity of this method is rather low.

Smaller amounts may be estimated by colorimetry of the thiocyanate complexes. The accuracy of this method is not very high, but as little as 10 μg of Nb_2O_5 may be estimated.

Tantalum. No satisfactory colorimetric method exists. A passable method is that involving pyrogallol in an oxalate solution.

(i) Estimation of niobium and titanium using hydrogen peroxide. The complexes formed by Nb(V) and Ti(IV) with hydrogen peroxide under suitable conditions absorb in the ultraviolet region.

The curves of optical density plotted as a function of the wavelength are shown in Figure 70 for solutions of niobium and titanium corresponding to 50 mg per litre of each of the oxides (50 ppm). The curve corresponding to a blank test involving all the reagents is also shown. It is evident that the most favourable conditions for the determination of niobium are at 340 mμ. The optical density due to the titanium is low. Titanium itself may be determined separately under conditions selected in such as way that the niobium interferes only slightly, *i.e.* in dilute sulphuric acid at 410 mμ.

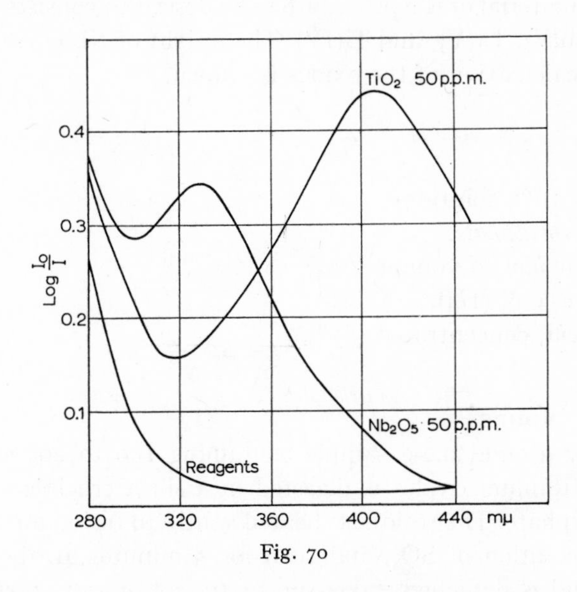

Fig. 70

It follows from the above that three separate calibrations are required: (a) determination of the absorption of niobium at 340 mμ in a medium containing 50% of H_2SO_4 and 10% of H_3PO_4; (b) determination of the absorption due to titanium under the same conditions; (c) determination of the absorption due to titanium in 1 *M* H_2SO_4 at 410 mμ.

(a) Niobium at 340 mμ. The molar extinction coefficient is 892 ± 2. 1 mg of Nb_2O_5 per litre corresponds to an optical density of 0.00671. 10 to 100 mg of Nb_2O_5 per litre can be determined to within 1%.

(b) Titanium at 340 mμ. The extinction coefficient is 238.5 ± 0.5. 1 mg of TiO_2 per litre corresponds to an optical density of 0.00298.

(c) Titanium at 410 mμ. The extinction coefficient is 745 ± 2. 1 mg of TiO_2 per litre corresponds to an optical density of 0.00933. 5 to 100 mg of TiO_2 per litre can be determined to within 0.5%.

As a result of this procedure, the niobium present may generally be estimated to within less than 1% (relative error).

Tantalum oxide is usually obtained by difference from the sum of the oxides. The absolute error in the determination of Ta_2O_5 does not exceed ± 1%. The method is lest satisfactory when the quantity of niobium involved is large and that of tantalum small.

Interfering ions. Interference is obtained from Fe(III), Mo(VI), U(VI), etc.; Nb(V), Ta(V) and Ti(IV) are usually separated from the other elements by methods described above. These three elements behave in a similar manner and, in particular, are coprecipitated by cupferron in acid solutions in the presence of tartaric acid.

The starting material of the procedure to be described consists of a mixture of the oxides of Nb(V), Ta(V), and Ti(IV). The weight of $Nb_2O_5 + Ta_2O_5 + TiO_2$ obtained after calcination of the oxides is known.

REAGENTS

Tartaric acid, 10% solution.
Potassium pyrosulphate.
Hydrogen peroxide, 20 volume (6%).
Sulphuric acid, 9 *N* (1/4).
Phosphoric acid, concentrated (85%).

OPERATING PROCEDURE

A very finely ground 0.2 g sample containing 100 to 200 mg of niobium, tantalum and titanium oxides is digested in a silica crucible with 3 g of potassium pyrosulphate. The residue is heated gently to fusion for 20–30 minutes, avoiding the evolution of SO_3, and then for 5 minutes at about 900°. After cooling, the solid is detached, taken up in 100 ml of 10% tartaric acid, and heated to complete disintegration. The volume is made up to 250 ml in a measuring flask.

Estimation of the titanium(IV). A suitable aliquot of the solution, for example 10 ml, is transferred to a 50 ml measuring flask. 10 ml of 9 *N* sulphuric acid and 2 ml of hydrogen peroxide are added, the volume is made up to 50 ml and colorimetry is performed at 410 mμ. This method is only suitable if the TiO_2/Nb_2O_5 ratio is not too low (several hundredths). For the estimation of traces of Ti(IV) in Nb(V), see Titanium.

Estimation of the niobium. About 25 ml of concentrated sulphuric acid are added to a 10 ml aliquot, the solution is cooled and transferred to a 50 ml measuring flask. 2 ml of hydrogen peroxide and then 5 ml of phosphoric acid are added, and the volume is made up to 50 ml. Colorimetry is performed at 340 mμ.

cf. G. TELEP AND B. F. BOLTZ, *Anal. Chem.*, 24 (1952) 163.
G. CHARLOT AND J. SAULNIER, *Chim. Anal.*, 35 (1953) 51.
H. SCHAFER AND F. SCHULTZE, *Z. Anal. Chem.*, 149 (1956) 73.

(ii) Colorimetry of niobium with thiocyanate. In highly acid solutions niobium (V) gives rise to yellow thiocyanate complexes which can be extracted with ethers and ketones. The reaction may also take place in tartaric acid solutions.

Interfering ions. Ta(V) interferes only slightly. U(VI), V(V), W(VI), and Mo(VI) interfere.

The Nb(V) complex exhibits an absorption maximum at 385 mμ, while Ti(IV) absorbs only slightly at this wavelength; a correction is nevertheless necessary. In the presence of even traces of Fe(III) some Sn(II) should be added.

Addition of acetone stabilizes the complex. If the reaction is carried out directly in an aqueous acetone medium, the decomposition of thiocyanic acid (which would otherwise give rise to coloured decomposition products) is prevented.

Sensitivity. The method is considerably more sensitive than that described above. $\varepsilon \sim 35{,}000$ at 385 mμ.

REAGENTS

Stannous chloride: 45 g of $SnCl_2 \cdot 2H_2O$ are dissolved in 50 ml of concentrated hydrochloric acid and made up to 100 ml.

Potassium thiocyanate: 29 g of KSCN are dissolved in 100 ml of water.

Acetone.

Tartaric acid, 10 g per litre.

OPERATING PROCEDURE

10 ml of concentrated hydrochloric acid, 1 ml of stannous chloride, 5 ml of water, and 10 ml of acetone are successively added to a 50 ml flask. The mixture is cooled, and 10 ml of the niobium solution in tartaric acid (10 g per litre) containing 10 to 40 μg of Nb(V), are added, followed by 10 ml of thiocyanate. The volume is made up to 50 ml, and the colorimetric determination is performed at 385 mμ after allowing the solution to stand for 5 minutes.

cf. G. W. C. MILNER AND A. A. SMALES, *Analyst,* 79 (1954) 315.
M. N. BUKHSH AND D. N. HUME, *Anal. Chem.,* 27 (1955) 116.
C. E. CROUTHAMEL, B. E. HJELTE AND C. E. JOHNSON, *Anal. Chem.,* 27 (1955) 507.
A. E. O. MARZYS, *Analyst,* 80 (1955) 199.
A. BACON AND G. W. C. MILNER, *Anal. Chim. Acta,* 15 (1956) 129.

Nb in pure tantalum
J. HASTINGS AND T. A. MCCLARITY, *Anal. Chem.,* 26 (1954) 683.
K. S. BERGSTRESSER, *Anal. Chem.,* 31 (1959) 1812.

Small amounts in minerals
G. W. C. MILNER AND A. A. SMALES, *Analyst,* 79 (1954) 315.
A. E. O. MARZYS, 79 (1954) 327.
F. N. WARD AND A. P. MARRANZINO, *Anal. Chem.,* 27 (1955) 1325.

In the presence of Ti(IV)
R. J. MUNDY, *Anal. Chem.*, 27 (1955) 1408.

In steels
G. W. C. MILNER AND A. A. SMALES, *Analyst*, 79 (1954) 425.

In rocks
F. S. GRIMALDI, *Anal. Chem.*, 32 (1960) 119.

(3) Other methods

(i) Colorimetry of the oxinate in chloroform. Ta(V) does not interfere. $\varepsilon \sim 10{,}000$ at 385 mμ.

J. L. KASSNER, A. GARCIA-PORRATA AND E. L. GROVE, *Anal. Chem.*, 27 (1955) 492.

With tiron. Ta(V) does not interfere.
H. FLASCHKA AND E. LASSNER, *Mikrochim. Acta*, (1956) 778.

(ii) Colorimetry of tantalum using pyrogallol. In acid solutions pyrogallol gives a yellow coloration with tantalum. Many ions interfere: V(V), Mo(VI), Sn(IV), W(VI), and U(VI). A correction can be made when Ti(IV) is present. The reaction takes place in a solution containing tartaric and oxalic acids; in this case, the effect of Nb(V) is negligible. The quantities of reagents added exert a great influence on the coloration.

Sensitivity. $\varepsilon \sim 5{,}000$ at 325 mμ.

REAGENTS

Hydrochloric acid-ammonium oxalate solution: 15 g of ammonium oxalate are dissolved in 150 ml of water, 760 ml of concentrated hydrochloric acid are added, and the volume is made up to 1 litre.

Pyrogallol: 50 g of pyrogallol are dissolved in a mixture made of 25 ml of concentrated hydrochloric acid and 10 ml of 1 *M* stannous chloride in concentrated hydrochloric acid solution.

Tartaric acid, 75 g per litre.

OPERATING PROCEDURE

50 ml of hydrochloric acid-ammonium oxalate solution and 20 ml of purogallol are added to 20 ml of a solution containing from 0.2 to 1 mg of Ta_2O_5 in tartaric acid at 75 g per litre. The volume is made up to 100 ml, the solution is left for half an hour, and colorimetry is then performed at 365 mμ.

In rocks
J. I. DINNIN, *Anal. Chem.*, 25 (1953) 1803.

Nb and Ta

A. E. O. Marzys, *Analyst*, 80 (1955) 194.

D. F. Wood and I. R. Scholes, *Anal. Chim. Acta*, 21 (1959) 121.

Nb in Ta

M. L. Theodore, *Anal. Chem.*, 30 (1958) 465.

(*iii*) *With gallic acid.* An oxalate medium is used; 2 to 40 ppm.

cf. H. Freund, K. H. Hammill and F. C. Bissonnette, Jr., *U.S. Bureau of Mines, Rept. Invest.*, 5 (1956) 242.

(*iv*) *Estimation of Ta(V) using phenylfluorone.* The following interfere: Ge, Te, Ti, Cr(VI), Sb(III), and large amounts of Nb(V). Ta(V) may be extracted with hexone.

In a hydrochloric acid-hydrofluoric acid medium. The interfering ions may be complexed with EDTA.

cf. C. L. Luke, *Anal. Chem.*, 31 (1959) 904.

Nitrogen and its compounds

N = 14.01

Ammonia – Ammonium ion – Nitrides

(1) Separation of trace amounts

(i) Evolution of NH_3. When an alkaline ammoniacal medium is distilled, free NH_3 is evolved. This operation can be carried out after any organic nitrogen or nitrides present have been converted into ammonium sulphate by Kjeldahl's method (digestion by concentrated sulphuric acid at the boil).

cf. P. L. KIRK, *Anal. Chem.*, 22 (1950) 354.

In steels

H. F. BEEGHLY, *Ind. Eng. Chem., Anal. Ed.*, 14 (1942) 137.

In biochemistry

W. HURKA, *Mikrochim. Acta*, 33 (1947) 11.
R. H. COTTON, *Ind. Eng. Chem., Anal. Ed.*, 17 (1945) 734.

(ii) Microdiffusion. Very minute traces of ammonia can be separated by this method.

cf. E. J. CONWAY, *Microdiffusion and Volumetric Analysis Error*, Crosby Lockwood and Son London, 1947.

(iii) Ammonium ion. Separation by ion-exchangers. This is of interest in particular cases and is also applicable to trace amounts.

(iv) Standard solutions. Many ammonium salts of guaranteed purity are available: NH_4Cl; $(NH_4)_2SO_4$; NH_4NO_3.

$$N/NH_4Cl = 0.2617$$
$$NH_3/NH_4Cl = 0.318.$$

(2) Colorimetry

(i) Nessler's reagent. With Nessler's reagent, ammonia and ammonium salts give an orange precipitate:

$$2HgI_4^{2-} + 2NH_3 \rightarrow 2NH_3HgI_2 + 4I^-$$
$$2NH_3HgI_2 \rightarrow \underset{\text{orange}}{NH_2Hg_2I_3\downarrow} + I^- + NH_4^+.$$

Gum arabic may be added to prevent flocculation. The intensity of coloration depends on the composition of the reagent used.

A number of ions, including Mg^{2+}, interfere but the addition of tartrate sometimes prevents this effect. In most cases, NH_3 is distilled off. Results are obtained either by a visual comparison with a series of standards or by colorimetry performed at 410 mμ.

Sensitivity: $\varepsilon = 6200$ at 400 mμ.

REAGENT

45.5 g of mercuric iodide and 35 g of potassium iodide are dissolved by shaking with a little water.

After dissolution, 112 g of potassium hydroxide are added, and the solution is made up to 1000 ml.

OPERATING PROCEDURE

The solution is diluted to about 1 mg of NH_3 per litre. A graduated 50 ml tube (Nessler tube) is filled and 2.5 ml of reagent are added. Colorimetry is performed after allowing the solution to stand for 10 minutes.

cf. E. GEIGER, *Helv. Chim. Acta*, 25 (1942) 1453.
R. F. REITEMEIER, *Ind. Eng. Chem., Anal. Ed.*, 15 (1943) 393.

In steels
A. GOTTA AND H. SEEHOF, *Z. Anal. Chem.*, 124 (1942) 216.

Artificial standards
R. DANET, *J. Pharm. Chim.*, 16 (1932) 68.
F. ALTEN AND E. HILLE, *Angew. Chem.*, 48 (1935) 137.

Colorimetry after Kjeldahl
G. L. MILLER AND E. E. MILLER, *Anal. Chem.*, 20 (1948) 481.

(ii) Phenoxide + hypochlorite. This method often obviates the necessity for a preliminary separation of NH_3.

Molar extinction coefficient: $\varepsilon = 3500$ at 625 mμ.

Principle. In an alkaline medium, in the presence of sodium hypochlorite, ammonia gives a blue coloration with phenol. Tartrate may be added if necessary to avoid precipitation at this pH.

REAGENTS

Concentrated hydrochloric acid, 12 *N.*

Purified sodium tartrate: 150 g of sodium tartrate are dissolved in 175 ml of water, and 5 ml of caustic soda solution containing 200 g per litre are added: the mixture is boiled for 15 min. Air is then bubbled through the hot solution for 5 minutes, and after cooling, the latter is made up to 250 ml. The point of this procedure is to eliminate any traces of ammonium salts present in the sodium tartrate.

Purified caustic soda solution, containing 200 g per litre.

200 g of caustic soda are dissolved in 750 ml of water: the solution is treated as in the case of the tartrate, and made up to 1000 ml.

Phenol solution: 25 g of pure crystalline phenol are dissolved in 100 ml of ethyl alcohol.

Pure sodium hypochlorite: commercial solution.

Standard solution: 0.47 g of 99.5–100% ammonium sulphate are dissolved in 1000 ml of water. 1 ml of this solution contains 100 μg of nitrogen. 25 ml of this solution are diluted to 1000 ml so that 1 ml contains 2.5 μg of nitrogen.

OPERATING PROCEDURE

An aliquot of 5 ml, containing from 1 to 15 μg of nitrogen, is placed in a 25 ml volumetric flask. Successive additions of 5 ml of tartrate if required, 2 ml of phenol, 4 ml of caustic soda, and 4 ml of hypochlorite are made. After standing for 5 minutes, the solution is made up to 25 ml, and estimated colorimetrically at 625 mμ after exactly 5 min.

NOTES. Volumes of the reagents used should be measured (graduated pipette or burette).

Since contamination is a frequent occurrence, frequent blank tests on all reagents should be carried out as a check.

If calcium is present, the tartrate should be replaced by purified citrate.

cf. J. P. RILEY, *Anal. Chim. Acta*, 9 (1954) 575.
U. BOHNSTEDT, *Z. Anal. Chem.*, 163 (1958) 415.
W. T. BOLLETER, C. J. BUSHMAN AND P. W. TIDWELL, *Anal. Chem.*, 33 (1961) 592.

Hydrazine $NH_2.NH_2$

Colorimetry

Using *p*-dimethylaminobenzaldehyde:

G. W. WATT AND D. CHRISP, *Anal. Chem.*, 24 (1952) 2006.

Using picryl chloride:

J. P. RILEY, *Analyst*, 79 (1954) 76.

Hydrazoic acid – Azides N_3^-

(1) Separation

The azeotropic mixture formed by HN_3 and water can be distilled from a strongly acid medium.

(2) Colorimetry

With Fe^{3+}, the red FeN_3^{2+} complex is formed. The determination may be carried out at 460 mμ.

SO_3^{2-}, $S_2O_3^{2-}$, and S^{2-} must be oxidized beforehand, CNO^- and CNS^- interfere.

cf. C. E. Roberson and C. M. Austin, *Anal. Chem.*, 29 (1957) 854.
A. Anton, J. G. Dodd and A. E. Harvey, Jr., *Anal. Chem.*, 32 (1960) 1209.

Nitrogen dioxide

(i) Photometry.

L. Harris and B. M. Siegel, *Ind. Eng. Chem., Anal. Ed.*, 14 (1942) 258.
J. K. Dixon, *J. Phys. Chem.*, 8 (1940) 157.

(ii) Use of the Griess reagent.

P. R. Averell, W. F. Hart, N. T. Woodberry and W. R. Bradley, *Ind. Eng. Chem., Anal. Ed.*, 19 (1947) 1040.

Mixtures of $N_2O + NO + NO_2$ (Visible and infrared)
H. Kienitz, *Chim. Anal.*, 24 (1952) 83.

Nitrites NO_2^-

(i) Standard solutions. For colorimetry, 98–100% sodium nitrite ($NaNO_2$) may be used.

Dilute solutions are stable if protected from bacteria by the addition of 1 ml of chloroform per litre, and if their acidification by carbon dioxide is avoided by the addition of 100 mg of caustic soda per litre.

$$N/NaNO_2 = 0.203$$

$$NO_2^-/NaNO_2 = 0.667$$

(ii) Colorimetry. Formation of an azo dye (Griess reagent). The most frequently used method consists in producing a red azo dye with sulphanilic acid and α-naphthylamine (Griess reagent).

Since the colour becomes more intense with time, the colorimetric estimation should be carried out after a predetermined period, the length of which depends on the pH and the temperature. The pH should be 1.7–3.0[1].

The absorption maximum is at 520 mμ.

Artificial standards can be used.

The method permits trace amounts to be estimated, but is not very accurate. Molar extinction coefficient: $\varepsilon \sim 40{,}000$ at 520 mμ.

Interfering ions. The coloration is specific.
Cr(III) at a concentration less than 40 ppm, Co(II) $<$ 100 ppm, $C_2O_4^{2-}$ $<$ 200 ppm, CO_3^{2-} $<$ 200 ppm, CN^- $<$ 100 ppm, Cr(VI) $<$ 80 ppm, SiO_3^{2-} $<$ 200 ppm, and W(VI) $<$ 10 ppm, do not interfere.

Fe(III), Sb(III), Bi(III), Ce(III), Au(III), Hg(II), and Ag(I) interfere; Fe^{3+} can be complexed by citrate[1].

REAGENTS

Sulphanilic acid, 0.6% in 2.4 *N* (1/5) hydrochloric acid.
α-Naphthylamine, colourless, 0.6 g in 1 ml of concentrated hydrochloric acid and 100 ml of water.
A stable mixture of the reagents can be made[1];
Crystalline sodium acetate in the form of a 25% (2 *M*) solution.

OPERATING PROCEDURE

1 ml of sulphanilic acid is added to the nitrite solution, and the latter allowed to stand for 3 to 4 minutes. 1 ml of α-naphthylamine and then 1 ml of acetate are now added, to give a pH of 2 to 2.5. This solution is made up to 50 ml, and allowed to stand for 10 minutes. Colorimetry is performed at 520 mμ.

cf. B. F. RIDER AND M. G. MELLON, *Ind. Eng. Chem., Anal. Ed.,* 18 (1946) 96.
(1) J. L. NELSON, L. T. KURTZ AND R. H. BRAY, *Anal. Chem.,* 26 (1954) 1081.

Artificial standards
R. DANET, *J. Pharm. Chim.,* 7 (1928) 113.
P. R. AVERELL, W. F. HART, N. T. WOODBERRY AND W. R. BRADLEY, *Anal. Chem.,* 19 (1947) 1040.

(iii) Azo dye with dimethyl-α-naphthylamine. A more intense and indefinitely stable coloration is obtained. The amine solution is prepared by dissolving 5 g of the reagent in 1 litre of 4 *N* acetic acid, obtained by the dilution of pure 17 *N* acid with 95% methanol.

The operating procedure is the same as that described above.

cf. F. G. GERMUTH, *Ind. Eng. Chem., Anal. Ed.,* 1 (1929) 28.

(iv) Other methods. Sulphanilamide and N-naphthyl-1-ethylenediamine hydrochloride.

cf. M. B. SHINN, *Ind. Eng. Chem., Anal. Ed.,* 13 (1941) 33.
H. BARNES AND A. R. FOLKHARD, *Analyst,* 76 (1951) 599.

Nitrates NO_3

(1) Separation of trace amounts

(i) General separation. Since silver and barium nitrates are soluble, all anions, the silver or barium salts of which are insoluble, can be separated. Chlorides may thus be separated by the addition of silver sulphate. A general separation can be effected by means of the following reagent: 8 ml of concentrated perchloric acid, 94 g of silver perchlorate, and 258 g of barium perchlorate in 1 litre of water.

OPERATING PROCEDURE

5 ml of the solution are treated with the reagent drop by drop, and the precipitate is allowed to settle after each 1 ml portion until no further precipitation takes place. The mixture is left to stand for at least 1 h, filtered and washed.

cf. A. DOLANGE AND P. W. HEALY, *Ind. Eng. Chem., Anal. Ed.*, 17 (1945) 718.

(ii) Destruction of nitrites. Nitrites cannot be separated by the above method, but may be decomposed by reduction with an azide, N_3^-, in acid medium.

cf. W. SEAMAN, A. R. NORTON, W. J. MADER AND J. J. HUGONET, *Ind. Eng. Chem., Anal. Ed.*, 14 (1942) 420.

(iii) Standard solutions. 99.8–100% KNO_3 of guaranteed purity can be weighed out after drying.

It is also possible to start with standardized nitric acid.

$$N/KNO_3 = 0.1385.$$

$$NO_3^-/KNO_3 = 0.613.$$

A 1.000 *N* solution of HNO_3 contains 14.01 mg of nitrogen or 62.0 mg of NO_3^- per millilitre.

(2) Colorimetry

(i) Using phenol-2,4-disulphonic acid. Nitrates give a yellow nitro derivative: the coloration is more sensitive in an alkaline medium. This method is normally capable of detecting from 10 to 400 μg to within about 5%.

Molar extinction coefficient: $\varepsilon \sim 10{,}000$ at 410 mμ.

Composition of the reagent and the working conditions are important factors.

Interfering ions. For accurate determination, NO_2^-, Cl^-, CO_3^{2-} and organic materials must be eliminated.

REAGENTS

25 g of pure phenol are dissolved in 150 ml of concentrated sulphuric acid, and 75 ml of oleum containing 13% of SO_3 are added. The solution is heated for 2 hours on a steam bath.

OPERATING PROCEDURE

The solution to be analyzed is evaporated to dryness and the whole of the residue is wetted with 2 ml of the reagent in one portion. After 10 minutes, 15 ml of water are added and the mixture stirred until solution is complete. After cooling, the solution is made alkaline by the addition of 4 N (1/3) ammonia, and filtered if necessary. Colorimetry is performed at 410 mμ.

cf. M. J. TARAS, *Anal. Chem.*, 22 (1950) 1020.

(ii) Using chromotropic acid.

P. W. WEST AND G. L. LYLES, *Anal. Chim. Acta*, 23 (1960) 227.

(iii) Absorption in the ultraviolet. The many interfering ions should all be separated before the operation. $\varepsilon = 5{,}800$ at 211 mμ; 23 at 355 mμ.

cf. A. P. ALTSHULLER AND A. F. WARTBURG, *Anal. Chem.*, 32 (1960) 174.
R. BASTIAN, R. WEBERLING AND F. PALILLA, *Anal. Chem.*, 29 (1957) 1795.

(iv) Reduction to ammonia and colorimetry of the NH_3.

R. F. REITEMEIER, *Ind. Eng. Chem., Anal. Ed.*, 15 (1943) 393.

(v) Use of redox indicators. Sulphonated diphenylamine, oxidized by NO_3^-, to a violet colour, can be used. By this method, 1 to 50 μg of nitrate in 10 ml of water may be estimated to within 5%. Certain oxidizing agents give the same reaction.

OPERATING PROCEDURE

0.1 g of potassium chloride are added to 10 ml of the solution, followed by 10 ml of concentrated sulphuric acid. The mixture is stirred, cooled, and then 0.1 ml of 0.2% sodium diphenylamine sulphonate are added. Colorimetry is performed directly.

cf. I. M. KOLTHOFF AND G. E. NOPONEN, *J. Am. Chem. Soc.*, 55 (1933) 1448.

(vi) Reduction to the nitrite and use of Griess' reagent. Nitrates may be reduced to nitrites with zinc powder and be then estimated colorimetrically.

REAGENTS

Sulphanilic acid, 0.6% in 2.4 N (1/5) hydrochloric acid.

α-Naphthylamine, colourless, 0.6 g in 1 ml of concentrated hydrochloric acid and 100 ml of water:

Crystalline sodium acetate in the form of a 25% solution (2 *M*).

OPERATING PROCEDURE

To 1 ml of the neutral solution containing 1–10 μg of NO_3^- are added, first 9 ml of 3.5 *M* (1/5) acetic acid, and then the reagent. The mixture is stirred for 50–60 seconds, centrifuged rapidly, and estimated colorimetrically as described on p. 323.

Limit of sensitivity: 0.5 μg of NO_3^-.

Accuracy: ± 5%.

cf. J. L. NELSON, L. T. KURTZ AND R. H. BRAY, *Anal. Chem.*, 26 (1954) 1081.

$NO_2^- + NO_3^-$

J. L. LAMBERT AND F. ZITOMER, *Anal. Chem.*, 32 (1960) 1684.

Osmium

Os = 190.2

(1) Separation

See The Platinum Metals p. 348.

(2) Colorimetry

With thiourea. Thiourea reacts with Os(VIII) to give a red complex. The concentration of chloride should be standardized.

Ru gives a blue-green coloration, which interferes. Pd(II) produces a yellow colour which interferes only slightly at 540 mμ.

Addition of Sn(II) promotes the reaction because the compound obtained corresponds to Os(III); in addition, Sn(II) reduces any ruthenium present, to a noninterfering form.

The osmium is usually first separated by volatilization or by the extraction of OsO_4.

cf. G. H. AYRES AND W. N. WELLS, *Anal. Chem.*, 22 (1950) 317.
W. J. ALLAN AND F. E. BEAMISH, *Anal. Chem.*, 24 (1952) 1608.

Separation and estimation

R. D. SAUERBRUNN AND E. B. SANDELL, *Anal. Chim. Acta*, 9 (1953) 86.

(3) Other methods

With 1-naphthylamine trisulphonic acid

E. L. STEELE AND J. H. YOE, *Anal. Chim. Acta*, 20 (1959) 205.

Traces by catalysis

R. D. SAUERBRUNN AND E. B. SANDELL, *Mikrochim. Acta*, 22 (1953).

Critical review of methods

F. E. BEAMISH AND W. A. E. MCBRYDE, *Anal. Chim. Acta*, 9 (1953) 349; 18 (1958) 551.
E. B. SANDELL, *Colorimetric Determination of Traces of Metals*, Interscience, 1959.

Oxygen and its compounds

O = 16.00

Oxygen

Colorimetry

The oxygen contents of various gases may be estimated colorimetrically by passing the latter continuously through appropriate solutions and estimating the oxygen dissolved. A number of automatic devices are in existence.

(i) Redox indicators. Indigo carmine. The pale yellow leuco derivative obtained by the reduction of indigo carmine with an exactly equivalent quantity of dithionite is slowly re-oxidized by oxygen.

REAGENTS

Indigo carmine, 0.1%. The reagent should be protected from air by means of a layer of mineral oil. Immediately prior to the experiment, a 5% solution of sodium dithionite should be prepared and added dropwise with a pipette to the coloured solution, until a faint green shade is obtained. The reagent should be used immediately.

cf. W. F. LOOMIS, *Anal. Chem.*, 26 (1954) 402; 28 (1956) 1347.
A. H. MEYLING AND G. H. FRANK, *Analyst*, 87 (1962) 60.

0 to 25 × 10^{-3} ppm
L. S. BUCHOFF, N. M. INGBÉR AND J. H. BRADY, *Anal. Chem.*, 27 (1955) 1401.

Using methylene blue
A. KLING, *Bull. Acad. Med.*, 119 (1938) 178.

Anthraquinone-β-sulphonic acid (Down to 0.001%)
L. J. BRADY, *Anal. Chem.*, 20 (1948) 1033.
C. STAFFORD, J. E. PUKETT, M. D. GRIMES AND B. J. HEINRICH, *Anal. Chem.*, 27 (1955) 2012.
F. W. KARASEK, R. J. LOYD, D. E. LUPFER AND E. A. HOUSER, *Anal. Chem.*, 28 (1956) 233.
L. SILVERMAN AND W. BRADSHAW, *Anal. Chim. Acta*, 14 (1956) 514.

3,3'-dimethylnaphthidine
J. BANKS, *Analyst*, 84 (1959) 700.

(ii) Oxidation of copper in the presence of ammonia. The oxygen contents of various gases and liquids can be estimated by means of absorption by copper moistened with $NH_3 + NH_4^+$, according to the reaction

$$2Cu\downarrow + O_2 + 8NH_3 + 4H^+ \rightarrow 2Cu(NH_3)_4{}^{2+} + 2H_2O.$$

The blue cuprammine complex formed may subsequently be determined colorimetrically. 0.2 to 2 ml of oxygen can be estimated in this way.

Alternatively, the Cu(II) formed may be estimated iodometrically.

cf. J. S. POWELL AND P. C. JOY, *Anal. Chem.*, 21 (1949) 296.
K. UHRIG, F. M. ROBERTS AND H. LEVIN, *Ind. Eng. Chem., Anal. Ed.*, 17 (1945) 31.
H. W. DEINUM AND J. W. DAM, *Anal. Chim. Acta*, 3 (1949) 353.

(iii) Oxidation of $Fe(OH)_2$. As in Winkler's method, it is possible, operating in the absence of air, to absorb oxygen by $Fe(OH)_2$. The latter is obtained by adding alkalis to ferrous salts. The solution is then re-acidified, after which the colorimetry of the Fe(III) formed can be carried out under the usual conditions.

cf. J. A. SHAW, *Ind. Eng. Chem., Anal. Ed.*, 14 (1942) 891.

(iv) Oxidation of $Mn(OH)_2$. It is similarly possible to oxidize $Mn(OH)_2$ and then, after acidification, to add an excess of iodide and estimate the iodine colorimetrically.

COMPARE
E. H. WINSLOW AND H. A. LIEBHAFSKY, *Ind. Eng. Chem., Anal. Ed.*, 18 (1946) 565.
S. BAIRSTOW, J. FRANCIS AND G. H. WYATT, *Analyst*, 72 (1947) 340.
K. WICKERT AND E. IPACH, *Z. Anal. Chem.*, 14 (1953) 350.

Traces down to 0.5 ppm
L. P. PEPKOWITZ AND E. L. SHIRLEY, *Anal. Chem.*, 25 (1953) 1718.

Traces down to 0.005 ppm
M. NEEDLEMAN, *Analyst*, 84 (1959) 720.

Review
H. A. J. PIETERS AND W. J. HANSSEN, *Anal. Chim. Acta*, 2 (1948) 712.

Ozone

Colorimetry

As in the case of oxygen, redox indicators are used.

(i) Oxidation of leuco-fluorescein. The leuco derivative of fluorescein is oxidized by ozone to fluorescein and the solution obtained is compared with standard solutions. The method is effective for very minute traces if the fluorescence is observed in ultraviolet light. As little as 2×10^{-7} volumes of ozone in air can be estimated in this way.

Interfering compounds. NO_2 and H_2O_2 do not interfere. Large amounts of both chlorine and CO_2 can interfere.

Reagent

1 mg of fluorescein is dissolved in 1 ml of 10% sodium hydroxide. 10 ml of saturated NaOH and 1 g of Zn are added, and the latter is separated off when the fluorescence has disappeared.

OPERATING PROCEDURE

1 drop of the reagent is added to 10 ml of 0.5% caustic soda and the gas to be analyzed is passed into the solution at the rate of 15 litres per hour. The fluorescence obtained is compared with, for example, that of a standard containing 0.001 μg of fluorescein per millilitre.

(ii) Oxidation of fluorescein. A solution of fluorescein is oxidized by ozone with the formation of a nonfluorescent, colourless compound. The amount of fluorescein oxidized may be determined by comparison with solutions containing known amounts of fluorescein.

In theory:

$$\frac{\text{mass of } O_3}{\text{mass of fluorescein oxidized}} = 0.29$$

This ratio is however, subject to variations governed by a number of factors.

cf. L. BENOIST, *Compt. Rend. (Paris)*, 168 (1919) 612.
A. MACHÉ, *Compt. Rend. (Paris)*, 200 (1935) 1760.
W. HELLER, *Compt. Rend. (Paris)*, 200 (1935) 1936.

(iii) Other methods. With indigo carmine. The limiting concentration is 10^{-6}%.

Y. DORTA-SCHAEPPI AND W. D. TREADWELL, *Helv. Chim. Acta*, 32 (1949) 356.

Absorptiometry in the ultraviolet. Amounts as low as 10^{-5} to 10^{-3} ppm in air can be estimated.

cf. R. STAIR, T. C. BAGG AND R. G. JOHNSON, *J. Res. Natl. Bur. Std.*, 52 (1954) 133.
N. A. RENZETTI, *Anal. Chem.*, 29 (1957) 869.

Hydrogen peroxide – H_2O_2; peroxides – per-salts

Colorimetry

(i) Oxidation of ferrous iron. In acid solutions Fe(II) is oxidized to Fe(III) by

hydrogen peroxide and its derivatives (perborates, perphosphates, sodium peroxide, etc.). The Fe(III) formed may then be estimated colorimetrically. A number of other oxidizing agents give the same reaction.

cf. J. S. REICHERT, S. A. MCNEIGHT AND H. W. RUDEL, *Ind. Eng. Chem., Anal. Ed.*, 11 (1939) 194.

(ii) Formation of complexes with Ti(IV). Hydrogen peroxide and its derivatives form orange-coloured complexes with acid solutions of Ti(IV). The colour is very stable. As little as 1 ppm of hydrogen peroxide may be detected by this method.

Interfering ions. See Titanium; Al(III), Ni(II), and BrO_3^- interfere.

REAGENT

Solution of Ti(IV) at 1 g per litre.

OPERATING PROCEDURE

The sample should contain 0.1 mg to 3 mg of hydrogen peroxide. The volume is made up to 75 ml, 10 ml of the reagent are added, and the volume is finally made up to 100 ml. Colorimetry is then performed.

cf. G. M. EISENBERG, *Ind. Eng. Chem., Anal. Ed.*, 15 (1934) 327.
P. BONET-MAURY, *Compt. Rend. (Paris)*, 218 (1944) 117.
G. JANICÉK AND POKORNY, *Chem. Listy*, 49 (1955) 1315.

(iii) Oxidation of iodides and colorimetric determination of the iodine. Catalyst: Mo(VI). Limit: 1 ppm.

cf. W. A. PATRICK AND H. B. WAGNER, *Anal. Chem.*, 21 (1949) 1279.

Palladium

Pd = 106.7

(1) Separations

The reader is referred to the methods of separation of the platinum metals, page 348.

(i) Precipitation of the hydroxide. $Pd(OH)_2$ can be precipitated in the presence of Pt(IV), using $Fe(OH)_3$ as the entraining agent. Rhodium and iridium precipitate as well.

cf. R. GILCHRIST AND E. WICHERS, *J. Am. Chem. Soc.*, 57 (1935) 2565.

(ii) Precipitation in the metallic state. Reduction with formic acid. The procedure used is similar to that applied in the case of platinum. The metal can then be weighed.

Traces: reduction with Sn(II). Te may be used as the entraining agent. Au, Ag, Pt, and Rh accompany the Pd.

cf. S. K. HAGEN, *Mikrochemie*, 20 (1936) 180.

(iii) Precipitation in the form of the iodide, PdI_2. Of the platinum metals, only rhodium can interfere.

cf. F. E. BEAMISH AND J. DALE, *Ind. Eng. Chem., Anal. Ed.*, 10 (1938) 697.

(iv) Precipitation with dimethylglyoxime. The yellow compound formed by Pd(II) with dimethylglyoxime is precipitated in 0.25 *N* hydrochloric acid. This method may be used with reasonably small quantities, but not with trace amounts. The extraction of the compound formed is, however, quantitative.

Interfering ions. The method is almost specific, since only Au(III), and to some extent Pt(IV), are reduced to the metallic state and are consequently coprecipitated. Although the precipitate may be redissolved in aqua regia and reprecipitated, it is preferable to carry out a previous separation of the Pt(IV) in the form of ammonium chloroplatinate. Rh should also be removed.

If Fe(III) is present, a large excess of dioxime should be added to reduce it to Fe(II). Cu(II) and Ni(II) do not interfere. The nickel can be precipitated from the filtrate. NO_3^- interferes slightly.

All interfering ions other than Au(III) can be eliminated by carrying out the precipitation in the presence of EDTA, even in neutral solutions.

REAGENT

Dimethylglyoxime, 1% in alcohol.

OPERATING PROCEDURE

The solution should be 0.2 *N* with respect to HCl and must not contain more than 100 mg of Pd(II). The precipitation is carried out at the boil and the mixture is left on a water bath until the precipitate separates out (30 minutes). The solution is then filtered through sintered glass in the hot, and the precipitate is washed with boiling water very slightly acidified with hydrochloric acid, and then with alcohol.

cf. G. H. AYRES AND E. W. BERG, *Anal. Chem.*, 25 (1953) 980.

Extraction

J. G. FRASER, F. E. BEAMISH, AND W. A. E. MCBRYDE, *Anal. Chem.*, 26 (1954) 495.
R. S. YOUNG, *Analyst*, 76 (1951) 49. (see also Colorimetry).
P. F. LOTT, R. K. VITCK AND K. L. CHENG, *Anal. Chim. Acta*, 19 (1958) 323.

Cyclohexane-1,2-dione dioxime

R. C. VOTER, C. V. BANKS AND H. DIEHL, *Anal. Chem.*, 20 (1948) 652.

(v) Precipitation by dithioxamide. At pH 3.0, Pd(II), Pt(IV), Au(III), Cr(III), Fe(III), and Cu(II) precipitate out. Palladium may thus be separated from Co(II), Ni(II), and from the other platinum metals.

cf. G. H. AYRES AND B. L. TUFFLY, *Anal. Chem.*, 24 (1952) 949.

Extraction in the presence of EDTA

J. XAVIER AND P. RAY, *Science and Culture*, 20 (1955) 609.

(vi) Precipitation and extraction with phenylthiourea.

cf. J. E. CURRAH, W. A. E. MCBRYDE, A. J. CRUIKSHANK AND F. E. BEAMISH, *Ind. Eng. Chem., Anal. Ed.*, 18 (1946) 120.
G. H. AYRES AND B. L. TUFFLY, *Anal. Chem.*, 24 (1952) 949.

(vii) Extraction of the dithizonate. Pd(II) dithizonate may be extracted from 1 *N* HCl with carbon tetrachloride. Palladium may in this way be separated from Pt(IV), Rh(III), Ir(III) and (IV), Os, and Ru. According to E. B. Sandell, the following are also extracted, at least in part: Au(III), Hg(II) and Cu(II).

(viii) Extraction with 1-nitroso-2-naphthol. See Colorimetry.

(ix) Concentration of small amounts by fire assay.

cf. J. G. FRASER AND F. E. BEAMISH, *Anal. Chem.*, 26 (1954) 1474.
M. E. V. PLUMMER AND F. E. BEAMISH, *Anal. Chem.*, 31 (1959) 1141.

See also Platinum, p. 346.

(2) Colorimetry

(i) With Sn(II)

G. H. AYRES AND J. H. ALSOP III, *Anal. Chem.*, 31 (1959) 1135.

(ii) Bromide complex. The method is not very selective.

G. H. AYRES AND B. L. TUFFLY, *Anal. Chem.*, 24 (1952) 949.

(iii) Iodide complex $\varepsilon \sim 10{,}000$.

J. G. FRASER, F. E. BEAMISH AND W. A. E. MCBRYDE, *Anal. Chem.*, 26 (1954) 495.

(iv) Nitrosodiphenylamine. Compounds containing the p-NO-$C_6H_4N=$ group form coloured complexes with Pd(II). Such reactions are extremely sensitive. The complex formed with nitrosodiphenylamine can be extracted with chloroform. Palladium may thus be separated from coloured ions, including Pt(IV), Ir, and Rh.

Pt(II) and Au(III) interfere.

cf. D. E. RYAN, *Analyst*, 76 (1951) 167.
J. H. YOE AND J. J. KIRKLAND, *Anal. Chem.*, 26 (1954) 1335.
R. A. MCALLISTER, *Analyst*, 86 (1961) 618.

(v) With dioximes in chloroform. The compounds formed with Pd(II) can be extracted (*e.g.* with chloroform) giving a yellow coloration. With dimethylglyoxime[1], $\varepsilon \sim 1{,}500$ at 375 mμ. Furyl dioxime[2] is more sensitive: $\varepsilon \sim 20{,}000$ at 380 mμ. The coloration is stable. Pt, Rh, Ir, Ru and Au do not interfere.

(1) W. NIELSCH, *Z. Anal. Chem.*, 142 (1954) 30.
(2) O. MENIS AND T. C. RAINS, *Anal. Chem.*, 27 (1955) 1932.

(vi) With 2-nitroso-1-naphthol in benzene. Extraction can be carried out with benzene or toluene in the presence of EDTA. $\varepsilon = 18{,}000$ at 370 mμ.

Pt, Ir, Rh, Ru, Os, Au, Zr, Fe, Co, Cu, Ni and Cr do not interfere.

cf. K. L. Cheng, *Anal. Chem.*, 26 (1954) 1894.

Review of methods

F. E. Beamish and W. A. E. McBryde, *Anal. Chim. Acta*, 9 (1953) 349 and 18 (1958) 551.

Phosphorus

P = 31.00

Phosphine, PH_3

Separation and colorimetry of gases

(1a) Gases containing phosphine may be bubbled through an oxidizing agent. A suitable apparatus incorporates three bubblers each containing 10 ml of 0,025% permanganate and 1 ml of N (1/40) sulphuric acid. The phosphoric acid formed may be estimated colorimetrically.

cf. W. L. MÜLLER, *Arch. Hyg. Bakt.*, 129 (1943) 286.

(1b) H_2S can be absorbed by dilute caustic soda. PH_3 is absorbed by solutions of Cu(II), Hg(II), Ag(I), etc.

cf. L. WOLF, W. DÜSING AND A. MARTOS, *Mikrochemie*, 18 (1935) 185.

(1c) Alternatively, colorimetry can be carried out on paper impregnated with a silver salt such as silver nitrate. A yellow spot appears (compare AsH_3).

PH_3 can be absorbed by a dispersion of silver nitrate on silica gel. A black spot is produced. The limit of sensitivity is 10^{-9} in air.

cf. J. P. NELSON AND A. J. MILUN, *Anal. Chem.*, 29 (1957) 1665.

Hypophosphorous acid, H_3PO_2

(1) Separation

Paper chromatography may be used for separation of hypophosphorous, phosphorous, pyrophosphorous, and hypophosphoric acids.

Y. VOLMAR, J. P. EBEL AND Y. FAWZI BASSILI, *Bul. Soc. Chim.*, (1953) 1085.

(2) Oxidation to orthophosphoric acid

Oxidation can be carried out with concentrated nitric acid. The solution is reduced to a very small volume on a water bath, taken up in fuming nitric acid, and concentrated again. The H_3PO_4 formed is then estimated.

Oxidation with bromine

G. L. JENKINS AND C. F. BRUENING, *J. Am. Pharm. Assoc.*, 25 (1936) 19.

(3) Colorimetry

Molybdenum blue may be used.

A. P. SCANZILLO, *Anal. Chem.*, 26 (1954) 411.

Standard solution. Sodium hypophosphite, NaH_2PO_2 can be used.

Orthophosphoric acid, H_3PO_4

(1) Separation of trace amounts

(i) Electrolysis with a mercury cathode. H_3PO_4 may be purified from a number of elements which deposit, and in particular from Mo(VI). The phosphate remains in solution.

(ii) Precipitation with Al(III). When aluminium hydroxide is precipitated from solution, any P(V) present is entrained in the form of $AlPO_4$. Ge(IV), Si(IV), and As(V) are then eliminated with HF + HCl + HBr, after which a suitable colorimetric method can be applied.

cf. H. LEVINE, J. J. ROWE AND F. S. GRIMALDI, *Anal. Chem.*, 27 (1955) 258.

(iii) Extraction of phosphomolybdate. Phosphorus may in this way be separated from Si(IV) and As(V), using a mixture of 25 parts by volume of amyl alcohol and 75 parts of ether as the solvent.

M. JEAN, *Anal. Chim. Acta*, 14 (1956) 172.

Butanol; butanol-ether; chloroform

N. S. GING, *Anal. Chem.*, 28 (1956) 1330.
E. RUF, *Z. Anal. Chem.*, 151 (1956) 169.
W. B. SILKER, *Anal. Chem.*, 28 (1956) 1782.
S. YOKOSUKA, *Bunseki Kagaku*, 5 (1956) 395.

(iv) Separation by means of ion-exchangers. In an acid medium H_3PO_4 may be freed from Fe^{3+}, Ca^{2+}, VO^{2+}, V^{3+}, etc., by fixing the latter on to a cation exchange resin – for example Amberlite IR 100 AG. V(V) can be separated after reduction to V^{3+} or VO^{2+}.

Preparation of the column. 20 g of the resin is left in contact with water for 2 hours and then transferred to a 50 ml burette. The resin is washed with 350 ml of 2 *N* (1/6) hydrochloric acid, and then with 200 ml of water. The rate of flow should be 3–5 ml per minute. The column is kept filled with water.

OPERATING PROCEDURE

The solution should be 0.1 *N* in HCl, a 50 ml portion is passed through the column. The latter is then washed four times with 50 ml of 0.1 *N* (1/100) hydrochloric acid and once with 50 ml of water. Up to 100 mg of Fe(III) and 100 mg of Al(III) can thus be separated.

Regeneration of the resin. 500 ml of 1 *N* hydrochloric acid are passed through the column, followed by 300 ml of water. The process should occupy about 20 minutes.

cf. K. HELRICH AND W. RIEMAN III, *Anal. Chem.*, 19 (1947) 651.
A. J. GOUDIE AND W. RIEMAN III, *Anal. Chem.*, 24 (1952) 1067.
B. H. KINDT, E. W. BALIS AND H. A. LIEBHAFSKY, *Anal. Chem.*, 24 (1952) 1501.

Estimation of P(V) in V(V)

S. HARTMANN, *Z. Anal. Chem.*, 151 (1956) 332.

(v) Standard solutions. A solution of phosphoric acid can be titrated acidimetrically.

In addition, guaranteed purity disodium phosphate $Na_2HPO_4 \cdot 12H_2O$ and sodium ammonium phosphate $Na(NH_4)HPO_4 \cdot 4H_2O$ are available.

1 ml of molar H_3PO_4 = 31.0 mg of P or 71.0 mg of P_2O_5.

$$P/Na_2HPO_4 \cdot 12H_2O = 0.0865$$
$$P_2O_5/2Na_2HPO_4 \cdot 12H_2O = 0.1981$$
$$P/Na(NH_4)HPO_4 \cdot 4H_2O = 0.1482$$
$$P_2O_5/2Na(NH_4)HPO_4 \cdot 4H_2O = 0.399$$

(2) Colorimetry

Trace amounts may be estimated using a method involving the phosphovanadomolybdate complex. Moreover, the method is accurate, yielding results accurate to within 0.1% with natural phosphates.

The 'molybdenum blue' method is less reliable, but can be used in the case of very minute traces.

The extraction of the phosphomolybdate complex is a selective and very sensitive method.

(i) Phosphovanadomolybdate complex. In the presence of V(V) and Mo(VI), phosphoric acid gives a yellow complex in acid solutions. As in the case of the phosphomolybdate complex, the pH exerts an important influence. It is thus necessary to work in a medium 0.5 to 0.9 *N* (or even up to 1.5 *N*) with respect

to nitric acid, HCl, $HClO_4$, or H_2SO_4. Under these conditions the coloration develops quite rapidly. In 1.7 *N* perchloric acid it is necessary to wait for 30 minutes for the colour to develop. Once formed, colour is stable for several weeks.

Concentrations of the reagents should be fixed exactly, these values are taken into account in the calibration curve, since they also absorb: Mo(VI) 0.02–0.06 *M*; V(V) 1 to 4×10^{-3} *M*.

The coloration is not very apparent to the eye. The absorption maximum is at 315 mμ, and at this value, $\varepsilon \sim 20{,}000$. At 400 mμ however, $\varepsilon \sim 2{,}500$. The reagent itself absorbs however at all wavelengths, and it is therefore necessary to plot a calibration curve point by point.

Interfering ions. Silica gives the yellow silicomolybdate complex. In 0.8 *N* nitric acid equal quantities of silicon and phosphorus produce satisfactory results, and, in a more strongly acidic medium, the silicon interferes even less. When large amounts of silica are present, the interference may be eliminated by boiling the solution with concentrated perchloric acid.

Large amounts of Fe(III) interfere. If however, the medium contains perchloric acid in the absence of Cl^- and SO_4^{2-}, Fe(III) is colourless and does not interfere. This is the case when the solution is boiled with concentrated perchloric acid. In insufficiently acidic solutions (even those containing perchloric acid) raising the temperature causes the degree of hydrolysis of Fe^{3+} to vary, thereby introducing large errors. Fe(III) can be complexed by the addition of F^-, but boric acid should be added to remove the excess of fluoride, which would otherwise interfere.

Compounds which precipitate interfere: Sn(IV), Nb(V), Ta(V), Ti(IV), Zr (IV), and large amounts of W(VI) and even of V(V); the precipitates formed entrain the phosphorus. Bi(III), Th(IV), As(V), Cl^-, and F^- retard the development of the coloration. When F^- is present, an excess of boric acid should be added to complex it. Cu(II) and Ni(II) interfere only slightly at 460 mμ. Reducing agents should be pre-oxidized. Cr(VI) interferes. As(V) gives a coloration 100 times less sensitive. Pyrophosphates do not interfere, so that orthophosphates may be estimated in the presence of these ions, provided that an excess of reagent is added. Citrate ions interfere.

The compound is soluble in many organic solvents, such as butyl, amyl and benzyl alcohols, cyclohexanol, and ethyl acetate. It can therefore be extracted, and thereby separated from numerous interfering ions, such as Cu(II), Ni(II), Cr(VI) etc.

REAGENTS

Ammonium molybdate, 5%.

Ammonium vanadate: 2.5 g are dissolved in 500 ml of hot water; after cooling 20 ml of nitric acid are added and the volume is made up to 1,000 ml.

OPERATING PROCEDURE

10 ml of 6 *N* (1/2) nitric acid, 10 ml of vanadate, and 10 ml of molybdate are added to the neutral solution containing 1 to 20 ppm of phosphorus: the volume is made up to 100 ml. After standing for 30 minutes, colorimetry is performed at 460 mμ.

cf. R. E. KITSON AND M. G. MELLON, *Ind. Eng. Chem., Anal. Ed.*, 16 (1944) 379.
K. P. QUINLAN AND M. A. DESESA, *Anal. Chem.*, 27 (1955) 1626.

In metals

Methods of Chemical Analysis of Metals, American Society for Testing Materials, A.S.T.M., Philadelphia, 1956.
H. K. LUTWAK, *Analyst*, 78 (1953) 661.

In biochemistry

P. FLEURY AND M. LECLERC, *Bull. Soc. Chim. Biol.*, 25 (1943) 201.

In alloys

W. T. ELWELL AND H. N. WILSON, *Analyst*, 81 (1956) 136.
O. B. MICHELSON, *Anal. Chem.*, 29 (1957) 60.

In copper alloys

H. K. LUTWACK, *Analyst*, 78 (1953) 661.

In steels

U. T. HILL, *Anal. Chem.*, 19 (1947) 318.
G. LINDLEY, *Anal. Chim. Acta*, 25 (1961) 334.

In minerals and rocks

J. A. BRABSON, J. H. KARCHMER AND M. S. KATZ, *Ind. Eng. Chem., Anal. Ed.*, 16 (1944) 553.
H. BAADSGAARD AND E. B. SANDELL, *Anal. Chim. Acta*, 11 (1954) 183.

In fertilisers

G. L. BRIDGER, D. R. BOYLON AND J. W. MARKEY, *Anal. Chem.*, 25 (1953) 336.

Precision colorimetry (Estimation in natural phosphates)

A. GEE AND V. R. DEITZ, *Anal. Chem.*, 25 (1953) 1320.

Extraction of the phosphomolybdate complex of safranine (ε = 190,000 at 532 mμ in acetophenone)

L. DUCRET AND M. DROUILLAS, *Anal. Chim. Acta*, 21 (1959) 86.

(ii) Reduction of the phosphomolybdate complex. Reduction of the phosphomolybdate complex gives rise to a blue coloration. The excess molybdate added should, as far as possible, be left unreduced. The pH is an important factor. A suitable acidity is 0.75 to 1.25 *N* H_2SO_4. The blue compound formed by the reduction of the phosphomolybdate is particularly absorbent at 830 mμ (ε =

26,800). In contrast, the 'molybdenum blue' produced by the reduction of Mo(VI) is particularly absorbent at about 630 mμ. It therefore follows that a wavelength of about 830 mμ should be used. The coloration is stable for 12 hours. Reduction may be effected by a number of suitable compounds. The best results are obtained with Fe(II), although hydrazine, 1-amino-2-naphthol-4-sulphonic acid (in the presence of sulphite), ascorbic acid[3], etc. are also suitable.

The phosphomolybdate complex may also be reduced after extraction in an organic solvent (butanol). The blue compound remains dissolved in the solvent. $\varepsilon \sim 25.000$ at 830 mμ.

Interfering ions. Analogous colorations are given by Si(IV) and Ge(IV). In a sufficiently acid solution Si(IV) does not interfere[1]. The addition of citric acid also eliminates interference from Si(IV)[2]. As(V) gives an analogous reaction, but may be reduced with sulphurous acid or with thiourea. Ions such as Nb(V) and Ta(V), Sn(IV), W(VI), Ti(IV), Zr(IV), and Bi(III) precipitate and entrain the phosphorus. Ti(IV) and Zr(IV) catalyze the reduction of Mo(VI)[3].

Ba(II), Sr(II), and Pb(II) precipitate in sulphuric acid. Large amounts of Cu(II), Ni(II), and Cr(III) interfere because of their colour. In the presence of large amounts of Cr(III), colorimetry should be performed at 690 mμ. V(V) involves the formation of the phosphovanadomolybdate complex, which is then reduced. NO_3^- and nitrous acid also interfere, but when the latter is present, a little sulphamic acid may be added to eliminate the effect.

Small amounts of $C_2O_4^{2-}$, F^-, and Fe(III) do not interfere. For 1 ppm of P(V), 1000 ppm of the following do not interfere, to within 2%: Al(III), Cd(II), Cr(III), Cu(II), Co(II), Ca(II), Mn(II), Ni(II), Zn(II), Cl^-, citrate, Si(IV), V(V), F^-, and B(III)[1].

COMPARE

(1) M. ROCKSTEIN AND P. W. HERRON, *Anal. Chem.*, 23 (1951) 1500.
(2) M. ZIMMERMANN, *Angew. Chem.*, 62 (1950) 291.
(3) M. JEAN, *Anal. Chim. Acta*, 14 (1956) 172.

Review:
N. S. GING, *Anal. Chem.*, 28 (1956) 1330.

Reduction with hydrazine.

REAGENTS

(1) *$Na_2MoO_4 \cdot 2H_2O$, 2.5% in 10 N (1/4) sulphuric acid.*
(2) *Hydrazine sulphate, 0.15% in water.*

Immediately before the test, 25 ml of (1) and 10 ml of (2) should be mixed and made up to 100 ml.

OPERATING PROCEDURE

20 ml of reagent are added to about 25 ml of neutralized solution, the mixture is made up to 50 ml, placed on a boiling water bath for 10 minutes and then cooled. Colorimetry is performed at 830 mμ.

COMPARE

D. F. BOLTZ AND M. G. MELLON, *Anal. Chem.*, 19 (1947) 873.
M. MACHEBOEUF AND J. DELSAL, *Bull. Soc. Chim. Biol.*, 25 (1943) 116.

In steels

H. L. KATZ AND K. L. PROCTOR, *Anal. Chem.*, 19 (1947) 612.

In germanium

C. L. LUKE AND M. E. CAMPBELL, *Anal. Chem.*, 25 (1953) 1588.
B. L. GRISWOLD, F. L. HUMOLLER AND A. R. MCINTIRE, *Anal. Chem.*, 23 (1951) 192.

Reduction with hydroquinone.

REAGENTS

Sulphuric acid, 0.2 N: 5.5 ml of concentrated sulphuric acid are added to 1 litre of water.
Ammonium molybdate, 5%.
Hydroquinone, 2%.
Sulphite-bisulphite mixture: 15 g of sodium bisulphite and 20 g of sodium sulphite are dissolved in 100 ml of water.

OPERATING PROCEDURE

5 ml of 0.2 *N* H_2SO_4, 2 ml of molybdate, and 1 ml of hydroquinone are added to 10 ml of a solution containing 25 to 125 μg of phosphorus. After standing for 15 to 20 minutes, 1 ml of the sulphite-bisulphite mixture is added, and the solution is left for a further 10 to 15 minutes. The volume is then made up to 25 ml and colorimetry is performed at 720 mμ.

COMPARE

R. E. KITSON AND M. G. MELLON, *Ind. Eng. Chem., Anal. Ed.*, 16 (1944) 466.
N. S. GING AND J. M. STURTEVANT, *J. Am. Chem. Soc.*, 76 (1954) 2087.
N. S. GING, *Anal. Chem.*, 28 (1956) 1330.

With ascorbic acid

P. S. CHEN, JR., T. Y. TORIBARA AND H. WARNER, *Anal. Chem.*, 28 (1956) 1756.
D. N. FOGG AND N. T. WILKINSON, *Analyst*, 83 (1958) 406.

In steels

M. JEAN, *Chim. Anal.*, 38 (1956) 37.

(iii) Extraction methods. (1) The phosphomolybdate complex can be extracted with a mixture of 20% of butan-1-ol and chloroform. Colorimetry can then be carried out at 310 mμ. The method is sensitive ($\varepsilon = 25{,}000$), highly selective, and rapid. 25 μg of P(V) can be estimated in the presence of 4 mg of As(V), 5 mg of Si(IV), and 1 mg of Ge(IV).

C. WADELIN AND M. G. MELLON, *Anal. Chem.*, 25 (1953) 1668.

(2) The phosphomolybdate complex may be extracted from aqueous solutions with *iso*-butanol, and then be reduced with stannous chloride. The blue coloration which remains in the solvent is stable.

In plants

W. A. PONS, JR. AND J. D. GUTHRIE, *Ind. Eng. Chem., Anal. Ed.*, 18 (1946) 184.
F. L. SCHAFFER, J. FONG AND P. KIRK, *Anal. Chem.*, 25 (1953) 343.

In steels

M. JEAN, *Anal. Chim. Acta*, 14 (1956) 172.
C. H. LUECK AND D. F. BOLTZ, *Anal. Chem.*, 28 (1956) 1168.

In bauxites

L. ERDEY AND V. FLEPS, *Acta Chim. Hung.*, 11 (1957) 195.

Meta-, pyro-, and polyphosphoric acids

The most important are: pyrophosphoric acid $H_4P_2O_7$, metaphosphoric acids $(HPO_3)_n$, (in particular trimeta- and tetrametaphosphoric acids,) and triphosphoric acid $H_5P_3O_{10}$.

(1) Hydrolysis

These acids may be converted to the stable H_3PO_4 by hydrolysis: an estimation may then be performed.

In principle, the above transformation may in each case be effected by heating for 1 hour at 100° in *N* (1/10) HNO_3.

In addition, triphosphoric acid can be hydrolyzed to ortho- and pyrophosphoric acids by boiling for 1 hour in 10 to 20% caustic soda.

$$P_3O_{10}^{5-} + 2HO^- \rightarrow P_2O_7^{4-} + PO_4^{3-} + H_2O.$$

It follows that by these successive colorimetric estimations of orthophosphoric acid, values may be obtained for: H_3PO_4, $H_3PO_4 + H_5P_3O_{10}$ (after alkaline hydrolysis) and the sum of the three acids (after acid hydrolysis).

L. E. NETHERTON, A. R. WREATH AND D. N. BERNHART, *Anal. Chem.*, 27 (1955) 860.

(2) Separations by chromatography

Among the methods proposed, separations of the acids by paper chromatography[1] and with ion-exchangers[2] are particularly important.

COMPARE

(1) J. P. EBEL, *Compt. Rend. (Paris)*, 234 (1952) 621.
J. P. EBEL AND Y. VOLMAR, *Compt. Rend. (Paris)* 233 (1951) 415.
J. P. EBEL, *Bull. Soc. Chim.*, (1953) 1089.
E. THILO AND H. GRUNZE, *Z. Anorg. Chem.*, 281 (1955) 262.
E. KARL-KROUPA, *Anal. Chem.*, 28 (1956) 1091.
H. HETTLER, *J. Chromatog.*, 1 (1958) 389.
O. PFRENGLE, *Z. Anal. Chem.*, 158 (1957) 81.
G. G. BERG, *Anal. Chem.*, 30 (1958) 213.
(2) S. LINDENBAUM, T. V. PETERS, JR., AND W. RIEMAN III, *Anal. Chim. Acta*, 11 (1954) 530.
J. A. GRANDE AND J. BEUKENKAMP, *Anal. Chem.*, 28 (1956) 1497.
T. V. PETERS, JR., AND W. REIMAN III, *Anal. Chim. Acta*, 10 (1956) 131.

Physicochemical methods of estimation

R. PARIS AND J. ROBERT, *Compt. Rend. (Paris)*, 223 (1946) 1135.

(3) Colorimetry

Many of the complexes formed by various ions, for example Fe(III), can be used. Thus, the red ferric thiocyanate complex is partially decolourized by the addition of metaphosphate or pyrophosphate. Any other ions which complex SCN^- or Fe^{3+} will interfere. The pH exerts an effect.

cf. H. E. WIRTH, *Ind. Eng. Chem., Anal. Ed.*, 14 (1942) 722.
H. W. THOENES AND G. KOSFELD, *Z. Anal. Chem.*, 159 (1953) 9.
R. H. KOLLOFF, H. K. WARD AND V. F. ZIEMBA, *Anal. Chem.*, 32 (1960) 1687.
J. MAURICE, *Bull. Soc. Chim.*, (1959) 819.

With o-phenanthroline + hydroxylamine + Fe(III)

W. B. CHESS AND D. N. BERNHART, *Anal. Chem.*, 30 (1958) 111.

Triphosphate of $Co(en)_3^{3+}$

H. J. WEISER, JR., *Anal. Chem.*, 28 (1956) 477.

Platinum

Pt = 195.2

(1) Separation of trace amounts

(i) Fire assay. This is still the one most frequently used, may be applied to platiniferous minerals and is capable of dealing with large samples.

cf. E. E. BUGBEE, *Textbook of Fire Assaying*, John Wiley and Sons, New York, 1940.
D. C. SHEPARD AND W. F. DIETRICH, *Fire Assaying*, McGraw Hill, New York, 1940.
E. A. SMITH, *Sampling and Assaying of the Precious Metals*, Griffin, London, 1947.
L. HOFFMAN AND F. E. BEAMISH, *Anal. Chem.*, 28 (1956) 1188.

(ii) Precipitation of traces by reduction with Sn(II) or SO_2. Te, which can be used as an entraining agent, is obtained by the reduction of tellurites.

Au, Ag, Pd and Rh accompany the platinum.

cf. S. K. HAGEN, *Mikrochemie*, 15 (1936) 180.
E. B. SANDELL, *Colorimetric Determination of Traces of Metals*, Interscience, 1959.

(iii) Ion-exchangers. In a chloride solution Pt(IV) is not collected by a cation-exchanger.

cf. H. G. COBURN, F. E. BEAMISH AND C. L. LEWIS, *Anal. Chem.*, 28 (1956) 1297.

(iv) Separation of the platinum metals. See p. 348.

(v) Standard solutions. These can be made from guaranteed purity hexachloroplatinic (IV) acid, $H_2PtCl_6 \cdot 6H_2O$ (99.5–100%)

$$Pt/H_2PtCl_6 \cdot 6H_2O = 0.377.$$

(2) Colorimetry

The colorimetric methods available for the estimation of platinum are not very selective, since the other platinum metals, particularly palladium, interfere. The method involving Sn(II) is rapid and quite selective.

(i) With stannous chloride. In 1.5 to 2.5 N hydrochloric acid, a specific,yellow to brown-black complex is obtained. The coloration develops immediately, remains stable, and can be extracted with ethers and esters.

Sensitivity. $\varepsilon \sim 10{,}000$ at 405 mμ.

Interfering ions. Pd(II) gives the same reaction, but its coloration can be removed by making the solution ammoniacal and subsequently re-acidifying. Ru gives the same coloration, but the reaction is ten times less sensitive and the colour is not extractable; Os is not extracted either. Rh and Ir give colorations of very low sensitivity which are however extractable. Au precipitates. Fe or Cu do not interfere in amounts lower than 200 ppm.

REAGENTS

Stannous chloride, crystalline, 10% in 2*N* (1/6) hydrochloric acid.

Standard solution of platinum, 0.01%.

OPERATING PROCEDURE

The sample should contain 25 to 250 μg of platinum. Hydrochloric acid is added until the final solution is 2 *N*. The volume is made up to 30 to 40 ml, 5 ml of stannous chloride are added, and the volume is adjusted to 50 ml. The colorimetric determination is performed at 406 mμ.

cf. G. H. AYRES AND A. S. MEYER, JR., *Anal. Chem.*, 23 (1951) 299.
A. S. MEYER, JR., AND G. H. AYRES, *J. Am. Chem. Soc.*, 77 (1955) 2671.
O. I. MILNER AND G. F. SHIPMAN, *Anal. Chem.*, 27 (1955) 1476.
I. MAZIEKIEN, L. ERMANIS AND T. J. WALSH, *Anal. Chem.*, 32 (1960) 645.

(ii) With iodide. Brownish pink PtI_6^{2-} is formed. The medium should not be too acidic.

Limit of sensitivity. About 0.5 ppm.

Interfering ions. The following give colorations: Pd(II), Au(III), Fe(III), Cu(II), Bi(III), Sb(III), etc. Rh gives a coloration which appears very slowly. SO_3^{2-}, $S_2O_3^{2-}$, and Hg(II) prevent the appearance of the coloration due to platinum, and NO_3^-, $CH_3CO_2^-$, and SO_4^{2-} delay it. Oxidizing and reducing agents interfere.

OPERATING PROCEDURE

The sample should contain 25–300 μg of platinum. 1 ml of *N* (1/12) hydrochloric acid and 1 ml of 5% potassium iodide are added, and the volume is made up to 50 ml. Colorimetry is performed after allowing the solution to stand for 1 hour in the dark. The colour is stable.

cf. E. B. SANDELL. *Colorimetric Determination of Traces of Metals,* Interscience Publ., New York, 1959.

(iii) Other methods. Using p-nitrosodimethylaniline. The method is not very selective but is very sensitive. $\varepsilon \sim 100{,}000$ at 525 mμ.

Review

F. E. BEAMISH AND W. A. E. MCBRYDE, *Anal. Chim. Acta*, 18 (1958) 551.

$PtCl_6^{2-}$ at 262 mμ. Very unselective.

J. J. KIRKLAND AND J. H. YOE, *Anal. Chim. Acta*, 9 (1953) 441.

Estimation of Au(III), Pd(II), Pt(IV) by means of dithizone. Os, Ir, Rh, and Ru do not interfere in a dilute acid medium.

R. S. YOUNG, *Analyst*, 76 (1951) 49.

Metals of the Platinum Group

(1) Methods of Separation

(i) Nitrous complexes. The platinum metals form complexes with nitrous acid, and other elements whose hydroxides are sparingly soluble may thus be precipitated at pH<10.

cf. R. GILCHRIST, *J. Res. Natl. Bur. Std.*, 20 (1938) 745 and 30 (1943) 89; *Chem. Rev.*, 32 (1943) 277.

(ii) By means of ion-exchangers (Using an anion-exchanger in HCl)

S. S. BERMAN AND W. A. E. MCBRYDE, *Can. J. Chem.*, 36 (1958) 835.

Rh and Ir are separated from Cu(II), Fe(III), and Ni(II) on cation-exchangers at pH 1.5. Rh and Ir are not collected.

A. G. MARKS AND F. E. BEAMISH, *Anal. Chem.*, 30 (1958) 1464.

Ir/Rh

M. L. CLUETT, S. S. BERMAN AND W. A. E. MCBRYDE, *Analyst*, 80 (1955) 204.

Separation of Rh from Pt, Pd and Ir. Rh(III) is collected by a cation-exchanger from a hydrochloric acid solution.

W. M. MACNEVIN AND E. S. MACKAY, *Anal. Chem.*, 29 (1957) 1220.

Separation of Pd from Pt and Ir. Ammine complexes of Pd(II) are collected by a cation-exchanger.

W. M. MACNEVIN AND W. B. CRUMMETT, *Anal. Chem.*, 25 (1953) 1628.

(iii) Fire assay. (See Platinum, p. 346)

Rh W. F. ALLEN AND F. E. BEAMISH, *Anal. Chem.*, 22 (1950) 451.
Ir R. R. BAREFOOT AND F. E. BEAMISH, *Anal. Chem.*, 24 (1952) 840.
Os W. J. ALLAN AND F. E. BEAMISH, *Anal. Chem.*, 24 (1952) 1569.

Review

F. E. Beamish, *Talanta*, 5 (1960) 1.

(iv) Separation of the platinum metals from one another. Osmium and ruthenium are usually separated by volatilization of the tetroxides[1]. The residual solution is adjusted to about pH 8, when the hydroxides of Pd(II), Rh(III), and Ir(III) should precipitate; a double precipitation is required if they are to be separated from platinum completely[2]. Fe(III) can if necessary be used as an entraining agent.

After redissolution, Pd can be precipitated by means of dimethylglyoxime. Rh is then precipitated by reduction with Ti(III); double precipitation is necessary. Ir remains in solution.

COMPARE

R. Gilchrist, *Applied Inorganic Analysis*, Eds Hillebrand, Lundell, Bright and Hoffman, Wiley, New York, 1953.
R. Thiers, W. Graydon and F. E. Beamish, *Anal. Chem.*, 20 (1948) 831.
W. F. Allen and F. E. Beamish, *Anal. Chem.*, 22 (1950) 451.
A. Hill and F. E. Beamish, *Anal. Chem.*, 22 (1950) 590.
R. Gilchrist, *Anal. Chem.*, 25 (1953) 1617.
(1) A. D. Westland and F. E. Beamish, *Anal. Chem.*, 26 (1954) 739.
J. M. Kavanagh and F. E. Beamish, *Anal. Chem.*, 32 (1960) 490.
C. V. Banks and J. W. O'Laughlin, *Anal. Chem.*, 29 (1957) 1412.
W. J. Allan and F. E. Beamish, *Anal. Chem.*, 24 (1952) 1608.
(2) R. Gilchrist, *Anal. Chem.*, 25 (1953) 1617.

(v) Other methods of separation

Rh R. Gilchrist and E. Wichers, *J. Am. Chem. Soc.*, 57 (1935) 2565.

From Ir by electrolysis

W. M. McNevin and S. M. Tuthill, *Anal. Chem.*, 21 (1949) 1052.

Extraction with 1-nitro-2-naphthol

K. Watanabe, *J. Chem. Soc. Japan*, 77 (1956) 1008.

Ru – Extraction of RuO_4

F. S. Martin, *J. Chem. Soc.*, (1954) 2564.

Os – Extraction of OsO_4

R. D. Sauerbrunn and E. B. Sandell, *Anal. Chim. Acta*, 9 (1953) 86.

Extraction of $OsCl_6^{2-}$ by means of tetraphenylarsonium. Osmium may be separated from Ru and Rh.

R. Neeb, *Z. Anal. Chem.*, 154 (1957) 23.

Review of methods

F. E. Beamish, *Talanta*, 5 (1960) 1 to 35.

Potassium

K = 39.10

(1) Separation of trace amounts

Colorimetric estimations of potassium generally require preliminary separations, which are often rather unsatisfactory. In general, it is preferable to estimate the potassium by flame spectrophotometry.

(i) Separation by means of ion-exchangers. Alkali-metal cations may in this way be separated easily from anions and molecules such as H_3PO_4, H_3BO_3, etc.

K(I) may be separated from Ca(II), Mg(II), V(V), Fe(III), Al(III), Cu(II), Ni(II), Co(II), Mn(II), and Zn(II) in an EDTA solution, and with the aid of an anionic resin[1].

(1) O. Samuelson and E. Sjöström, *Anal. Chem.*, 26 (1954) 1908.
O. Samuelson and S. Forsblom, *Z. Anal. Chem.*, 144 (1955) 323.

(ii) Electrolysis at a mercury cathode. Many elements may be removed by deposition at the cathode; the voltage should be controlled.

cf. T. D. Parks, H. O. Johnson and L. Lykken, *Anal. Chem.*, 20 (1948) 148.

(iii) Standard solutions. These can be made from a standardized solution of guaranteed purity potassium hydroxide.

Guaranteed purity KCl, KNO_3 and K_2SO_4 are also available.

1 ml of 1.000 *N* KOH = 39.1 mg of K or 47.1 mg of K_2O.

K/KCl = 0.525	$K_2O/2KCl$ = 0.632
K/KNO_3 = 0.387	$K_2O/2KNO_3$ = 0.466
$2K/K_2SO_4$ = 0.2243	K_2O/K_2SO_4 = 0.270

(2) Colorimetry

The various colorimetric methods of estimating potassium involve the previous precipitation of a sparingly soluble compound: chloroplatinate (VI), cobaltinitrite, dipicrylaminate. This compound is then redissolved in order to carry out the colorimetric determination. Such methods are therefore unsuitable for use with minute traces. Being the least soluble, cobaltinitrite is the compound most frequently used, although its composition is ill-defined and the method is not accurate.

The precipitation of potassium should always be made as complete as possible; the solution used should therefore be concentrated and may, for this reason, be reduced to a volume as low as 1 ml. The precipitate obtained is separated and washed by centrifuging.

Microvolumetric determination of cobaltinitrite and in particular flame spectrophotometry are preferable to colorimetry.

(i) Potassium chloroplatinate. The sample used should contain 0.1 to 10 mg of K^+. The precipitate obtained is redissolved, and colorimetry is then performed, either directly using the orange-yellow chloroplatinate (IV) complex, or with respect to the platinum content of the solution.

A number of ions interfere, particularly NH_4^+.

cf. M. F. ADAMS AND J. L. ST JOHN, *Ind. Eng. Chem., Anal. Ed.*, 17 (1945) 435.

In biochemistry

R. M. TENERY AND C. E. ANDERSON, *J. Biol. Chem.*, 135 (1940) 659.
P. W. SALIT, *J. Biol. Chem.*, 136 (1940) 191.
R. E. ECKEL, *J. Biol. Chem.*, 195 (1952) 191.

(ii) Cobaltinitrite. The potassium is precipitated in the form of potassium sodium cobaltinitrite, $K_2Na[Co(NO_2)_6]$. The composition of this precipitate varies however with concentration of the various ions present, particularly K^+ and Na^+, with pH, with temperature, and also with the operating procedure. The complex may be redissolved and colorimetry of the cobalt or the nitrite can then be carried out. A number of ions interfere, particularly NH_4^+.

Limit of sensitivity. 3 ppm. Under normal conditions, 50 to 500 μg of K^+ may be estimated.

The precipitation of the very sparingly soluble potassium silver cobaltinitrite is also sometimes used[1].

cf. J. F. REED, A. MEHLICH AND J. R. PILAND, *Proc. Soil Soc. Am. Proc.*, 9 (1944) 56.
F. A. UHL, *Z. Anal. Chem.*, 123 (1942) 322.
(1) E. M. CHENERY, *Analyst*, 77 (1952) 102.

Lithium cobaltinitrite reagent

T. DUPUIS, *Compt. Rend. (Paris)*, 237 (1953) 256.

Turbidimetry

L. BURKHART, *Plant. Physiol.*, 16 (1941) 411.

(3) Other methods

Absorptiometry in the ultraviolet. A solution of the tetraphenylborate in a mixture of acetonitrile and water is used; colorimetry is performed at 266 or 274 mμ.

R. T. Pflaum and L. C. Howich, *Anal. Chem.*, 28 (1956) 1542.

Rare earth metals

(1) Separation of trace amounts

(i) Precipitation of the fluorides. The fluorides of the rare earth metals can be precipitated from acid solutions with an entraining agent consisting of either one of the rare earth metals in the pure state, e.g. Y(III), or Ce(III)[1] or of Ca(II) or Sr(II).

(1) B. HELGER AND R. RYNNINGER, *Svensk Kemisk. Tid.*, 64 (1952) 224.

(ii) Extraction of the oxinates. The oxinates of the rare earth metals are soluble in chloroform. Thus Ce(III) oxinate can be extracted from a citrate medium at pH 9.9–10.5.

cf. W. WESTWOOD AND A. MAYER, *Analyst*, 73 (1948) 275.

(iii) With ion-exchangers. Separation from Fe(III), Al(III), etc. can be achieved[1]. Separation from Zr(IV) is carried out in an F^- medium[2].

(1) E. SCHUMACHER, *Helv. Chim. Acta*, 39 (1956) 531.
(2) H. J. HETTEL AND V. A. FASSEL, *Anal. Chem.*, 27 (1955) 1311.

(iv) Electrolysis at a mercury cathode. The rare earth metals remain in solution.

(v) Separation from cerium. See Cerium.

(vi) Separation of the rare earth metals from one another. This presents a particularly difficult problem, and is currently performed by means of ion-exchangers.

cf. D. H. HARRIS AND E. R. TOMPKINS, *J. Am. Chem. Soc.*, 69 (1947) 2792.
L. HOLLECK AND L. HARTINGER, *Angew. Chem.*, 68 (1956) 411.

Chromatography on a cellulose column
F. H. POLLARD, J. F. W. MAC OMIE AND H. M. STEVENS, *J. Chem. Soc.*, (1952) 4730.

Chromatography on paper
M. LEDERER, *Anal. Chim. Acta*, 15 (1956) 122.

(2) Colorimetry

No generally satisfactory colour reactions exist, since the reagents which have been proposed often give rise to adsorption compounds and are not selective. Such reagents are alizarin S[1], aluminon[3], and naphthazarine[4].

Alternatively, colorimetric estimations of the oxinates[2] or the compound formed with 'arsenazo' [3-(2-arsonophenylazo)-4,5-dihydroxynaphthalene-2,7-disulphonic acid] may be performed.

Y(III) may also be estimated by means of pyrocatechol violet[5].

(1) R. W. Rinehart, *Anal. Chem.*, 26 (1954) 1820.
(2) K. S. Bergstresser, *Anal. Chem.*, 30 (1958) 1630.
(3) L. Holleck, D. Eckardt and L. Hartinger, *Z. Anal. Chem.*, 146 (1955) 103.
(4) T. Moeller and M. Tecotzky, *J. Am. Chem. Soc.*, 77 (1955) 2649.
(5) J. P. Young, J. C. White and R. G. Ball, *Anal. Chem.*, 32 (1960) 928.

With 'arsenazo'. This is the most accurate method. The pH should be fixed at 8.0. *Sensitivity*: $\varepsilon = 29{,}000$ at 570 mμ.

Interfering ions. Interference is obtained with many elements such as U(VI), Th(IV), Ti(IV), Zr(IV), Fe(III), Cu(II), Ca(II), Mg(II), etc. Small amounts of aluminium can be complexed with sulphosalicylic acid. Iron can be separated in the form of its thiocyanate complexes, by collection on anion-exchangers.

When working at pH 5, the reaction is less sensitive but small amounts of Ca(II) and Mg(II) do not interfere.

Reagents

'Arsenazo' [sodium 3-(2-arsonophenylazo)-4,5-dihydroxynaphthalene-2,7-disulphonate: 150 mg of reagent are dissolved in 100 ml of water.

Buffer solution: equal portions of 1 *M* triethylamine and 0.5 *M* nitric acid are mixed together.

Operating procedure

25 ml of solution containing 10 to 100 μg of the rare earths are mixed with 2 ml of arsenazo followed by 5 ml of triethylamine. The volume is made up to about 40 ml, and the pH adjusted to 8.0. The volume is then made up to 50 ml and the colorimetric estimation is performed at 570 mμ.

cf. V. I. Kuznetsov, *Zh. Analit. Khim.*, 7 (1952) 226.
J. S. Fritz, M. J. Richard and W. J. Lane, *Anal. Chem.*, 30 (1958) 1776.
C. V. Banks, J. A. Thompson and J. W. O'Laughlin, *Anal. Chem.*, 30 (1958) 1792.

(3) Spectrophotometry

Most rare earth metals give characteristic absorption spectra consisting of narrow bands in the visible, ultraviolet, and near infrared regions. In this way,

Pr(III), Nd(III), Sm(III), Eu(III), Tu(III), and Yb(III) can be estimated to within 1% in mixtures.

The sensitivity of the method is low: $\varepsilon \sim 2$ to 10 (chlorides).

cf. W. PRANTL AND K. SCHEINER, *Z. Anorg. Chem.*, 220 (1934) 107.
Spectra of the Rare Earths, Office of Technical Services, U.S. Dept. of Commerce, Washington, 1961.
M. M. WOYSKI AND R. E. HARRIS, in *Treatise on Analytical Chemistry*, eds. I. M. KOLTHOFF AND P. J. ELVING, Interscience, III, 8, 1963.
L. HOLLECK AND L. HARTINGER, *Angew. Chem.*, 67 (1955) 648.
C. V. BANKS AND D. W. KLINGMAN, *Anal. Chim. Acta*, 15 (1956) 356.
LINDSAY, *Bulletin R-4*, Chemical Division, American Potash and Chemical Corporation, Chicago; *Anal. Chem.*, 32 (1960) 19A.

11 elements in the state of chlorides
D. C. STEWART AND D. KATO, *Anal. Chem.*, 30 (1958) 164.

Samarium
R. RASIN-STREDEN, W. DAUSCHAN AND O. ZEMEK, *Mikrochim. Acta*, (1956) 512.

Terbium
E. I. ONSTOTT AND C. J. BROWN, *Anal. Chem.*, 30 (1958) 172.

Dy(III), Ho(III), and Er(III)
C. K. JORGENSEN, *Acta. Chem. Scand.*, 11 (1957) 981.

Differential spectrophotometry – yttrium
C. V. BANKS, J. L. SPOONER AND J. W. O'LAUGHLIN, *Anal. Chem.*, 28 (1956) 1894.

Nd/Er – Pr/Nd/Sm
C. V. BANKS, J. L. SPOONER AND J. W. O'LAUGHLIN, *Anal. Chem.*, 30 (1958) 458.

(4) Fluorescence and Spectrofluorometry

(4a) In fused salts. Ce, Eu, Sm, Gd, Dy, Tm, Ho, (U).

cf. H. HABERLANDT, *Mikrochim. Acta*, 36/37 (1951) 1075.

Samarium
C. G. PEATTIE AND L. B. ROGERS, *Spectrochim. Acta*, 7 (1956) 321.

(4b) In aqueous solution. Limit of sensitivity. 0.01 ppm for Tb(III), 0.1 ppm for Ce(III), 1 ppm for Eu(III).

Tb V. A. FASSEL AND R. H. HEIDEL, *Anal. Chem.*, 26 (1954) 1134.

Rhenium

Re = 186.3

(1) Method of separation

(i) Volatilization of Re_2O_7. $HReO_4$ and Re_2O_7 are lost, particularly when the solutions are evaporated to render silica insoluble. This property can be utilized for the separation of Re if arrangements are made to recover the distillate.

W. Geilmann and H. Bode, *Z. Anal. Chem.*, 130 (1950) 323.

(ii) The dry method. After digestion with fused carbonate or caustic soda, ReO_4^- passes into the aqueous solution.

(iii) Extraction of the cupferrates. Re(VII) is not extracted.

R. J. Meyer and C. L. Rulfs, *Anal. Chem.*, 27 (1955) 1387.

(iv) Extraction of the oxinates. Re(VII) remains in the aqueous phase, and may thus be separated from Mo(VI).

W. Geilmann and H. Bode, *Z. Anal. Chem.*, 133 (1951) 177.

(v) Extraction of perrhenic acid. The solvent is butyl alcohol or *iso*-amyl alcohol.

S. Tribalat, *Ann. Chim.*, 8 (1953) 642.

(vi) Extraction of tetraphenylarsonium perrhenate. In the example quoted, the rhenium is considered to be present as an impurity in a molybdenite.

Reagent

2% solution of tetraphenylarsonium chloride.

Operating procedure

1 g of the molybdenite is weighed out and digested with about 10 millilitres of fuming nitric acid in a 200 ml conical flask on a water bath. After digestion, the solution is transferred to a capsule and, while still on the water bath, the nitric acid is expelled with hydrochloric acid, without ever evaporating to dryness. The solution is reduced to 2 ml, neutralized approximately with 10 *N* caustic soda, adjusted to pH 8–9 with sodium bicarbonate and made up to 20 ml in a graduated tube.

After shaking, a 5 ml portion is placed in a separating funnel. 0.5 ml of tetraphenylarsonium chloride solution is added, followed by 10 to 15 ml of chloroform. The mixture is shaken for several minutes and separated. The chloroform phase is then shaken with anhydrous calcium chloride and filtered through a filter moistened with chloroform. This operation is repeated in order to free the solvent from any remaining traces of aqueous molybdate solution. The filter is rinsed lightly with a jet of chloroform.

cf. S. TRIBALAT, *Anal. Chim. Acta*, 3 (1949) 113.
J. M. BEESTON AND J. R. LEWIS, *Anal. Chem.*, 25 (1953) 651.

In molybdenite
S. TRIBALAT, I. PAMM AND M. L. JUNGFLEISCH, *Anal. Chim. Acta*, 6 (1952) 142.

Extraction of Mo(VI) and W(VI)
I. R. ANDREW AND C. H. R. GENTRY, *Analyst*, 82 (1957) 372.

(vi) With ion-exchangers. ReO_4^- may easily be separated from cations and molecules.

Using an anionic resin in an alkaline medium such as 0.05 *N* caustic soda, Re(VII) is collected in favour of Mo(VI). The ReO_4^- is subsequently eluted with perchloric acid.

Separation from Mo(VI)
V. W. MELOCHE AND A. F. PREUSS, *Anal. Chem.*, 26 (1954) 1911.
M. PIRS AND R. J. MAGEE, *Talanta*, 8 (1961) 395.

(vii) Standard solutions. These can be made from the pure metal. The latter is oxidized with a current of chlorine in the presence of several millilitres of water. The resulting solution of perrhenic acid is then gently boiled and adjusted to a known volume.

Alternatively, about 100 mg of potassium perrhenate may be dissolved in 50 ml of 6 *N* (1/6) sulphuric acid, and the volume made up to 1 litre.

$$Re/KReO_4 = 0.644$$

(2) Colorimetry

(i) With thiocyanate (after S. TRIBALAT). The colorimetric estimation is analogous to that of Mo(VI). Under reducing conditions rhenium and SCN^- give rise to a yellow complex which may be extracted with ethers and esters.

Sensitivity. $\varepsilon = 35{,}000$ at 430 mμ.

Interfering ions. W(VI), Mo(VI), Cu, Se, Te, Au, Pt, and V all interfere, as also do large amounts of Fe(III).

REAGENTS

Ammonium thiocyanate, 20% solution.

$SnCl_2$, aqueous, 20% solution in concentrated hydrochloric acid.

Standard solution of rhenium.

OPERATING PROCEDURE

(1) Re(VII) is dissolved in concentrated HCl (1–2 ml).

(2) If the separation has been performed by means of tetraphenylarsonium, the chloroform is evaporated on a water bath to a volume 0.5 to 1 ml and poured into a test-tube, in which it is shaken with 1 to 2 ml of concentrated hydrochloric acid.

0.2 ml of ammonium thiocyanate solution and 0.2 ml of stannous chloride solution are then added, shaking the mixture after each addition. The coloration due to the rhenium is extracted with 1 ml of *iso*-amyl alcohol, which forms a homogeneous phase with the chloroform present.

The coloration is compared with that of standards obtained from known amounts of rhenium similarly treated with thiocyanate and stannous chloride in concentrated hydrochloric acid and extracted with the same volume of chloroform-*iso*-amylol.

If no coloration is observed at the end of the treatment of the 5 ml sample, it may be deduced that the latter contains less than 1 μg of rhenium, which corresponds to a rhenium content in the mineral of less than 4×10^{-6}. It is then advisable to repeat the procedure with the remaining 15 ml, which is extracted with 10 to 15 ml of chloroform in the presence of an amount of reagent which may be three times greater than that used previously.

The limit of sensitivity corresponds to 1 to 2×10^{-6} of rhenium in the mineral. The extraction of rhenium is always quantitative and this limit is therefore determined only by the sensitivity of the reaction involving the thiocyanate. The latter does not allow less than 1 to 0.5 ppm of rhenium to be detected with certainty.

With special precautions, it is possible to detect 2×10^{-7}.

cf. W. GEILMANN AND H. BODE, *Z. Anal. Chem.*, 128 (1948) 489.
V. PATROVSKY, *Chem. Listy*, 51 (1957) 1295.
S. TRIBALAT, *Anal. Chim. Acta*, 3 (1949) 113.

(3) Other methods

Absorptiometry of $ReCl_6^{2-}$ in the ultraviolet. $\varepsilon = 12{,}000$ at 281.5 mμ.

R. J. MEYER AND C. L. RULFS, *Anal. Chem.*, 27 (1955) 1387.
V. W. MELOCHE AND R. L. MARTIN, *Anal. Chem.*, 28 (1956) 1671.

Absorptiometry of tetraphenylarsonium perrhenate. The chloroform solution possesses an absorption band between 245 and 265 mμ. It is therefore possible to carry out the estimation immediately after the separation by extraction.

The method is not very sensitive. $\varepsilon \sim 4{,}000$. 20 mg of Mo(VI) and W(VI) interfere.

T. R. ANDREW AND C. H. R. GENTRY, *Analyst*, 82 (1957) 372.

With 2,4-diphenylthiosemicarbazide. The medium is 5–7 *M* HCl; a red compound which can be extracted with chloroform is produced. Mo(IV) interferes.

The sensitivity is the same as that of the thiocyanate method.

W. GEILMANN AND R. NEEB, *Z. Anal. Chem.*, 151 (1956) 401.

General reference

S. TRIBALAT, *Rhénium et Technétium*, Gauthier-Villars, 1957.

Rhodium

Rh = 102.9

(1) Separation

See the Platinum Metals, p. 348 .

(2) Colorimetry

(i) With stannous chloride. The method is sensitive, but the other platinum metals interfere.

(ii) In the presence of Pt at two wavelengths.

G. H. AYRES, B. L. TUFFLY AND I. S. FORRESTER, *Anal. Chem.*, 27 (1955) 1742.
F. PANTANI AND G. PICCARDI, *Anal. Chim. Acta*, 22 (1960) 231.

(iii) In the presence of Ir.

A. D. MAYNES AND W. A. E. McBRYDE, *Analyst*, 79 (1954) 230.

(iv) In U-Rh.

R. D. GARDNER AND A. D. HUES, *Anal. Chem.*, 31 (1959) 1488.

(v) With 2-mercaptobenzoxazole.

D. E. RYAN, *Anal. Chem.*, 22 (1950) 599.

(vi) With 2-mercapto-4,5-dimethylthiazole.

D. E. RYAN, *Analyst*, 75 (1950) 557.

Rubidium
Rb = 85.5

Cesium
Cs = 132.9

These elements accompany potassium owing to their mutual chemical similarity. The methods of extraction and estimation are therefore those described on p. 350 for K^+.

The separation of K^+, Rb^+, and Cs^+ from one another presents a difficult problem. These elements are usually estimated by flame spectrophotometry.

Separation

(i) Solubility differences of the chlorides. The difference in the respective solubilities of the chlorides in alcohol saturated with gaseous hydrogen chloride can be utilized, potassium chloride being the least soluble.

cf. R. C. Wells and R. E. Stevens, *Ind. Eng. Chem., Anal. Ed.*, 6 (1934) 439.
J. C. Hillyer, *Ind. Eng. Chem., Anal. Ed.*, 9 (1937) 236.
D. Meier and W. D. Treadwell, *Helv. Chim. Acta*, 34 (1951) 805.

Review of methods

W. J. O'Leary and J. Papish, *Ind. Eng. Chem., Anal. Ed.*, 6 (1934) 107.

(ii) Extraction of the tetraphenylborates with nitrobenzene

R. C. Fix and J. W. Irvine, Jr., *Mass. Inst. Techn., Lab. Nuclear Science, Ann. Progr Rept.*, May 1956.

(iii) Chromatography

G. Kayas, *J. Chim. phys.*, 47 (1950) 408.
M. J. Cabell and A. A. Smales, *Analyst*, 82 (1957) 390.

On paper and on cellulose

J. Fouarge and G. Duyckaerts, *Anal. Chim. Acta*, 14 (1956) 527.

Cs – Separation from the alkali metals by ion-exchangers

S. A. Ring, *Anal. Chem.*, 28 (1956) 1200.

Cs – Review of colorimetric methods

C. Duval and M. Doan, *Mikrochim. Acta*, (1953) 200.

Ruthenium

Ru = 101.7

(1) Separation

See the Platinum Metals, p. 348.

Extraction of RuO_4.

J. H. Forsythe, R. J. Magee and C. L. Wilson, *Talanta*, 3 (1960) 324.

(2) Colorimetry

(i) Ruthenate

E. D. Marshall and R. Richard, *Anal. Chem.*, 22 (1950) 795.

(ii) Perruthenate

G. A. Stoner, *Anal. Chem.*, 27 (1955) 1186.
R. P. Larsen and L. E. Ross, *Anal. Chem.*, 31 (1959) 176.

(iii) Thiourea. This is one of the most widely used methods: $\varepsilon \sim 4{,}000$ at 650 mμ. Os and Pd interfere.

G. H. Ayres and H. F. Young, *Anal. Chem.*, 22 (1950) 1277.
R. P. Yaffe and A. F. Voigt, *J. Am. Chem. Soc.*, 74 (1952) 2503.

Scandium

Sc = 45.0

(1) Separation

Most methods of separation described for the rare earth elements are also suitable for scandium (*cf.* Rare Earth Metals).

(i) Precipitation of the hydroxide. Sc is precipitated in an acetate buffer solution at pH 5.4 as $Sc(OH)_2Cl$, and is thus separated from the other rare earths[1].

The hydroxide does not precipitate quantitatively in ammoniacal solution.

(1) I. P. ALIMARIN AND T. YUNG-SCHAING, *Talanta*, 8 (1961) 317.

(ii) Precipitation of scandium ammonium tartrate. The solubility of scandium ammonium tartrate is fairly low (6.7 mg of Sc_2O_3 per litre). Scandium is separated in this way from Al(III), Fe(III), Zr(IV) and Th(IV).

W. FISCHER AND R. BOCK, *Z. Anorg. u. Allgem. Chem.*, 249 (1942) 146.
W. FISCHER, O. STEINHAUSER, E. HOHMANN, E. BOCK AND P. BORCHERS, *Z. Anal. Chem.*, 133 (1951) 57.

(iii) Precipitation by mandelic acid. Scandium is separated from Th(IV) and from the rare earths at pH 2–3.

I. P. ALIMARIN AND S. HAN-SI, *Talanta*, 9 (1962) 1.

(iv) Extraction of the thiocyanate. The complex thiocyanate may be extracted with ether. This gives separation from the rare earths, Th(IV), Zr(IV), Mn(II), Mg(II) and Ca(II).

W. FISCHER, O. STEINHAUSER, E. HOHMANN, E. BOCK AND P. BORCHERS, *Z. Anal. Chem.*, 133 (1951) 57.

(v) Extraction with N-benzoylphenylhydroxylamine. The separation with cupferron is poor, but is better with N-benzoylphenylhydroxylamine. Zr(IV) and Ti(IV) are extracted by *iso*-amyl alcohol in 5–8 *N* HCl.

I. P. ALIMARIN AND T. YUNG-SCHAING, *Talanta*, 8 (1961) 317.

The following methods of separation by extraction may be used, but are less efficient.

(vi) Extraction of the chloride. Scandium may be separated from the rare earths by extraction with butyl phosphate in HCl solution. It is separated in this way

from Al(III) and Be(II), and, in presence of H_2O_2, from Ce(IV) and Ti(IV). Zr(IV) interferes with the separation. U(IV) and Th(IV) are extracted.

A. R. Eberle and M. W. Lerner, *Anal. Chem.*, 27 (1955) 1551.

(vii) Extraction of the nitrate.

R. Bock and E. Bock, *Z. Anorg. Allgem. Chem.*, 263 (1950) 146.

(viii) Extraction of the oxinate. The oxinate is extracted with benzene at pH 9.7–10.5. Scandium can then be determined colorimetrically at 378 mμ. $\varepsilon \sim 6{,}900$.

F. Umland and H. Puchelt, *Anal. Chim. Acta*, 16 (1957) 334.

(ix) By means of ion-exchangers. (a) Cation exchangers. Sc(III) is separated from the rare earths by elution with citric acid or with EDTA. Sc(III) is eluted first.

cf. P. Radhakrishna, *Anal. Chim. Acta*, 8 (1953) 140.
V. K. Iya, *Compt. Rend.*, 236 (1953) 608.
V. K. Iya and J. Loriers, *Compt. Rend.*, 237 (1953) 1413.

In ores

R. C. Vickery, *J. Chem. Soc.*, (1955) 245.

(b) Anion exchangers. Sc(III) is retained and thus separated from Th(IV), Al(III) and the rare earths.

cf. K. A. Kraus, F. Nelson and G. W. Smith, *J. Phys. Chem.*, 58 (1954) 11.

(x) By fixation on cellulose. Sc(III) is the first to be eluted by HNO_3 + ether, and is thus separated from the other rare earths. Zr(IV) may be fixed in the presence of tartaric acid.

cf. F. H. Burstall, G. R. Davies, R. P. Linstead and R. A. Wells, *Nature*, 163 (1949) 64.

Paper chromatography.

O. H. Johnson and H. H. Krause, *Anal. Chim. Acta*, 11 (1954) 128.

Review of the methods of separation.

R. C. Vickery, *J. Chem. Soc.*, (1956) 3113.

(2) Colorimetry

Quinalizarin. Quinalizarin gives a blue coloration which can be extracted with *iso*-amyl alcohol. The rare earths do not interfere.

G. Beck, *Mikrochim Acta*, 34 (1949) 282.

Alizarin. Alizarin S may be used.

A. R. Eberle and M. W. Lerner, *Anal. Chem.*, 27 (1955) 1551.

Oxinate. Oxinate in benzene gives $\varepsilon \sim 7{,}500$.

F. Umland and H. Puchelt, *Anal. Chim. Acta*, 16 (1957) 334.

'Tiron'. At pH 6, $\varepsilon \sim 8{,}000$ at 310 mμ, As, Ti, Fe, Ge, Mo, Th, Zr, oxalate, citrate, and tartarate interfere.

cf. H. Hamaguchi, N. Onuma, R. Kuroda and R. Sugisita, *Talanta*, 9 (1962) 563.

# Selenium
Se = 79.0

# Tellurium
Te = 127.6

(1) Separations

When the volatile selenious and tellurous acids are calcined, some loss of selenium and tellurium will inevitably occur. The volatile chlorides also evaporate when a hydrochloric acid solution is heated.

(i) Precipitation of Se and Te. Selenious and tellurous acids may be reduced in acid solutions by many reducing agents, with the precipitation of Se and Te (see p. 366).

In 9 *N* HCl solution, SO_2 reduces Se(IV) but not Te(IV), Cu(II) etc. Both Se(IV) and Te(IV) are reduced by SO_2 in 2–4 *N* HCl.

Sn(II) reduces both selenium and tellurium, and is more effective for quantities smaller than 1 mg.

With SO_2.

cf. H. GOTO AND T. OGAWA, *Sci. Repts. Res. Inst. Tohuku Univ., Ser. A*, 4 (1952) 121.
H. GOTO AND Y. KAKITA, *Sci. Repts. Res. Inst. Tohuku Univ., Ser. A*, 4 (1952) 28.

Precipitation of Te.

H. BODE AND E. HETTEVER, *Z. Anal. Chem.*, 173 (1960) 285.

With Sn(II).

H. GOTO AND Y. KAKITA, *Sci. Repts. Res. Inst. Tohuku Univ., Ser. A*, 7 (1955) 365.
R. R. DE MEIO, *Anal. Chem.*, 20 (1948) 488.

In presence of Cu(II) and NO_3^-
F. D. L. NOAKES, *Analyst*, 76 (1951) 542.

In Steels
S. E. WIBERLEY, L. BASSETT, A. M. BURRILL AND H. LYNG, *Anal. Chem.*, 25 (1953) 1586.

With Cr(II). Selenium is precipitated in 3.5 *N* HCl or 6 *N* H_2SO_4. Tellurium is precipitated in HCl < 3.5 *N*.

cf. H. GOTO, Y. KAKITA AND S. SUKUJI, *Nippon Kinzoku Gakkai-Shi*, B15 (1951) 617.

With Fe(II). In presence of EDTA, selenium is precipitated at pH 3, and tellurium at pH 10.

cf. V. SIMON AND V. GRIM, *Chem. Listy*, 48 (1954) 1415.

With hypophosphite. Te/Pb, Sb.

B. S. Evans, *Analyst*, 67 (1942) 387.

Traces

C. L. Luke, *Anal. Chem.*, 31 (1959) 572.
N. Leontovitch, *Chim. Anal.*, 42 (1960) 329.

(ii) Separation by distillation. $SeOCl_2$, $TeOCl_2$, $SeOBr_2$, and $TeOBr_2$ are volatile in > 6 *N* HBr or HCl above 100° C.

Distillation can be carried out in the presence of concentrated hydrobromic acid, which forms the corresponding volatile derivative with Se(IV) or Te(IV). When Se(VI) or Te(VI) are present, heating with concentrated hydrobromic acid first reduces them to Se(IV) and Te(IV). Selenium and tellurium may in this way be separated from all elements except As(III), Sb(III) and Ge(IV).

Operating procedure

The solution is placed in a distillation flask, 100 ml of 40% of hydrobromic acid containing 1 ml of bromine are added, and 75 ml of liquid are distilled over.

cf. H. C. Dudley and H. G. Byers, *Ind. Eng. Chem., Anal. Ed.*, 7 (1935) 3.
J. S. McNulty, E. J. Center and R. M. McIntosh, *Anal. Chem.*, 23 (1951) 123.
J. L. Lambert, P. Arthur and T. E. Moore, *Anal. Chem.*, 23 (1951) 1101.

In an HCl medium
R. Dolique and S. Perahia, *Bull. Soc. Chim.*, 13 (1946) 44.

(iii) Separation of Se from Te. (a) Tetraphenylarsonium chloride precipitates only Te (VI)in 5 *N* hydrochloric acid. The following interfere: Br^-, I^-, F^-, NO_3^-, Mo(VI), W(VI), Hg(II), Sn(IV), Bi(III), Tl, Zn(II), Cd(II), and Fe(III).

cf. H. Bode, *Z. Anal. Chem.*, 134 (1951) 100.

(b) Extraction of the diethyldithiocarbamate of Te(IV).

H. Bode, *Z. Anal. Chem.*, 144 (1955) 90.

(c) Extraction of Te(IV) from hydriodic acid. The solvent is *iso*-amyl alcohol.

In urine.
C. K. Hanson, *Anal. Chem.*, 29 (1957) 1204.

(d) Extraction of Se with 3,3'-diaminobenzidinium in HCl. The solvent used is toluene. This enables separation from Cu(II), Mo(VI), Fe(III), Ti(IV), Cr(III), Ni(II), Co(II), Te and As.

K. L. Cheng, *Anal. Chem.*, 28 (1956) 1738.

(e) With ion-exchangers. Se(IV) can be separated from Te(IV) with Amberlite IR 120 in 0.3 *N* HCl[1]. Te(IV) can be separated from Sb and Sn on the anionic resin[2] Dowex 1.

(1) F. AOKI, *Bull. Chem. Soc. Japan*, 26 (1953) 480.
(2) G. W. SMITH AND S. A. REYNOLDS, *Anal. Chim. Acta*, 12 (1955) 151.

(2) Colorimetry

(i) Se in the elementary state. When reduced with SO_2, I^-, Sn(II), hydroxylamine, Hg_2Cl_2, etc., Se is precipitated in orange-yellow to red form. 1 ml of 5% gum arabic per 10 ml of solution should be added to the sample to avoid flocculation of the colloidal selenium produced. 0.5 g of hydroxylamine hydrochloride are then added, and the mixture is left overnight.

10 µg can be estimated satisfactorily.

cf. in biochemistry
R. DOLIQUE, J. GIROUX AND S. PERAHIA, *Bull. Soc. Chim.*, 13 (1946) 48.

(a) Te. Reduction with Sn(II).

cf. R. H. DE MEIO, *Anal. Chem.*, 20 (1948) 488.
W. KLEMM, K. GEIERSBERGER, B. SCHAELFER AND H. MINDT, *Z. Anorg. Allgem. Chem.*, 255 (1948) 287.
K. GEIERSBERGER, *Z. Anal. Chem.*, 135 (1952) 15.

(b) Reduction with hypophosphite.

cf. R. A. JOHNSON, F. P. KWAN AND DON WESTLAKE, *Anal. Chem.*, 35 (1953) 1017.

In steels
N. LEONTOVITCH, *Chim. Anal.*, 42 (1960) 329.

(ii) Se by means of 3,3'-diaminobenzidine. With Se, diaminobenzidine gives a yellow coloration in 0.1 *N* hydrochloric acid. Te and SO_4^{2-} do not interfere.

Cr(III) interferes owing to its colour. When this metal is present, the selenium compound should be separated by extraction.

Fe(III) should be complexed with fluoride, and Cu(II) with oxalate.

Sensitivity. $\varepsilon \sim 20{,}000$ at 340 mµ in toluene.

REAGENTS

Diaminobenzidine hydrochloride, 0.1% in 0.1 *N* hydrochloric acid.

Hydrochloric acid, 1 N.

OPERATING PROCEDURE

10 ml of neutral solution containing 10 to 100 μg of Se are mixed with 5 ml of *N* hydrochloric acid and then 25 ml of diaminobenzidine. The volume is made up to 50 ml, and colorimetry is performed at 350 mμ after waiting for 1 hour.

cf. J. HOSTE AND J. GILLIS, *Anal. Chim. Acta*, 12 (1955) 158.

Extraction of 0.1 ppm
K. L. CHENG, *Anal. Chem.*, 28 (1956) 1738.

In the presence of EDTA – in steels
K. L. CHENG, *Chemist-Analyst*, 45 (1956) 67.

5 μg/1g H_2SO_4
T. DANZUKA AND K. UENO, *Anal. Chem.*, 30 (1958) 1370.

In copper
C. L. LUKE, *Anal. Chem.*, 31 (1959) 572.

Fluorimetry (0.02 μg)
J. H. WATKINSON, *Anal. Chem.*, 32 (1960) 980.

(iii) Estimation of Te(IV) by means of diethyldithiocarbamate. The compound formed can be extracted with chloroform or carbon tetrachloride.

Interfering ions. Se(IV) does not interfere at pH 8.5. Bi(III), Sb(III), and Tl(III) interfere.

Small amounts of Cu(II), Hg(II), etc., can be complexed with cyanide and EDTA.

REAGENTS

EDTA: disodium salt of ethylenediaminetetraacetic acid, 10%.
Potassium cyanide, 10%.
Sodium diethyldithiocarbamate, 0.2%.

OPERATING PROCEDURE

1 ml of EDTA and 1 ml of cyanide are added to 25 ml of solution containing from 10 to 50 μg of Te. The pH is adjusted to 8.5 and 10 ml of diethyldithiocarbamate are added. The mixture is extracted three times with 7 ml portions of CCl_4, shaking for 2 minutes. The extracts are made up to 25 ml with CCl_4 and colorimetry is performed at 428 mμ.

cf. H. BODE, *Z. Anal. Chem.*, 144 (1955) 90 and 176.
C. L. LUKE, *Anal. Chem.*, 31 (1959) 572.

(iv) Iodometry

H_2SeO_3

J. L. LAMBERT, P. ARTHUR AND T. E. MOORE, *Anal. Chem.*, 23 (1951) 1101.

Se in Te

K. GEIERSBERGER, *Z. Anal. Chem.*, 135 (1952) 15.

TeI_6^{2-} 0.2 μg. At 335 mμ. Bi(III), Fe(III), Cu(II), and Se(IV) interfere.

cf. R. JOHNSON AND F. P. KWAN, *Anal. Chem.*, 23 (1951) 651.
K. GEIERSBERGER AND A. DURST, *Z. Anal. Chem.*, 135 (1952) 11.
E. G. BROWN, *Analyst*, 79 (1954) 50.

$TeCl_6^{2-}$

(In concentrated HCl at 376 mμ)
M. W. HANSON, W. C. BRADBURY AND J. K. CARLTON, *Anal. Chem.*, 29 (1957) 490.

$TeBr_6^{2-}$

N. W. FLETCHER AND R. WARDLE, *Analyst*, 82 (1957) 743.

(3) Other methods

(i) Dissolution of Se and Te in concentrated sulphuric acid.
$Se + H_2SO_4 \rightarrow SeSO_3 + H_2O$. The $SeSO_3$ is estimated at 350 mμ and the $TeSO_3$ at 520 mμ.

cf. S. E. WIBERLEY, L. G. BASSETT, A. M. BURRILL AND H. LYNG, *Anal. Chem.*, 25 (1953) 1586.

(ii) Absorptiometry of telluric acid in the ultraviolet

L. W. SCOTT AND G. W. LEONARD, JR., *Anal. Chem.*, 26 (1954) 445.

(iii) Te – By means of thiourea

W. NIELSCH, *Z. Anal. Chem.*, 144 (1955) 191.
W. NIELSCH AND L. GIEFER, *Z. Anal. Chem.*, 145 (1955) 347; 155 (1957) 401.

(iv) Extraction of the complex with 'bismuthiol II'

J. JANKOVSKY AND O. KSIR, *Talanta*, 5 (1960) 238.

Silicon

Si = 28.06

In solution, silica may exist in one of the following forms: as an anion in alkaline media, as a colloid, or as the complex SiF_6^{2-}.

(1) Separation

Silicon can be separated from a large number of elements by digestion with fused carbonates or preferably with fused caustic soda; P(V) and As(V) do not follow the silicon.

Any interfering ions present may best be separated when the silicon is in the form of SiF_6^{2-}, by extraction of their cupferrates, oxinates, etc., or by electrolysis at a mercury cathode.

Some silicon may be lost when a fluoride solution of silica is heated above 100° C, owing to the volatilization of the azeotropic $H_2SiF_6 - H_2O$ mixture.

Standard solutions. Pure silica is calcined to constant weight, digested with fused carbonates and the product brought to a known volume. The solution should be stored in polythene vessels. During storage, the compositions undergo a gradual change owing to the formation of a colloidal hydroxide, some of which react only very slowly with the molybdate reagent after 12 hours.

The disadvantages mentioned may also be avoided by digesting SiO_2 with hydrofluoric acid in the cold, in a platinum crucible. The solution should again be stored in polythene containers.

The hydrofluoric acid solutions are stable.

(2) Colorimetry

Two important methods are available: (1) formation of the yellow silicomolybdate complex; (2) reduction of this complex and colorimetry of the blue compound formed.

If appropriate precautions are observed, the first method may be made accurate to within $\pm$ 0.5%, and may then be used for the rapid estimation of large amounts of silica. The more common gravimetric method is lengthy, and does not allow silica to be estimated in the presence of fluorides. In contrast, the colorimetric method can be used when fluorides are present, and indeed is the only method available in this important instance.

A certain number of general precautions should be observed: thus, the use of glass vessels is not recommended, since they are attacked, particularly by alkaline solutions. The reagents used may also contain silica and it is consequently essential to perform a blank test.

Interfering ions. Phosphates, arsenates and germanates give analogous reactions. The separation of these three elements may indeed present a major problem. A certain number of their compounds can be extracted with organic solvents: phosphomolybdate, phosphovanadomolybdate, and the various 'molybdenum blues' obtained by reducing complexes of Mo(VI) with phosphorus(V), silica, etc. The pH of the solution plays an important role in such extractions. Interfering elements of this type may also be removed by operating under suitable conditions of acidity, or again, by the formation of complexes.

(i) Separation of phosphorus(V) and arsenic(V) by extracting their molybdate complexes. At the pH of 1.2, ethyl acetate dissolves phosphomolybdate and arsenomolybdate, whilst the silicomolybdate remains in aqueous solution (Figure 71). Although the separation is not accurate, it possesses the advantage of being rapid. It cannot however be applied to small amounts, since the absolute error remains substantially constant.

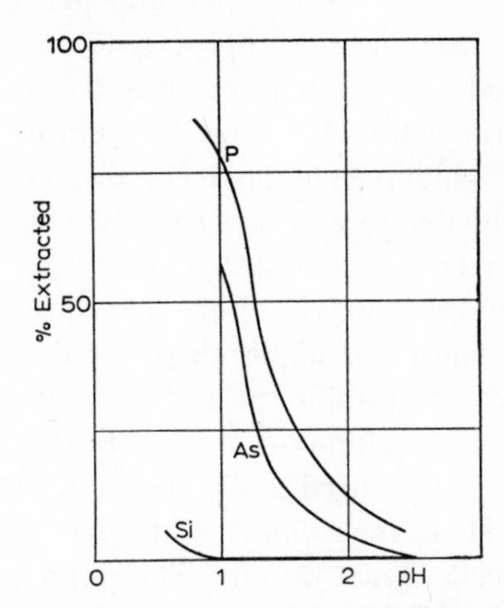

Figure 71. Extraction curve of the molybdate complexes

OPERATING PROCEDURE

1 ml of 10% persulphate is added to the solution in order to oxidize reducing agents if necessary, followed by 2.5 ml of 2% ammonium molybdate. The solution itself may contain up to 100 μg of silicon. The pH is then adjusted to 1.2 by means of nitric acid (pH-meter) and the volume is brought to 10 ml. The solution is heated for 20 minutes at about 40°, cooled, and the complexes are extracted with 10 ml portions of ethyl acetate. The extracts are washed with nitric acid at pH 1.2. 1 mg of phosphorus and 0.4 mg of arsenic may be separated by this method.

The silicon present in the aqueous solution obtained may be estimated directly at this pH; it is however more reliable to perform the colorimetry after adjusting the pH to 1.7.

cf. J. HURÉ AND T. ORTIS, *Bull. Soc. Chim.*, 16 (1949) 834.
J. PAUL AND W. F. R. POVER, *Anal. Chim. Acta*, 22 (1960) 185.

Review of the methods
C. WALDEN AND M. G. MELLON, *Anal. Chem.*, 26 (1954) 1668.

(ii) Colorimetry of the silicomolybdate complex. A yellow silicomolybdate complex, $Si(Mo_{12}O_{40})^{4-}$, is formed.

It has been shown that two complexes, α and β exist. The α complex is stable, while the β complex gradually changes into α over a period of time. The α complex can however be formed directly in slightly acidic (pH 2.3–3.9) hot solutions. With the β complex, the pH is kept low and the conditions are rigorously defined. Operating procedures for both instances will however be described.

(a) Formation of the β complex. Both time and pH of the solution are very important factors. Formation of the complex is slower in more acidic solutions. The pH is generally fixed at 1.5–1.7, or even up to 2.0, and it is consequently necessary to let the solution stand for at least 20 minutes. After 45 minutes, the transformation of the less highly coloured α complex is fairly appreciable.

The results are reproducible to within 0.2–0.5%, but errors of up to 5% may occur if the pH is not precisely fixed—1.25 to 2.2.

The colour develops more slowly in dilute solutions.

The silicomolybdate complex is highly dissociated, and an excess of molybdate is therefore required. A suitable ratio consists of 5 ml of 10% molybdate per 10 mg of silica, the total volume of the solution being 100 ml. The effect of temperature is slight and need only be controlled if high accuracy is required. Silica often exists in the colloidal state, even in alkaline solutions, and only sufficiently fine micelles react rapidly. The estimation should therefore be carried out before the silica has had time to age. Freshly prepared solutions of silicate should thus be used and, if this is not possible, the silica may be complexed by fluorides and then decomplexed in the presence of the molybdate reagent by means of boric acid or Al(III).

Sensitivity. If the available apparatus makes no provision for ultraviolet colorimetry, the estimation may be performed at about 420–430 mμ, and the result compared against those yielded by artificial standards. The absorption of the complex is however much greater in the ultraviolet and, if this is used, very minute traces may be estimated. In this case however, many substances interfere—for example traces of nitrite, etc.

$\varepsilon \sim$ 20,000 at 300 mμ, $\sim$ 7,000 at 350 mμ.

Interfering ions. P(V), As(V), and Ge(IV) give analogous colorations. The silicomolybdate appears at an appreciable rate in 0.1–0.25 *N* sulphuric acid, particularly between pH 1.5–1.7. The phosphomolybdate forms best in 0.2–1 *N* H_2SO_4, particularly in the range 0.5–0.7 *N* H_2SO_4. At pH 1.5–1.7 and 470 mμ, the colorations due to phosphorus and arsenic are less sensitive than that due to silicon: 6.90 mg of P(V) = 1.00 mg of Si(IV) = 187 mg of As(V). A satisfactory but approximate result may be obtained by making this correction after the phosphorus and the arsenic have been estimated separately. Once the silicomolybdate complex has been formed under the normal conditions, the solution is adjusted to 2 to 3.5 *N* H_2SO_4, when the phosphomolybdate and arsenomolybdate complexes are destroyed; the silicomolybdate complex remains stable for a further 20 minutes. It is then preferable to reduce the latter immediately to the blue compound.

In the presence of oxalic acid (3 ml of 10% solution), citric acid, or tartaric acid (4 ml of 10% solution), the coloration due to the phosphorus is still less sensitive, being negligible up to 5 parts of phosphorus per 1 part of silicon (approximate colorimetry).

If F^- is present, some silicofluoride complex is formed and the coloration due to the silicon is thus diminished. The addition of boric acid (20 parts of H_3BO_3 per 1 part of F^-) restores the colour quantitatively. Alternatively, Al(III) may be added. These methods allow silicon to be estimated easily and accurately in the presence of fluorides—*e.g.* in hydrofluosilicic acid—and also after the digestion of silicon-bearing alloys or minerals with hydrofluoric acid; the latter instance may offer great advantages.

Ions which precipitate will interfere: Zr(IV), Sn(IV), Ti(IV), W(VI), etc. They may however be complexed with fluoride, after which a calculated quantity of boric acid is added, sufficient to decomplex the silicon (unstable complex) but insufficient to decomplex the Ti(IV), Zr(IV), etc. completely (considerably more stable complexes). Tests should be carried out in each particular instance to determine the amounts of F^- and H_3BO_3 required for a given amount of these elements. Any interference due to Fe(III) can be eliminated in the same way.

In the presence of V(V), which gives yellow vanadomolybdates, the silicomolybdate can be extracted with butanol from *N* H_2SO_4; 25 mg of V(V) does not interfere. Since the silicomolybdate complex is not very stable, ions which complex Mo(VI) can interfere. Among these are large amounts of Cl^-, citrate and tartrate ions, etc. Reducing agents should be oxidized by the addition of a little persulphate. Organic substances may be destroyed with permanganate, the slight excess of which can be removed by adding one drop of oxalate. Coloured ions such as e.g. Cr(III) may interfere; the silicomolybdate complex should be extracted when they are present.

NOTE. Silicate solutions should never be stored in glass, and paraffin-waxed glass, plastic, platinum, or even quartz containers are used instead. It is however permissible to adjust solutions to a definite volume in a measuring flask, or to pipette a known volume without giving rise to error, provided such manipulation is rapid. Special apparatus is required in the case of hydrofluoric acid solutions.

Artificial standards. Artificial standards of indefinite stability can be made up with chromate. If, however, the accuracy need not be very high, the preparation of standard solutions may be dispensed with by comparing the colour against that of a chromate solution buffered with borax.

Blank test. The reagents used, and particularly the caustic soda of guaranteed purity, contain small amounts of silica. It is therefore essential to determine their silica contents by a blank test and to make the necessary corrections.

Fixation of the pH. The pH should be fixed even in moderately accurate estimations. A pH meter may be used, or, if there is no suitable buffer for the solution, the latter may be adjusted to about pH 4.5 (reagent paper) and 9.5 ml of 1.0 N sulphuric acid added per 100 ml of total volume.

REAGENTS

Sodium molybdate, 10%, or ammonium molybdate neutralized with caustic soda to phthalein.

Sulphuric acid of pH 1.6.

Ammonium sulphate, 10% solution, freshly prepared.

OPERATING PROCEDURE

To 10 ml of solution, containing 0.4 to 1 mg of silica, are added 1 ml of 9 N (1/4) sulphuric acid, 5 ml of persulphate, and 5 ml of molybdate. The pH is adjusted to 1.6 ± 0.1 and the volume is made up to 50 ml with the sulphuric acid solution of pH 1.6. Colorimetry is performed after 20 minutes.

(b) Formation of the α complex[1]. This occurs at pH 2.5 to 3.9, at the boiling point of the solution.

The colorimetric technique is a little less sensitive, but the coloration is stable and an accuracy of $\pm 0.1\%$ (relative error) may be obtained. The temperature of the measurement is an important factor, involving the coefficient of 0.4% per degree.

The estimation may be performed in the presence of $< 0.01\ M$ EDTA.

REAGENTS

Disodium salt of ethylenediaminetetraacetic acid, 0.05 M.

Chloroacetate buffer: 2 M monochloroacetic acid + 2 M ammonium chloroacetate; ammonium molybdate at 35.3 g per litre, 500 ml.

OPERATING PROCEDURE

As an example, a sample of rock containing 20–70% SiO_2 may be considered. A 100–300 mg portion of the sample is digested with 10 g of fused caustic soda in a nickel crucible for 5–10 minutes. The product is taken up in 40 ml of EDTA solution and a little water, and the volume is made up to 500 ml (509.4 g at 20°) in a polyethylene vessel. 25 ml of this solution are mixed with 20 ml of buffer solution; the pH of the resulting solution should lie between 3.5 and 3.7. The mixture is placed on a boiling water bath for 5 to 10 minutes, cooled, and made up to 50 ml. The colorimetric determination is then performed.

Complexes of Si(IV) and Mo(VI)

J. D. H. STRICKLAND, *J. Am. Chem. Soc.*, 74 (1952) 862.

Rate of formation of the complex

G. B. ALEXANDER, *J. Am. Chem. Soc.*, 75 (1953) 5655.
M. JEAN, *Chim. Anal.*, 38 (1956) 37.
S. LACROIX AND M. LABALADE, *Anal. Chim. Acta*, 3 (1949) 383.
H. W. KNUDSEN, C. JUDAY AND V. W. MELOCHE, *Ind. Eng. Chem., Anal. Ed.*, 12 (1940) 270.
W. H. HADLEY, *Analyst*, 70 (1945) 43.

In the presence of P(V)

M. C. SCHWARTZ, *Ind. Eng. Chem., Anal. Ed.*, 14 (1942) 893.

Chromate standards

H. W. SWANK AND M. G. MELLON, *Ind. Eng. Chem., Anal. Ed.*, 6 (1934) 348.

In alloys

1956 book of A.S.T.M. Methods of Chemical Analysis of Metals, Philadelphia.
O. P. CASE, *Ind. Eng. Chem., Anal. Ed.*, 16 (1944) 309.
S. LACROIX AND M. LABALADE, *Anal. Chim. Acta*, 3 (1949) 383.

In steels

D. ROZENTAL AND H. C. CAMPBELL, *Ind. Eng. Chem., Anal. Ed.*, 17 (1945) 222.

In bauxites, ilmenites, fluorine derivatives, silicates, etc.

S. LACROIX AND M. LABALADE, *Anal. Chim. Acta*, 3 (1949) 383.

In sea water (Influence of chlorides)

G. S. BIEN, *Anal. Chem.*, 30 (1958) 1525.
M. A. DE SESA AND L. B. ROGERS, *Anal. Chem.*, 26 (1954) 1278.
(1) A. RINGBOM, P. E. AHLERS AND S. SIITONEN, *Anal. Chim. Acta*, 20 (1959) 78.

(iii) Colorimetry by reduction of the silicomolybdate complex. Under suitable conditions, the yellow silicomolybdate complexes may be reduced to give a blue coloration, while the excess of molybdate remains unchanged.

One of the yellow complexes should first be found under favourable conditions such as those described above.

Of the numerous reducing agents available, 1-amino-2-naphthol-4-sulphonic acid in the presence of sulphurous acid has often been recommended. Alternatively, sulphurous acid may be used on its own. The most reliable results are however obtained with Fe(II).

The colorimetry is carried out at 815 mμ. The colour is stable for twelve hours.

Interfering ions. Ge(IV) and P(V) exhibit the same reaction. The sensitivity of the reaction due to phosphates can however be diminished by adding oxalate, citrate, or tartrate. If after the formation of the yellow complexes the medium is made sufficiently acid (2 *N* H_2SO_4) P(V) and As(V) no longer interfere. In this case the reduction should be carried out immediately after acidification.

Sensitivity. $\varepsilon \sim 20{,}000$ at 810–820 mμ.

(a) Extraction of the blue compound. If a quantity smaller than 1 μg per 25 ml is to be estimated, the blue coloration, which is soluble in alcohols such as butyl or *iso*-amyl, should be extracted. This is only possible however, if the medium is sufficiently acid: 1 to 2 *N* H_2SO_4.

(b) Reduction by Fe(II). Application to the determination of silicon in aluminium alloys.

REAGENTS

Oxalic acid, 4% solution.
Sulphuric acid, 9 *N* (1/4).
Ammonium molybdate, 10% solution.
Mohr's salt, 2.5% in 0.4 *N* (1/100) sulphuric acid.
Caustic soda solution stored in a suitable container.

OPERATING PROCEDURE

The sample— for example, 0.5 g—is digested in a nickel crucible, using 5 ml of caustic soda, and warming gently if necessary. 5 ml of 3% (10 volume) hydrogen peroxide are added, and the mixture is boiled for 10 minutes. The solution is diluted to 50 ml, filtered into 15 ml of 9 *N* sulphuric acid, and the residue is washed. The volume is made up to 500 ml in a measuring flask, and a 50 ml aliquot is transferred into a 100 ml measuring flask, mixed with 5 ml of molybdate, shaken and left for 5 minutes. 15 ml of sulphuric acid, 10 ml of oxalic acid, and 10 ml of Mohr's salt are added, and the mixture is diluted to 100 ml. Colorimetry is performed at 820 mμ The coloration is stable for at least 1 hour. A blank test is carried out to determine the silica present in the reagents.

cf. D. F. BOLTZ AND M. G. MELLON, *Ind. Eng. Chem., Anal. Ed.*, 19 (1947) 873.
R. F. MILTON, *Analyst,* 76 (1951) 431.

A. B. Carlson and C. V. Banks, *Anal. Chem.*, 24 (1952) 472.
O. A. Kenyon and H. A. Bewick, *Anal. Chem.*, 25 (1953) 145.

In aluminium products
A. L. Olsen, E. A. Gee, V. McLendon and D. D. Blue, *Ind. Eng. Chem., Anal. Ed.*, 16 (1944) 462.
J. A. Brabson, I. W. Harvey, G. E. Maxwell and O. A. Schaeffer, *Ind. Eng. Chem., Anal. Ed.*, 16 (1944) 705.

Study of the blue and yellow silicomolybdate complexes
J. D. H. Strickland. *J. Am. Chem. Soc.*, 74 (1952) 862.
M. Jean, *Chim. Anal.*, 38 (1956) 41.

Other methods of reduction
M. Jean, *Chim. Anal.*, 38 (1956) 37.

In steels
T. R. Andrew and C. H. R. Gentry, *Analyst*, 81 (1956) 339.
W. F. Sanders and C. H. Cramer, *Anal. Chem.*, 29 (1957) 1139.

In alloys
C. L. Luke, *Anal. Chem.*, 25 (1953) 148.

In the presence of Ni and W
T. R. Andrew, *Analyst*, 82 (1957) 423.

In Ni
W. Gann, *Z. Anal. Chem.*, 150 (1956) 254.

In ThO_2
O. Menis and D. L. Manning, *Anal. Chim. Acta*, 16 (1957) 67.

In Ti
M. Codell and G. Norwitz, *Anal. Chim. Acta*, 16 (1957) 327.

In tungsten, titanium and other metallic carbides
B. Bagshawe and R. J. Truman, *Analyst*, 79 (1954) 17.

In Zr, Be, Al, Ca
A. B. Carlson and C. V. Banks, *Anal. Chem.*, 24 (1952) 472.

In fluorine compounds
J. A. Brabson, H. C. Mattraw, G. E. Maxwell, A. Darro and M. F. Needham, *Anal. Chem.*, 20 (1948) 505.
A. K. Babko and L. M. Etushenko, *Zavodsk. Lab.*, 23 (1957) 423.

In plants
R. J. Volk and R. L. Weintraub, *Anal. Chem.*, 30 (1958) 1011.

In silicate rocks
J. P. Riley, *Anal. Chim. Acta*, 19 (1958) 413.

In the presence of F^- and P(V)

S. GREENFIELD, *Analyst*, 84 (1959) 380.

Extraction

J. PAUL, *Anal. Chim. Acta*, 23 (1960) 178.

Fluosilicic acid, H_2SiF_6

With water, H_2SiF_6 gives an azeotropic mixture which can be distilled.

(2) Colorimetry

The fluosilicate complex is relatively unstable.
(2a) The colorimetry of silica can be carried out in the presence of substances which complex F^-, such as B(III), Al(III), etc.[1].
(2b) Alternatively, the F^- can be estimated.

(1) P. R. GRAFF AND F. J. LANGMYRH, *Anal. Chim. Acta*, 21 (1959) 429.

Silver

Ag = 107.9

(1) Separation of trace amounts

(i) Precipitation of the chloride or the bromide. Silver chloride and particularly silver bromide, are suitable for precipitation by entrainment with the corresponding mercurous salt, the latter being subsequently driven off by calcination.

cf. D. G. HIGGS, *Analyst*, 69 (1944) 270.

(ii) Precipitation of the sulphide. Traces of silver may be separated by the well-known method of precipitating silver sulphide with hydrogen sulphide in 0.1 *N* (1/400) sulphuric acid, using Cu(II) or Hg(II) as an entraining agent.

(iii) Precipitation by p-dimethylaminobenzylidenerhodanine in the presence of EDTA (see Colorimetry).

(iv) Extraction by dithizone (see Colorimetry).

(v) Electrolysis. In a cyanide medium. The solution contains the complex ion $Ag(CN)_2^-$. Cu(II) and Pt(IV), which give extremely stable complexes, do not interfere.

OPERATING PROCEDURE

The solution is neutralized with dilute NaOH to incipient precipitation. The precipitate is then just redissolved with 10% potassium cyanide, and the solution is electrolysed.

In an acetate medium. 25 g of sodium or ammonium acetate are added to 100 ml of the solution after weakly acidifying with nitric acid (less than 1 ml of concentrated acid). Electrolysis is performed at the boiling-point.

Numerous metals including copper[1] may be separated.

cf. A. LASSIEUR, *Electroanalyse rapide*, Presses Universitaires, Paris, 1927.

(1) A. SCHLEICHER, *Elektroanalytische Schnellmethoden*, F. Enke, Stuttgart, 1947.

Microelectrolysis

A. FRIEDRICH AND S. RAPAPORT, *Mikrochemie*, 18 (1935) 227.

F. HERNLER AND R. PFENINGBERGER, *Mikrochemie*, 25 (1938) 208.

In an ammoniacal medium. Silver and copper are electrodeposited according to the following equations, at different potentials.

$$Ag(NH_3)_2^+ + e \rightarrow Ag\downarrow + 2NH_3 \qquad E_0 = 0.37 \text{ V.}$$
$$Cu(NH_3)_4^{2+} + 2e \rightarrow Cu\downarrow + 4NH_3 \qquad E_0 = 0.05 \text{ V.}$$

Ag may therefore be separated from Cu(II) by controlling the potential.

Ag-Cu alloys

W. L. MILLER, *Ind. Eng. Chem., Anal. Ed.*, 8 (1936) 431.
H. DIEHL, *Electrochemical Analysis with Graded Cathode Potential Control*, G. F. SMITH Chemical Co., Columbus, 1948.

(vi) Internal electrolysis

cf. J. G. FIFE, *Analyst*, 62 (1937) 723.

(vii) Standard solutions. The pure metal and 99.9–100% silver nitrate, $AgNO_3$, are usually available.

Dilute silver solutions are not stable in glass (24 hours) because of adsorption phenomena, but silica vessels are suitable.

$$Ag/AgNO_3 = 0.635.$$

A 0.1000 *N* silver nitrate solution contains 10.8 mg of silver per millilitre.

(2) Colorimetry

The methods described above also allow the estimation of small amounts of silver in pyrites, copper, lead etc., concurrently with the dry method.

With certain precautions, the dithizone method is accurate to within 1–2%, and may in fact be used for quite large amounts of silver.

The rhodanine method is less accurate.

(i) The dithizone method. In an acidic medium, the yellow silver dithizonate, which can be extracted by carbon tetrachloride, is formed. It is not however, possible to work in neutral or alkaline solutions, since the violet-red silver dithizonate (enolic form) appearing under such conditions, cannot be extracted.

A method involving extraction titrations may also be used.

Sensitivity. Molar extinction coefficient $\varepsilon = 27{,}000$ at 462 mμ.

Interfering ions. Though the extraction from 0.5 *N* acid is very selective, the following ions still interfere: Hg(II), Pd(II), Au(III), and Pt(II), and also large amounts of Cu(II). In the presence of EDTA at pH 4.7 large amounts of Cu(II), Bi(III), and Pb(II) do not interfere even when present at a ratio of 100,000 to 1.

In the presence of large amounts of Cu(II), copper dithizonate may be used as the reagent; the variation of colour indicates the replacement of Cu(II) by Ag(I). Calibration curves may then be used with accurate results.

In the presence of Hg(II) and Cu(II), it is possible to extract all three elements, and then pass back only the Ag(I) into an aqueous phase containing hydrochloric acid and sodium chloride. The silver dithizonate is then re-extracted after diluting the solution[1].

(a) In the absence of large amounts of Cu(II). Silver dithizonate does not absorb light at 620 mμ, and the dithizone may therefore be estimated at this wavelength. The amount of dithizone taken up by the reaction is deduced by comparison with the initial solution.

REAGENTS

Dithizone, 0.001% in carbon tetrachloride.

Standard silver solution, 2 mg per litre in 0.1 N sulphuric acid: 9 *N* (1/4) sulphuric acid.

OPERATING PROCEDURE

25 ml of a solution containing 2 to 10 μg of silver are mixed with 2 ml of sulphuric acid and accurately measured 5 ml of dithizone. The mixture is shaken for 1 minute, allowed to settle, and transferred slowly into the spectrophotometer cell. This is compared with the pure dithizone solution at 620 mμ.

(b) In the presence of large amounts of Cu(II).

REAGENTS

Copper dithizonate: a 0.001% solution of dithizone in carbon tetrachloride is shaken with an excess of copper sulphate solution in 0.05 *N* (1/1000) sulphuric acid. The dithizonate formed is washed twice with 0.01 *N* (1/4000) sulphuric acid.

OPERATING PROCEDURE

The conditions used are the same as those described earlier, with the dithizone replaced by the copper dithizone. The solution is shaken for 2 minutes, and a mixture of the red copper dithizonate and the yellow silver dithizonate is finally obtained. The estimation is carried out at 465 mμ.

In zinc

H. FISCHER AND G. LEOPOLDI, *Metall. Erz*, 35 (1938) 86.

In minerals

M. SHIMA, *Japan Analyst*, 2 (1953) 96.

In lead

P. D. JONES AND E. J. NEWMAN, *Analyst*, 87 (1962) 66.

(1) H. FRIEDEBERG, *Anal. Chem.*, 27 (1955) 305.

(ii) With rhodanine. In 0.05 *N* acid, *p*-dimethylaminobenzylidenerhodanine gives a red-violet colloidal precipitate with Ag^+ ions.

The conditions of the experiment should be fixed exactly; these are: acidity, concentration of the reagent, time after which the measurement is carried out, nature and the time of shaking, concentration of inert salts, and temperature.

The colloid may be stabilized with a mixture of sucrose + acetone.

The pH can be fixed by means of a mixture of sodium sulphate (200 g) and sodium bisulphate (90 g).

An accuracy of about 3% may be attained.

Sensitivity: $\varepsilon \sim 20{,}000$.

Interfering ions. Large concentrations of salts promote flocculation of the colloid. Au(III), Pd(II), Hg(II), and Cu(I) rise to colorations. EDTA may also be present during the estimation[1].

REAGENTS

Rhodanine, 0.05% in alcohol.

Standard solution of Ag^+, 0.01% or 0.001% in 0.1 *N* (1/100) nitric acid.

OPERATING PROCEDURE

There should be less than 40 μg of silver in 20 ml of solution, and a 0.05 *N* solution of nitric acid should be used. To this are added 0.5 ml of reagent, and the solution is made up to 25 ml. Colorimetry should be performed after 5 minutes for 1 ppm, and after 20 minutes for amounts from 1 to 0.5 ppm.

cf. G. C. B. CAVE AND D. N. HUME, *Anal. Chem.*, 24 (1952) 1503.
(1) A. RINGBOM AND E. LINKO, *Anal. Chim. Acta*, 9 (1953) 80.

(iii) Estimation of the precious metals by the dry method. Only the principle of this method shall be given below, as full descriptions may be found in many classical works.

Silver, gold, and platinum present in alloys and minerals may be estimated by this method, which is particularly useful where wet methods are less satisfactory; this is the case with the estimation of small amounts in a material difficult to attack.

In principle, the precious metals are alloyed with lead; in the molten state, this alloy separates, particularly from the silicates, because of its high density. The lead is then separated by atmospheric oxidation and fusion of the oxide formed; the latter is absorbed by the wall of a porous crucible (cupellation). A 'button' containing Au, Pt, and Ag remains, which is weighed to within 0.01 mg or 0.05 mg (assayer's balance or semimicrobalance).

Au and Ag can be separated by attack with nitric acid in which gold remains insoluble, provided that the Ag/Au ratio is higher than 3. Consequently it is often necessary to add a known quantity of silver (inquartation).

The gold remaining after the nitric acid treatment is weighed, and the silver is calculated by the difference.

When Ag and Pt are present, the former is dissolved by 30 *N* (5/6) sulphuric acid at 240° C. When Au, Ag and Pt are present, and where the ratio $\frac{Ag}{Au + Pt}$ is greater than 3, the platinum passes into solution at the same time as the silver on prolonged treatment with nitric acid.

On the dry method

N. H. FURMAN, *Scott's Standard Methods of Chemical Analysis*, van Nostrand, New York, 1939.

F. P. TREADWELL, *Analyse Quantitative*, Dunod.

E. E. BUGBEE, *Textbook of Fire Assaying*, Willey, New York, 1940.

C. M. HOKE, *Testing Precious Metals*, Jewellers' Techn. Assoc., New York, 1946.

E. A. SMITH, *The Sampling and Assay of the Precious Metals*, London, Griffin, 1947.

Dry micro method

J. DONAU, *Mikrochemie*, 17 (1935) 174; 19 (1936) 108.

Sodium

Na = 23.00

(1) Separation of trace amounts

The best method of separation involves the precipitation of a triple acetate such as zinc uranyl sodium acetate. If any interfering ions are present, the separations appropriate to each particular case should be performed.

Electrolysis at a mercury cathode may be of use.

(i) Precipitation of Na(I) in the form of a triple acetate. Sodium zinc uranyl acetate — $NaCH_3COO \cdot Zn(CH_3COO)_2 \cdot 3UO_2(CH_3COO)_2 \cdot 6H_2O$ — which is sparingly soluble between pH 4.5 and pH 6.0, can be precipitated. The solubility of the compound is not negligible (1 mg of Na^+ per millilitre at 20°), so that the concentration of reagent saturated with the triple acetate should be high, and the washing should be performed with reagents saturated with the salt in question.

REAGENTS

(1) *Acetic acid, 5 M (30%)* 6 ml; *$UO_2(CH_3CO_2)_2 \cdot 2H_2O$* 10 g; *Water* 50 ml } A

Acetic acid, 5 M (30%) 30 g; *$Zn(CH_3CO_2)_2 \cdot 3H_2O$* 3 ml; *Water* 50 ml } B

A and B are mixed and warmed gently to complete the dissolution. A little sodium chloride is added, and the solution is filtered after 48 hours.

(2) *Alcohol saturated with the triple acetate.*

OPERATING PROCEDURE

The solution is concentrated and placed in a centrifuge tube. Ten times its own volume of reagent are added, and the mixture is centrifuged after 1 hour. The precipitate is washed five times with a few millilitres of reagent and then five more times with a few millilitres of alcohol saturated with the triple acetate. If very little sodium is present, it may be necessary to wait 12 hours for complete precipitation.

Interfering ions. K^+ may precipitate in the form of sulphate if its concentration exceeds 25 g per litre; no interference is however occasioned by this effect in the case of volumetric or colorimetric determinations. Li^+ precipitates similarly to Na^+, and the two should be separated or a different reagent used[1]. Sr(II) must also be separated.

Phosphate and arsenate ions precipitate the uranium; if however, the triple acetate is redissolved in water to carry out the colorimetry, uranyl phosphate and arsenate remain insoluble, and may be removed by filtering or centrifuging the solution. Similarly, Mo(IV) precipitates in the form or uranyl molybdate, but can be complexed by means of citrate or tartrate ions. Oxalate ions interfere. F^- does not interfere but, when Al(III) is present, $Na_3[AlF_6]$ may precipitate. Mg(II), Co(II), Ni(II), and Mn(II) may replace Zn(II) in the triple acetate; in this case, colorimetric and volumetric determinations still give correct results, but gravimetric determinations are falsified. A certain number of hydroxides precipitate.

cf. H. H. BARBER AND I. M. KOLTHOFF, *J. Am. Chem. Soc.*, 50 (1928) 1625 and 51 (1929) 3233.

In the presence of P(V), Li(I) and K(I)
H. R. SHELL, *Anal. Chem.*, 22 (1950) 575.

In the presence of Mo(VI)
C. H. HALE, *Ind. Eng. Chem., Anal. Ed.*, 15 (1943) 516.

In the presence of Li^+ *and* K^+
E. R. CALEY AND L. B. ROGERS, *Ind. Eng. Chem., Anal. Ed.*, 15 (1943) 32.

In plants
W. M. BROADFOOT AND G. M. BROWNING. *J. Assoc. Off. Agr. Chem.*, 24 (1941) 916.

In potassium hydroxide
D. WILLIAMS AND G. S. HAINES, *Ind. Eng. Chem., Anal. Ed.*, 16 (1944) 157.
(1) E. R. CALEY AND L. B. ROGERS, *Ind. Eng. Chem., Anal. Ed.*, 15 (1943) 32.

(ii) Standard solutions. These can be made from standardized caustic soda of guaranteed purity. Many alternative compounds exist in grades of guaranteed purity, *e.g.* Na_2CO_3 (which should be dried to constant weight at 270–300°), NaCl, $NaNO_3$.

1 ml of 1.000 *N* caustic soda = 23.00 mg of Na or 31.00 mg of Na_2O.

$$Na/NaCl = 0.393 \qquad Na_2O/2NaCl = 0.530$$

$$Na/NaNO_3 = 0.2706 \qquad Na_2O/2NaNO_3 = 0.365$$

(2) Colorimetry

The colorimetric estimation of sodium is a method of limited importance, since it entails a preliminary separation involving the precipitation of the triple acetate, and consequently presupposes that large amounts of sodium are present. If this is in fact the case, the triple acetate may be estimated satisfactorily by a volumetric method. Flame spectrophotometry is the most widely recommended method of estimation.

Besides the sodium zinc uranyl acetate, sodium magnesium uranyl, or sodium manganese uranyl triple acetates may be used. After the precipitate has been redissolved in water, the uranium can be estimated in the first two cases, and the manganese in the last case. The triple acetate may be precipitated by the technique described on p. 384.

With very small amounts, and if the accuracy need not be very high, a simpler technique may be resorted to, consisting of the nephelometric determination of the triple acetate.

cf. W. S. HOFFMAN AND B. OSGOOD, *J. Biol. Chem.*, 124 (1938) 347.
L. JENDRASSIK AND M. HOLASZ, *Biochem. Z.*, 298 (1938) 74.
D. R. MCCORMICK AND W. E. CARLSON, *Chemist-Analyst*, 31 (1942) 15.

Triple acetate of Mn
W. C. WOELFEL, *J. Biol. Chem.*, 125 (1938) 219.
E. LEVA, *J. Biol. Chem.*, 132 (1940) 487.

Strontium

Sr = 87.6

1) Separation

Strontium may only be estimated colorimetrically when other elements have previously been separated. The separations involved are particularly laborious when other alkaline earth metals are present. In consequence of the above, estimation by flame spectrophotometry is generally preferred.

(i) Separation of the alkaline earth metals. See Calcium.

$SrSO_4$ should be precipitated in a 50% alcohol medium and SrC_2O_4 in an 80% alcohol medium.

The separation from Ca(II) is difficult. Calcium nitrate is soluble in amyl alcohol, anhydrous acetone[1], butyl cellosolve[2], and a mixture of absolute alcohol and ether, while strontium nitrate remains insoluble. Alternatively, strontium nitrate can be precipitated from an 80% nitric acid solution; Ba(II) and Pb(II) also precipitate[3].

The alkaline earth metals can be separated by means of ion-exchangers[4].

(1) R. N. Shreve, C. H. Watkins and J. C. Browning, *Ind. Eng. Chem., Anal. Ed.*, 1 (1939) 215.
(2) H. H. Barber, *Ind. Eng. Chem., Anal. Ed.*, 13 (1941) 572.
(3) H. H. Willard and E. W. Goodspeed, *Ind. Eng. Chem., Anal. Ed.*, 8 (1936) 414.
(4) M. Lerner and W. Rieman III, *Anal. Chem.*, 26 (1954) 610.

(ii) Extraction of the oxinate. This is performed at pH 11.3.

D. Dryssen, *Svensk Kem. Tidskr.*, 67 (1955) 311.

(2) Colorimetry

(i) With chloranilic acid. Sr(II) chloranilate is sparingly soluble. The excess of reagent is estimated colorimetrically. The reaction is not at all specific.

cf. P. J. Lucchesi, S. Z. Lewin and J. E. Vance, *Anal. Chem.*, 26 (1954) 521.

(ii) With murexide and o-cresolphthalein complexone.

F. H. Pollard and J. V. Martin, *Analyst*, 81 (1956) 348.
D. S. Russell, J. B. Campbell and S. S. Berman, *Anal. Chim. Acta*, 25 (1961) 81.

Sulphur and its compounds

S = 32.07

Sulphur

(1) Separation

Extraction. Sulphur may be dissolved in suitable solvents: crystalline sulphur is soluble in carbon disulphide, and all forms of sulphur are soluble in pyridine[1]. The element may be separated from hydrocarbons by acetone[2], alcohols, etc.

(1) H. SOMMER, *Ind. Eng. Chem., Anal. Ed.*, 12 (1940) 368.

In fertilisers

Official Methods of Analysis of A.O.A.C., 8th ed., 1955.

(2) J. K. BARTLETT AND D. A. SKOOG, *Anal. Chem.*, 26 (1954) 1008.

Reduction. Sulphur may be reduced to sulphide or to hydrogen sulphide.

In acetone solution, sulphur can be reduced by copper to cuprous sulphide[1], and may also be reduced by zinc in hydrochloric acid[2].

(1) A. F. HARDMAN AND H. E. BARBEHENN, *Ind. Eng. Chem., Anal. Ed.*, 7 (1935) 103.
H. LEVIN AND E. STEHR, *Ind. Eng. Chem., Anal. Ed.*, 14 (1942) 107.

(2) M. MAURICE, *Anal. Chim. Acta*, 16 (1957) 578.

(2) Colorimetry

Thiocyanate. In acetone solution sulphur reacts with cyanide; the resulting thiocyanate is estimated.

cf. J. K. BARTLETT AND D. A. SKOOG, *Anal. Chem.*, 26 (1954) 1008.

In the ultraviolet region. A solution of sulphur in aqueous alcohol absorbs in the ultraviolet region. H_2S also absorbs in this region[1].

Colorimetry of sulphur solutions in hexane has also been proposed[2].

(1) N. G. HEATLEY AND E. J. PAGE, *Anal. Chem.*, 24 (1952) 1854.

(2) M. J. MAURICE, *Anal. Chim. Acta*, 16 (1957) 574.

Turbidimetry.

G. L. MACK AND J. M. HAMILTON, *Ind. Eng. Chem., Anal. Ed.*, 14 (1942) 604.
L. PEYRON, *Compt. Rend.*, 222 (1946) 740.

Sulphates; sulphuric acid

(i) Separation by ion-exchangers. The anion SO_4^{2-} can be separated from the accompanying cations by means of a cation-exchanger. This procedure should be carried out prior to estimation by the majority of methods.

Preparation of the column. The dimensions of the column are as follows: height 30 to 35 cm, diameter 1 cm; the column is filled with Amberlite IR 120 resin in the H^+ form, possessing a grain size of about 50 mesh. The resin is pretreated in the usual way.

OPERATING PROCEDURE

25 ml of water are added to 25 ml of the solution, in which the cations should have a concentration of 0.2 *N* at most. The resulting solution is passed through the column at the rate of 5 ml per minute. The column is then washed with 100 ml of water.

A certain number of metals which give sulphate complexes, such as *e.g.* Cr(III) or Zr(IV), may not be completely exchanged.

cf. O. SAMUELSON, *Z. Anal. Chem.*, 116 (1939) 328.

(ii) Adsorption chromatography. Many anions can be separated by adsorption on alumina, frequently with greater efficiency than by means of ion-exchangers. The residue may be as small as 0.5 ppm. For example, the residue of Cl^- separated by the two methods differ by a factor of 4,500 in favour of the chromatographic method.

The latter is carried out in a medium 5 *M* in $HClO_4$ and 2 *M* in HCl. The SO_4^{2-} collected is eluted with ammonia or with caustic soda.

cf. S. S. FRITZ, S. S. YAMAMURA AND J. RICHARD, *Anal. Chem.*, 29 (1957) 158.

(2) Colorimetry

(i) Photometry of the rhodizonate

cf. R. N. WALTER, *Anal. Chem.*, 22 (1950) 1332.

(ii) Turbidimetry of $BaSO_4$. This method permits the estimation of small amounts of S^{2-}, SO_3^{2-}, and $S_2O_3^{2-}$ after oxidation to SO_4^{2-}, and of course a direct estimation of SO_4^{2-}. 0.1 to 3 μg of S may be estimated in this way.

The precipitate may be kept in the dispersed state with the aid of products such as polyoxyethylenesorbitol ('Tween 20')[1].

cf. W. VOLMER AND F. FRÖHLICH, *Z. Anal. Chem.*, 126 (1944) 401.
J. F. TREON AND W. E. CRUTCHFIELD, JR., *Ind. Eng. Chem., Anal. Ed.*, 14 (1942) 119.

In cements

L. BLONDIAU, *Ann. Chim. Anal.*, 26 (1944) 4 and 26.
(1) P. BLANC, P. BERTRAND AND L. LIANDIER, *Chim. Anal.*, 37 (1955) 305.

Nephelometry of $BaSO_4$. 20 to 200 μg of sulphur, in the form of SO_4^{2-} in 25 ml of solution can be estimated by this method.

5 ml of 1% barium chloride are added to the solution, which is then stirred and left for fifteen minutes. A comparison is then performed against standards prepared at the same time.

cf. E. CANALS AND A. CHARRA, *Bull. Soc. Chim.*, 12 (1945) 89.
H. J. KELLY AND L. B. ROGERS, *Anal. Chem.*, 27 (1955) 759.

(3) Other methods

(i) Using barium chloranilate

R. J. BERTOLACINI AND J. E. BARNEY II, *Anal. Chem.*, 29 (1957) 281.

(ii) With methylene blue thiocyanate. The solution is first passed through an SCN^- resin.

L. DUCRET AND M. RATONIS, *Anal. Chim. Acta*, 21 (1959) 91.

(iii) Reduction to S(II). Sulphur may be reduced to the sulphide in a medium containing Sn(II) and conc. H_3PO_4. The sulphide is then estimated colorimetrically.

T. KIBA AND I. KISHI, *Bull. Chem. Soc. Japan*, 30 (1957) 44.
L. GUSTAFSSON, *Talanta*, 4 (1960) 227, 236.
G. HEINEMANN AND H. W. RAHN, *Ind. Eng. Chem., Anal. Ed.*, 9 (1937) 458.

(iv) Estimation of concentrated H_2SO_4 *(96–100%). Colorimetry with quinalizarin*

E. ZIMMERMAN AND W. W. BRANDT, *Talanta*, 1 (1958) 374.

(v) Standard solutions. Standardized sulphuric acid may be used. K_2SO_4 and $(NH_4)_2SO_4$ of guaranteed purity are also available.

$$1 \text{ ml of } 1.000\ N\ H_2SO_4 = 32.1 \text{ mg of S or } 80.2 \text{ mg of } SO_3.$$

$$S/K_2SO_4 = 0.1842 \qquad SO_3/K_2SO_4 = 0.460$$

$$S/(NH_4)_2SO_4 = 0.243 \qquad SO_3/(NH_4)_2SO_4 = 0.607$$

Sulphides

(1) Separation

(i) H_2S is generally liberated when sulphides are attacked by strong acids. The gas is collected in zinc or cadmium acetate and ZnS or CdS are precipitated.

Sulphur in the form of the sulphide can be separated from SO_3^{2-}, $S_2O_3^{2-}$, and from many other anions by precipitation with zinc or cadmium acetate. $S_2O_3^{2-}$ is partially adsorbed.

cf. A. KURTENACKER AND R. WOLLAK, *Z. Anorg. Allgem. Chem.*, 161 (1927) 201.
R. POMEROY, *Anal. Chem.*, 26 (1954) 570.

S in titanium
M. CODELL, G. NORWITZ AND C. CLEMENCY, *Anal. Chem.*, 29 (1957) 1496.

Separation of traces in the form of CdS (10^{-9} in air)
M. B. JACOBS, M. M. BRAVERMAN AND S. HOCHEISER, *Anal. Chem.*, 29 (1957) 1349.

(ii) Oxidation by heating in a current of oxygen. The substance to be analyzed is heated in a current of oxygen or air at a suitable temperature.

The method is applicable to natural sulphides, steels, etc.

In slags
C. J. B. FINCHAM AND F. D. RICHARDSON, *Iron and Steel Inst.*, 172 (1952) 53.

In minerals
W. G. RICE-JONES, *Anal. Chem.*, 25 (1953) 1383.

In minerals and slags
R. BOULIN, R. DESGUIN AND E. JAUDON, *Chim. Anal.*, 36 (1954) 123.

In copper alloys
Centre technique de la fonderie [Technical Centre of the Foundry Industry] S-1, 1957.

In pyrites
L. M. JOLSON, E. I. DJADITSCHEWA AND L. B. GINZBURG, *Z. Anal. Chem.*, 110 (1937) 184.
A. MENDELOWITZ, *J. S. African Chem. Inst.*, 10 (1957) 36.

In steels, causes of error
J. W. FULTON AND R. E. FRYXELL, *Anal. Chem.*, 31 (1959) 401.

(iii) Standard solutions. $Na_2S \cdot 9H_2O$ of guaranteed purity is dissolved in cold boiled water.

The solution oxidizes rapidly in the air.

$$S/Na_2S \cdot 9H_2O = 0.1336.$$

(2) Colorimetry

(i) Formation of methylene blue. In acid solutions sulphides react with *p*-aminodimethylaniline in the presence of ferric iron to give methylene blue.

REAGENTS

Solution of the amine: 27.2 g of *p*-aminodimethylaniline sulphate are dissolved in 100 ml of 1/2 sulphuric acid.

Diluted solution of the amine: 25 ml of the preceding solution are diluted to 1 l with 1/2 sulphuric acid.

Ferric chloride: 100 g of $FeCl_3 \cdot 6H_2O$ are dissolved in 100 ml of water.

OPERATING PROCEDURE

1.5 ml of diluted amine solution, and then 5 drops of ferric chloride are added to 35 ml of solution containing 10 to 40 μg of sulphur in the form of sulphide. The volume is made up to 50 ml and the solution is left to stand for 15 minutes. The colorimetric measurement is performed at 680 mμ.

cf. J. K. FOGO AND M. POPOWSKY, *Anal. Chem.*, 21 (1949) 734.
M. S. BUDD AND H. A. BEWICK, *Anal. Chem.*, 24 (1952) 1536.
L. GUSTAFSSON, *Talanta*, 4 (1960) 227.

In steels
O. H. KRIEGE AND A. L. WOLFE, *Talanta*, 9 (1962) 673.

(ii) With nitroprusside. This gives a violet coloration in a slightly alkaline solution.

(iii) Nephelometry of sparingly soluble sulphides. H_2S is absorbed in caustic soda and a coloured sulphide is precipitated in the colloidal state, this may be compared with standards.

COMPARE

Bi_2S_3
H. KOREN AND W. GIERLINGER, *Mikrochim. Acta*, (1953) 220.

Other sulphides
E. TREIBER, H. KOREN AND W. GIERLINGER, *Mikrochim. Acta*, 40 (1952) 32.

Review of colorimetric estimations
H. KOREN AND W. GIERLINGER, *Mikrochim. Acta*, (1953) 220.

Sulphites, sulphurous acid, sulphur dioxide

(1) Separation

If sulphites are present in a highly diluted acidic medium, SO_2 is liberated. This gas may then be removed from the solution by boiling or by entrainment with an inert gas. Thiosulphates also give SO_2.

Gases may be purified from SO_2 by dissolving the latter in water, absorbing it in an alkaline solution, oxidizing it with iodine or hydrogen peroxide, or adsorbing it on silica gel.

Traces can be collected by mercuric chloride, and may subsequently be estimated colorimetrically[1].

(1) P. W. WEST AND G. C. GAEKE, *Anal. Chem.*, 28 (1956) 1816.
H. STRATMANN, *Mikrochim. Acta*, (1954) 668.

(2) Colorimetry

(i) With fuchsin in the presence of formaldehyde. 0.1 to 10 ppm of SO_2 may be estimated in this way.

cf. A. STEIGMANN, *Anal. Chem.*, 22 (1950) 492.
S. ATKIN, *Anal. Chem.* 22 (1950) 947.
P. W. WEST AND G. C. AND M. KATZ, *J. Air Pollution Control Assoc.*, 7 (1957) 25.

Continuously
H. L. HELWIG AND C. L. GORDON, *Anal. Chem.*, 30 (1958) 1810.

(ii) With the iodine-iodide-starch paste reagent

M. KATZ, *Anal. Chem.*, 22 (1950) 1040.

(iii) Decoloration of dichromate

S. SUSSMAN AND I. L. PORTNOY, *Anal. Chem.*, 24 (1952) 1652.

Thiosulphates

Colorimetry

(i) In the form of thiocyanate. In the presence of CN^- and Cu^{2+}, SCN^- is obtained, which can then be estimated colorimetrically by means of Fe^{3+}. Limit of sensitivity: 10^{-5} *M*.

cf. B. H. SÖRBO, *Biochim. Biophys. Acta*, 23 (1957) 412.

Dithionites (hydrosulphites)

Colorimetry

This is performed by the reduction of dyes (redox indicators).

cf. *Naphthol yellow S*
T. P. WHALEY AND J. A. GYAN, *Anal. Chem.*, 29 (1957) 1499.

Tantalum

See Niobium and Tantalum, p. 310.

Technetium

Tc = 99

The methods of separation and colorimetric estimation of rhenium are generally also applicable to the case of technetium. With less than a few micrograms of technetium however, the determination of the radioactivity of the substance is the only feasible method.

(i) Separation by extraction of tetraphenylarsonium pertechnetate.

S. TRIBALAT AND J. BEYDON, *Anal. Chim. Acta*, 8 (1953) 22.

(ii) Separation of Mo(VI) by ion-exchangers.

N. F. HALL AND D. H. JOHNS, *J. Am. Chem. Soc.*, 75 (1953) 5787.
E. H. HUFFMAN, R. L. OSWALT AND L. A. WILLIAMS, *J. Inorg. Nucl. Chem.*, 3 (1956) 49.

(iii) Colorimetry with SCN^-. $\varepsilon = 52{,}500$ at 513 mμ.

C. E. CROUTHAMEL, *Anal. Chem.*, 29 (1957) 1756.
O. H. HOWARD AND C. W. WEBER, *Anal. Chem.*, 34 (1962) 530.

(iv) Colorimetry with thioglycollic acid.

F. J. MILLER AND P. F. THOMASON, *Anal. Chem.*, 32 (1960) 1429.

(v) Separation and estimation of traces.

G. E. BOYD AND Q. V. LARSON, *J. Phys. Chem.*, 60 (1956) 707.

(vi) General documentation.

S. TRIBALAT, *Rhénium et Technétium*, Gauthier-Villars, Paris, 1957.

Tellurium

See Selenium and Tellurium, p. 365.

Thallium

Tl = 204.4

(1) Separation of trace amounts

(i) Precipitation of the sulphide. This method is only of interest in the absence of large amounts of precipitable elements and only for the purpose of concsn-trating traces of Tl.

An ammoniacal medium should be used, together with entraining agente: Ag(I), Pb(II), or Hg(II).

(ii) Precipitation of the chromate. Separation from Zn(II), Ni(II), Co(II), and Se(IV) may be effected in ammoniacal solutions.

Thallium may also be separated from Cd(II), Cu(II), Hg(II), and Ag(I) in CN^- media, and from As(III) and Sb(III) in a medium containing NH_3 and H_2O_2.

Trace amounts are entrained by Ba(II) (Sandell).

(iii) Precipitation of metallic Tl. Zinc is used in an acid medium.

H. B. Knowles, *Anal. Chem.*, 21 (1949) 1539.

(iv) Precipitation by thionalide. Thallium may be separated from all other elements in the presence of tartrate and cyanide in alkaline solutions.

Reagents

Thionalide, 5% in acetone, freshly prepared.
Potassium cyanide, 10% in water.
Caustic soda, 2 *N* (1/5).
Acetone.
Sulphuric acid, *N* (1/40).

Operating procedure

The solution should be neutral or slightly acidic. To 3–5 ml in a centrifuge tube are added 0.5 ml of 2 *N* caustic soda and 0.5 ml of cyanide, followed by 5–6 drops of thionalide. The mixture is brought to 90° C on the water bath for 5 minutes with shaking, in order to crystallize the precipitate. After cooling, the mixture is centrifuged and the precipitate washed three times with 3 ml of acetone. The precipitate can then be dissolved in 2 drops of sulphuric acid and 1 ml of alcohol, with warming.

cf. R. BERG, E. S. FAHRENKAMP AND W. ROEBLING, *Mikrochemie*, (1936) 42.
Z. Anal. Chem., 109 (1937) 305.

(v) Extraction of the chloride. Tl(III) can be extracted from a 6 *N* (1/2) HCl solution in the form of $HTlCl_4$ and $TlCl_3$, with ethers.

This is the best method for separating Tl(III) from Pb(II), but many other elements are also extracted: Au(III), Fe(III), Sb(V), Sn(IV), Ga(III), etc.

OPERATING PROCEDURE

The solution, which should contain 0.5 to 2 mg of Tl(III), is brought to 100 ml, and made 6 *N* with respect to hydrochloric acid; the chloride is then extracted three times with 20 ml of *iso*-propyl ether.

cf. K. LOUNAMAA, *Z. Anal. Chem.*, 147 (1955) 196.

(vi) Extraction of the bromide and the iodide.

H. M. IRVING AND F. J. C. ROSSOTTI, *Analyst*, 77 (1952) 801.

Cl^-, Br^-, I^-

F. A. POHL AND K. KOBES, *Mikrochim. Acta*, (1957) 318.

(vii) Extraction of the dithizonate. Tl(I) dithizonate can be extracted between pH 9 and pH 12. The extraction is slow.

The interfering ions can be complexed with CN^- and citrate ions. Under these conditions, only Pb(II), Sn(II) and Bi(III) accompany Tl(I). Large amounts of Ni(II), Co(II), Hg(II), and Zn(II) retard the extraction.

R. S. CLARKE, Jr. AND CUTTITTA, *Anal. Chim. Acta*, 19 (1958) 555.

(viii) Standard solutions. The pure metal is dissolved in dilute nitric acid.

(2) Colorimetry

(i) Oxidation of I^-. If Tl(I) is present, it should be oxidized with bromine, and the excess of the latter removed by reaction with phenol or by brief boiling. The following reaction is used:

$$Tl^{3+} + 3I^- \rightarrow TlI\downarrow + I_2.$$

The iodine is then estimated colorimetrically in CCl_4. 500 μg may be estimated to within 1–5% by this method.

Interfering ions. Oxidizing agents such as Cu(II), As(V), W(VI), and Mo(VI) interfere. Traces of Fe(III) do not interfere, and the influence of larger amounts of this ion can be eliminated by the addition of phosphate.

REAGENTS

10 ml of concentrated *hydrochloric acid,* 10 g of *disodium phosphate,* and 90 ml of *saturated bromine water.*

Phenol, 25% in pure acetic acid.

Potassium iodide, 0.2%.

OPERATING PROCEDURE

1 g of ammonium chloride is added to the neutral solution and the volume is made up to about 20 ml. 25 ml of bromine reagent are added, and the mixture is boiled for not longer than 3 minutes. 0.25 ml of phenol are added, and after standing for 5 minutes, the solution is made up to about 60 ml. 5 ml of iodide are added and the excess iodine is extracted with carbon tetrachloride, and estimated colorimetrically.

cf. P. A. SHAW, *Ind. Eng. Chem., Anal. Ed.,* 5 (1933) 92.
L. A. HADDOCK, *Analyst,* 60 (1935) 394.

(ii) Absorptiometry of $TlCl_6^{3-}$ at 250 mμ. Limit 4 μg.

cf. D. PESCHANSKI, *Bull. Soc. Chim.,* (1956) 1574.

(iii) With rhodamine B. In 2 *M* hydrochloric acid, Tl (III) reacts with rhodamine B to give a red-violet coloration which can be extracted with benzene. Sb(V), Au(III), Fe(III), Hg(II), and Ga(III) give an analogous reaction.

REAGENTS

Hydrochloric acid, 2 *N* (1/6).

Bromine water, saturated.

Rhodamine B: 0.2 g of rhodamine B is dissolved in 100 ml of 2 *N* hydrochloric acid.

Benzene.

OPERATING PROCEDURE

0.5 ml of bromine water are added to the sample to oxidize the thallium to Tl(III); the sample itself consists of 5 ml of solution made 2 *N* with respect to hydrochloric acid and containing 0.5 to 10 μg of thallium. The mixture is heated at about 95° (until the excess of bromine has disappeared), cooled and made up to 10 ml with 2 *M* hydrochloric acid, and 1 ml of rhodamine B is added, followed by 5 ml of benzene. The mixture is shaken for 1 minute, the benzene phase is separated and colorimetry is performed at 560 mμ.

(3) Other methods

(i) Extraction of $TlCl_6^{3-}$ *and* $TlBr_6^{3-}$ *+ rhodamine B.*

H. ONISHI, *Bull. Chem. Soc. Japan*, 29 (1956) 945; 30 (1957) 567.
J. F. WOOLLEY, *Analyst*, 83 (1958) 477.
C. L. LUKE, *Anal. Chem.*, 31 (1959) 1680.

With methyl violet and brilliant green
N. T. VOSKRESENSKAYA, *Zh. Analit. Khim.*, 11 (1956) 585.

Extraction
M. KOVAŘIK AND M. MOUČKA, *Anal. Chim. Acta*, 16 (1957) 249.

In minerals
I. A. BLYUM AND I. A. UL'YANOVA, *Zavodsk. Lab.*, 23 (1957) 283.

Thorium

Th = 232.1

(1) Separation of trace amounts

(i) Precipitation of the fluoride. Trace amounts of thorium are most commonly separated by precipitating the fluoride with La(III) as an entraining agent. Thorium may in this way be easily separated from U(VI), Nb(V), Ta(V), Ti(IV), Zr(IV), and Fe(III).

cf. H. H. WILLARD, A. W. MOSEN AND R. D. GARDNER, *Anal. Chem.*, 30 (1958) 1614.

(ii) Precipitation of the hypophosphate ThP_2O_6. This method is also used principally with traces. Bi(III) can be added as an entraining agent. The medium employed should be 2 *N* with respect to $HClO_4$.

cf. L. KOSTA, *Energia nucleare (Milan)*, 4 (1957) 37.

(iii) By means of arsonic acids. The precipitation is generally only complete in buffered acetate solutions. Ti(IV), Zr(IV), Ce(IV), U(IV), Sn(IV), Nb(V), and Ta(V) precipitate as well in more acidic media.

cf. A. C. RICE, H. C. FOGG, AND C. JAMES, *J. Am. Chem. Soc.*, 48 (1926) 895.
H. GRUNDMANN, *Aluminium*, 24 (1942) 105.

(iv) Extraction of the nitrate. Thorium nitrate can be extracted with numerous solvents, particularly with mesityl oxide, and may thus be separated from almost other elements, especially from the rare earth metals. Only Zr, V, and U interfere.

Alternatively, the extraction may be performed with a mixture of hexone and tributyl phosphate. Zr(IV) can be complexed with mesotartaric acid[1].

The extraction is facilitated by the addition of large amounts of aluminium nitrate, lithium nitrate etc. to the aqueous phase.

(v) With mesityl oxide

REAGENTS

Washing solution: 380 g of aluminium nitrate, $Al(NO_3)_3 \cdot 9H_2O$, are dissolved in 170 ml of water and 30 ml of concentrated nitric acid are added.

Mesityl oxide.

OPERATING PROCEDURE

(1) To 15 ml of nitrate solution are added 2 ml of concentrated nitric acid and 19 g of aluminium nitrate. The nitrate is extracted with 20 ml of mesityl oxide, shaking for 20 seconds. The extraction is repeated once, and the two organic portions are combined and washed three times with the washing solution.

(2) Instead of aluminium nitrate, 16 g of lithium nitrate may be used, and the thorium nitrate extracted under identical conditions.

The thorium can be reclaimed into the aqueous phase by shaking with water.

cf. C. V. BANKS AND C. H. BYRD, *Anal. Chem.*, 25 (1953) 416.
H. LEVINE AND F. S. GRIMALDI, *U.S. Atomic Energy Commission Rept.* AECD, 3186, 1950.
C. V. BANKS AND R. E. EDWARDS, *Anal. Chem.*, 27 (1955) 947.

With hexone
R. BOCK AND E. BOCK, *Z. Anorg. Allgem. Chem.*, 263 (1950) 152.
(1) D. A. EVEREST AND J. V. MARTIN, *Analyst*, 84 (1959) 312.

Partition chromatography; cellulose
N. F. KEMBER, *Analyst*, 77 (1952) 78.

*(vi) With tri-*n*-octylphosphine oxide.* Thorium may be separated from the rare earth metals in cyclohexane solutions with the aid of this reagent. Sulphates and phosphates do not interfere.

cf. W. J. ROSS AND J. C. WHITE, *Anal. Chem.*, 31 (1959) 1847.
R. F. APPLE AND J. C. WHITE, *Chemist-Analyst*, 49 (1960) 421.

(vii) With ion-exchangers. An appropriate cationic or anionic resin allows the separation of thorium from a large number of elements.

(1) *Cationic resin.* Thorium may be separated from anions such as phosphates, but not from sulphates.

In 2 *M*[1] or 4 *M*[2] hydrochloric acid, Th(IV) is collected while the divalent elements, U(VI), Fe(III), Mo(VI), the rare earth metals, and Al(III) pass through. Thorium is then eluted with 3 *M* H_2SO_4[5].

(2) *Anionic resin.* In 9 *M* hydrochloric acid, thorium is not collected by the resin, while Fe(III) and U(VI) are retained[1,3].

Thorium can also be collected as a sulphate complex at about pH 2, and may thus be separated from the rare earth metals. Al(III), Fe(III), and U(VI) accompany Th(IV)[4].

COMPARE

(1) O. A. NIETZEL, B. W. WESSLING AND M. A. DE SESA, *Anal. Chem.*, 30 (1958) 1182.
F. W. E. STRELOW, *Anal. Chem.*, 33 (1961) 1648.
(2) F. W. E. STRELOW, *Anal. Chem.*, 31 (1959) 1201.

(3) K. A. Kraus, G. E. Moore and F. Nelson, *J. Am. Chem. Soc.*, 78 (1956) 2692.
(4) R. A. Nagle and T. R. S. Murthy, *Analyst*, 84 (1959) 37.
(5) K. S. Chung and J. Riley, *Anal. Chim. Acta*, 28 (1963) 1.

(viii) Standard solutions. These can be made from pure thorium metal or ThO_2.

$$Th/ThO_2 = 0.879.$$

(2) Colorimetry

(i) With 'thoron'. 'Thoron', 'thorin', or APANS, [disodium 2-(2-hydroxy-3,6-disulpho-1-naphthylazo)-benzenearsonate] reacts with Th(IV) to give a coloration which is stable at pH 1.

Interfering ions. A number of ions interfere. Zr(IV), Fe(III), U(IV), the rare earth metals, and oxidizing agents (including NO_3^-) should be separated. However, the addition of tartaric acid often enables Th(IV) to be estimated without previous separations[1].

Sensitivity. $\varepsilon \sim 13{,}000$ at 545 mμ.

Reagents

Hydrochloric acid, 3 *M* (1/4)

'Thoron': 100 mg are dissolved in 100 ml of water.

Operating procedure

4 ml of slightly acidic solution containing from 20 to 100 μg of thorium are mixed with 1 ml of 3 *M* hydrochloric acid and then 1 ml of reagent. The volume is made up to 10 ml and the colorimetric determination is performed at 545 mμ in comparison with a blank test containing exactly the same volume of reagent.

cf. C. V. Banks and C. H. Byrd, *Anal. Chem.*, 25 (1953) 416.
J. Clinch, *Anal. Chim. Acta*, 14 (1956) 162.
M. H. Fletcher, F. S. Grimaldi and L. B. Jenkins, *Anal. Chem.*, 29 (1957) 963.
F. S. Grimaldi, L. B. Jenkins and M. H. Fletcher, *Anal. Chem.*, 29 (1957) 848.

(1) *In the presence of Zr(IV)*
F. S. Grimaldi and M. H. Fletcher, *Anal. Chem.*, 28 (1956) 812.
D. A. Everest and J. V. Martin, *Analyst*, 84 (1959) 312.

(3) Other methods

(i) With 'arsenazo'. 2-(1,8-dihydroxy-3,6-disulpho-2-naphthylazo)-benzenearsonic acid.

cf. J. S. Fritz and E. C. Bradford, *Anal. Chem.*, 30 (1958) 1021.
F. V. Zaikovskü and L. I. Gerkhardt, *Zh. Analit. Khim.*, 13 (1958) 274.

M. ISHIBASHI AND S. HIGASHI, *Japan Analyst*, 4 (1955) 14.
A. L. ARNFELT AND I. EDMUNDSSON, *Talanta*, 8 (1961) 473.

(ii) With SPADNS (2-*p*-sulphophenylazo-1,8-dihydroxynaphthalene-3,6-disulphonic acid).

$\varepsilon = 20{,}000$ at 575 mμ. SO_4^{2-} and Zr(IV) interfere; U does not interfere.

cf. G. BANERJEE, *Anal. Chim. Acta*, 16 (1957) 56.
J. A. COOPER AND M. J. VERNON, *Anal. Chim. Acta*, 23 (1960) 351.

(iii) With SNADNS (*o*-arsonophenylazochromotropic acid). Zr(IV), Ti(IV), Ce(III), U(VI), Fe(III), sulphates, and phosphates interfere.

T. TAKAHASHI AND S. MIYAKE, *Talanta*, 3 (1959) 155.

(iv) With quercetin (3,5,7,3′,4′-pentahydroxyflavone). $\varepsilon = 33{,}000$ at 422 mμ.

O. MENIS, D. L. MANNING AND G. GOLDSTEIN, *Anal. Chem.*, 29 (1957) 1426.
E. TOMIC AND H. KHALIFA, *Mikrochim.*, (1957) 668.

(v) With Eriochrome Black T in the presence of EDTA; $\varepsilon = 35{,}000$.

P. F. LOTT, K. L. CHENG AND B. C. H. KWAN, *Anal. Chem.*, 32 (1960) 1702.

Tin

Sn = 118.7

(1) Separation of trace amounts

It should be borne in mind that $SnCl_4$ is volatile and can be lost when a hydrochloric acid solution is evaporated during the course of a separation.

(i) Precipitation of the hydroxide. $Sn(OH)_4$ is sparingly soluble in nitric and perchloric acids, and the precipitate formed adsorbs numerous impurities. Some Sb(V) is present in the precipitate, as also are As(V) and P(V). The precipitate is soluble in sodium sulphide.

When only traces of tin are present, $Sn(OH)_4$ can be entrained by $Mn(OH)_4$[1].

In the presence of P(V)

O. Gates and L. Silverman. *Ind. Eng. Chem., Anal. Ed.*, 11 (1939) 370.

Purification of SnO_2

E. R. Caley and M. G. Burford, *Ind. Eng. Chem., Anal. Ed.*, 8 (1936) 114.

(1) S. Kühnel-Hagen, N. Hohman-Bang and P. Gjertsen, *Acta Chem. Scand.*, 2 (1948) 343.

(ii) Precipitation of the sulphide. SnS_2 can be precipitated from an acid medium (4 *N* H_2SO_4). When only traces of tin are present Cu(II) may be used as an entraining agent. Tin can be separated from W(VI), V(V), and Ti(IV) by this method, using a tartaric acid medium.

cf. F. R. Williams and J. Whitehead, *J. Applied Chem.*, 2 (1952) 213.

(iii) Distillation of the chlorides. As(III) and Sb(III) can be separated by distilling off their respective chlorides (see p. 172). HCl and HBr are then added, and the tin, which passes over at 140–150° C, is collected. Ti(IV) interferes if present in large amounts.

(iv) Distillation of $SnBr_4$.

cf. W. D. Mogerman, *J. Res. Natl. Bur. Std.*, 33 (1944) 307.

H. Onishi and E. B. Sandell, *Anal. Chim. Acta*, 14 (1956) 153.

(v) Precipitation by means of phenylarsonic acid. In 0.6 *N* (1/20) hydrochloric acid or 3 *N* (1/12) sulphuric acid, Sn(IV), Th(IV), and Zr(IV) are precipitated

and may thus be separated from Cu(II), Zn(II), Pb(II), As(III), Sb(III), Ni(II), Fe(III), Cd(II), and Al(III).

cf. J. S. KNAPPER, K. A. CRAIG AND C. G. CHANDLEE, *J. Am. Chem. Soc.*, 55 (1933) 3945.

(vi) Extraction of the iodide with ether, benzene, etc.

5 ml of solution containing 6.9 *M* HI are treated with 20 ml of ether.

The following are completely extracted: Sb(III), Hg(II), Cd(II), and Au(III); the following pass across partially: Bi(III), Zn(II), Mo(VI), Te(IV), and In(III)[1]. Sn(II) may also be extracted[2].

(1) S. KITAHARA, *Bull. Inst. Phys. Chem. Res. (Tokyo)*, 24 (1948) 454.
(2) H. M. IRVING AND F. J. C. ROSSOTTI, *Analyst*, 77 (1952) 801.
M. MALAT, *Z. Anal. Chem.*, 187 (1962) 404.

(vii) Extraction by means of diethyldithiocarbamate. The compound formed between Sn(II) and diethylammonium diethyldithiocarbamate can be extracted.

cf. P. F. WYATT, *Analyst*, 80 (1955) 368.

(viii) With anion-exchangers. Sn(IV) is collected from an HCl solution[1] or from oxalic acid[2].

M. ARIEL AND E. KIROWA, *Talanta*, 8 (1961) 214.
(1) A. EVEREST AND J. H. HARRISON, *J. Chem. Soc.*, (1957) 1439.
(2) G. W. SMITH AND S. A. REYNOLDS, *Anal. Chim. Acta*, 12 (1955) 151.

(ix) Standard solutions. These can be made from pure tin, 1g of the metal is dissolved in 100 ml of 6 *N* (1/2) hydrochloric acid and the volume is made up to 1 litre with hydrochloric acid.

98–100% $SnCl_2 \cdot 2H_2O$ can also be used; this is dissolved in dilute hydrochloric acid.

$$Sn/SnCl_2 \cdot 2H_2O = 0.526.$$

(2) Colorimetry

In a certain number of methods, Sn(II) is used to reduce substances which act as redox indicators. Such methods do not permit the estimation of very slight traces, are not very accurate, and necessitate preliminary separations.

Dithiol gives a coloured compound with tin. This reagent is however unstable and the reaction is not very selective.

Phenylfluorone gives a coloration and can therefore be used under all conditions after suitable preliminary purifications have been performed.

(i) With dithiol. (Toluene-3,4-dithiol). Sn(II) gives a red precipitate which may be estimated colorimetrically in the colloidal state. Even small traces may be detected by this method.

Limit of sensitivity. Under normal conditions, 1 to 6 μg of tin can be estimated, and as little as 0.1 ppm can be detected.

The coloration can be extracted with ethers and alcohols.

Interfering ions. Many ions interfere with the results. Thus Bi(III) gives a brick red precipitate, Cu(II), Ni(II), and Co(II) give black precipitates, Ag(I), Hg(II), Pb(II), Cd(II), and As(III), etc, give yellow precipitates. The NO_2^- ion interferes. It is however possible to operate in the presence of 20 times more Fe and Pb than Sn. Mn(II) and Zn(II) give rise to low results.

The acidity should be accurately fixed.

Sn(IV) is reduced slowly by the reagent and then undergoes reaction. The reagent is usually stabilized by the addition of thioglycollic acid, which rapidly reduces Sn(IV) to Sn(II).

REAGENTS

10 ml of thioglycollic acid are added to 0.25 ml of *melted dithiol*. The solution is diluted to 200 ml with alcohol and stored in small stoppered bottles protected from light.

Gum arabic in 10% solution.

OPERATING PROCEDURE

5 ml of the solution, containing less than 50 μg of Sn(IV), are neutralized with ammonia and then 0.5 ml of concentrated hydrochloric acid are added, followed by 0.5 ml of the reagent. The mixture is placed on a boiling water bath for 1 minute and, after cooling, shaken with 2 ml of gum arabic. Colorimetry is performed at 530 mμ.

The colour is stable for 1 month in a sealed tube in the dark.

cf. M. FARNSWORTH AND J. PEKOLA, *Anal. Chem.*, 26 (1954) 735.
T. C. J. OVENSTON AND C. KENYON, *Analyst*, 80 (1955) 566.

In beer
I. STONE, *Ind. Eng. Chem., Anal. Ed.*, 13 (1941) 791.

In foodstuffs
R. DE GIACOMI, *Analyst*, 65 (1940) 216.
N. H. LAW, *Analyst*, 67 (1942) 283.
H. CHEFTEL, F. CUSTOT AND M. NOWAK, *Bull. Soc. Chim.*, 16 (1949) 441.

In biochemistry
J. SCHWAIBOLD, W. BORCHERS AND G. NAGEL, *Biochem. Z.*, 306 (1940) 113.

In minerals

H. Onishi and E. B. Sandell, *Anal. Chim. Acta*, 14 (1956) 153.

(ii) With phenylfluorone (2,3,7-Trihydroxy-9-phenyl-6-fluorone).

The coloration becomes specific when Sn(IV) has been separated by carbamate-chloroform extraction.

Interfering ions. A number of ions interfere. Ge(IV) gives the same reaction. Zr (IV), Ga(III), Fe(III), Ta(V), and P(V) interfere.

The limit of sensitivity is 0.02 ppm.

Small amounts of Mo(VI), Ti(IV), Nb(V), and Ta(V) may be complexed with hydrogen peroxide.

Reagents

Phenylfluorone: 50 mg of reagent are dissolved in 50 ml of methanol and 1 ml of concentrated hydrochloric acid. The volume is made up to 500 ml with methanol.

Gum arabic: 0.5 g of powder is dissolved in 50 ml of water. A fresh solution should be prepared every day.

Dilute sulphuric acid: 80 ml of water are mixed with 20 ml of concentrated sulphuric acid.

Dilute - N (1/12) - hydrochloric acid.

Buffer solution: 450 g of sodium acetate are dissolved in 350 ml of water; 240 ml of concentrated acetic acid are added, and the volume made up to 1 litre.

Hydrogen peroxide, 10 volume (3%)

Standard tin solution: 1 g of guaranteed purity tin is dissolved in 100 ml of 1/2 hydrochloric acid with gentle warming. The solution is made up to 1 litre. A solution containing 20 μg of tin per ml is prepared by dilution with 0.2 *N* (1/25) hydrochloric acid.

Operating procedure

To 10 ml of solution containing 10 to 60 μg of tin are added 5 ml of dilute sulphuric acid, 1 ml of hydrogen peroxide, 10 ml of buffer solution, 1 ml of gum arabic, and finally 10 ml of phenylfluorone. The mixture should be stirred after the addition of each reagent. After standing for 5 minutes, the volume is made up to 50 ml with dilute hydrochloric acid, with stirring. Colorimetry is performed immediately at 510 mμ.

cf. C. Luke, *Anal. Chem.*, 28 (1956) 1276.
L. B. Ginzburg and E. P. Shkrobot, *Zavodsk. Lab.*, 23 (1957) 527.
R. L. Bennett and H. A. Smith, *Anal. Chem.*, 31 (1959) 1441.

In Cu and Pb alloys

C. L. Luke, *Anal. Chem.*, 31 (1959) 1803.

(3) Other methods

Colorimetry of Sn (IV) oxinate in chloroform

P. F. Wyatt, *Analyst*, 80 (1955) 374.

A. R. Eberle and M. W. Lerner, *Anal. Chem.*, 34 (1962) 627.

With hematoxylin

H. Teicher and L. Gordon, *Anal. Chem.*, 25 (1953) 1182.

Sn(II)–Cl⁻–crystal violet ($\varepsilon \sim$ 85,000 at 595 mμ in butyrone)

L. Ducret and H. Maurel, *Anal. Chim. Acta*, 21 (1959) 79.

Nephelometry with cupferron

M. Jean, *Ann. Chim.*, 3 (1948) 516.

Turbidimetry with arsonic acids

P. Karsten, H. L. Kies and J. J. Walraven, *Anal. Chim. Acta*, 7 (1952) 355.

C. M. Dozinel and H. Gill, *Chemist-Analyst*, 45 (1956) 105.

Copper alloys

H. J. G. Challis and J. T. Jones, *Anal. Chim. Acta.*, 21 (1959) 58.

C. M. Dozinel, *Modern Methods of Analysis of Copper and its Alloys*, Elsevier, Amsterdam, 1963.

Titanium

Ti = 47.90

(1) Separation of trace amounts

Of the more familiar methods, the following can be used:

(i) Precipitation of the hydroxide with ammonia. Al(III) is used as the entraining agent.

(ii) Precipitation of the hydroxide with caustic soda. Fe(III) is used as the entraining agent. Titanium may thus be separated from V(V), Mo(VI), P(V), and Cr(VI). A double precipitation should be performed.

(iii) Digestion with fused sodium carbonate. The product is dissolved in water. Ti(IV) is retained in the insoluble matter.

(iv) With cupferron. Ti(IV) precipitates in acid solutions and may thus be separated from Al(III) and P(V). Zr(IV), Mo(VI), Fe(III), V(V), etc., also precipitate. Co(II) and Ni(II) are coprecipitated if present in high concentrations.

In the presence of EDTA and at pH 4.3 to 7, the precipitation of Zr(IV), Th(IV), etc. is prevented[1].
The compound can be extracted under identical conditions[2].

(1) A. K. Majumdar and J. B. R. Chowdhury, *Anal. Chim. Acta*, 15 (1956) 105.

Extraction

T. C. J. Ovenstone, C. A. Parker and C. G. Hatchard, *Anal. Chim. Acta*, 6 (1952) 7.

Separation of Nb(V) and Ta(V)

I. P. Alimarin and I. M. Gibalo, *Dokl. Akad. Nauk SSSR, Otd. Khim. Nauk*, 109 (1956) 511.

(2) K. L. Cheng, *Anal. Chem.*, 30 (1958) 1941.

(v) With arsonic acids. Ti(IV) and Zr(IV) precipitate.

cf. Hydroxybenzenearsonic acid

F. Richter, *Z. Anal. Chem.*, 121 (1941) 1.

Precipitation of the arsenate (Using Zr(IV) as the entraining agent)

In steels

J. R. Simmler, K. H. Roberts and S. M. Tuthill, *Anal. Chem.*, 26 (1954) 1902.

(vi) With oxine.

cf. A. CLAASSEN AND J. VISSER, *Rec. Trav. Chim.*, 60 (1941) 715.

(vii) Electrolysis at a mercury cathode. Titanium may be separated from a number of other elements, the latter being deposited while titanium remains in solution.

(viii) Standard solutions. (1) These can be prepared from the pure metal.

(2) 1.7 g of TiO_2 of guaranteed purity, calcined to constant weight, are digested with 10 g of fused pyrosulphate. The product is taken up in 200 ml of hot 4 *N* (1/9) sulphuric acid, and the volume is made up to 2 litres by means of 4 *N* H_2SO_4.

(3) Alternatively, TiO_2 can be heated to gentle boiling with a mixture of 10 parts of concentrated sulphuric acid and 1 part of ammonium sulphate. TiO_2 passes into solution.

(4) A known weight of TiO_2 is digested with fused Na_2O_2 and taken up in 4 *N* (1/9) sulphuric acid.

(5) A commercial solution of titanous chloride of approximately molar strength can be used. This is diluted and titrated by precipitating $Ti(OH)_4$ by the addition of ammonia and hydrogen peroxide at the boil. The TiO_2 produced is weighed.

(2) Colorimetry

The most important method involves the formation of complexes between Ti(IV) and hydrogen peroxide. This procedure is selective and very accurate, but of low sensitivity.

For trace amounts, the chromotropic acid method which can be made specific, or the thymol method which is not accurate but is sensitive and practically specific, can be used; vanadium does not interfere under these conditions.

(i) With hydrogen peroxide. With hydrogen peroxide, Ti(IV) gives orange-coloured complex ions. A 1.5–5.0 *N* sulphuric acid medium is generally used. The colorimetry is carried out at 410 mμ, and the coloration is stable indefinitely.

Molar extinction coefficient: $\varepsilon = 550$ at 436 mμ.

Interfering ions. Ions which complex Ti(IV) interfere to some extent: F^- should be absent; $C_2O_4^{2-}$ attenuates the coloration; large amounts of SO_4^{2-} and H_3PO_4 also have an appreciable influence; citrate ions interfere, but tartrate ions do so to a much smaller extent. The ionic strength of the solution is a factor of some importance, since the complex is fairly unstable.

Yellow complexes are given by a certain number of other ions such as U(VI), Mo(VI), and Nb(V); the colorations formed are however, weak. On the other hand V(V) interferes greatly, and must be estimated at the same time as Ti(IV) by working at two wavelengths (see below).

Coloured ions may interfere: Ni(II) only slightly and Fe(III), U(VI) and Cr(III) to a far greater extent. Fe(III) can be complexed by the addition of phosphoric acid, an identical amount of which should be added to the standards.

In addition, solutions with and without hydrogen peroxide may be compared.

REAGENT

Hydrogen peroxide, 3% (10 volume).

OPERATING PROCEDURE

The solution should contain 10 to 50 ppm of Ti(IV). The acidity is adjusted to 2 *N* H_2SO_4, 3 ml of hydrogen peroxide are added, and the volume is made up to 100 ml.

cf. A. WEISSLER, *Ind. Eng. Chem., Anal. Ed.*, 17 (1945) 695 and 775.

In metals

1956 Book of A.S.T.M. Methods of Chemical Analysis of Metals, American Society for Testing Materials, Philadelphia.

In steels

A. WEISSLER, *Ind. Eng. Chem., Anal. Ed.*, 17 (1945) 775.
L. MAILLARD AND J. ETTORI, *Compt. Rend. (Paris)*, 202 (1936) 594.
W. T. L. NEAL, *Analyst*, 79 (1954) 403.
R. A. PAPUCCI, *Anal. Chem.*, 27 (1955) 1175.

In Ta

J. HASTINGS, T. A. MCCLARITY AND E. J. BRODERICK, *Anal. Chem.*, 26 (1954) 381.

In U-Ti, precision colorimetry

G. W. C. MILNER AND P. J. PHENNAH, *Analyst*, 79 (1954) 414.

(ii) Estimation of Ti(IV) and V(V) at two wavelengths. The wavelengths used are 410 mμ and 460 mμ. The two colours differ only slightly (Figure 72), and the method is not accurate. A nomogram allows the Ti and V contents to be obtained as functions of the optical densities read at the two wavelengths.

cf. A. WEISSLER, *Ind. Eng. Chem., Anal. Ed.*, 17 (1945) 775.

In aluminium

H. GINSBERG, *Angew. Chem.*, 51 (1938) 663.

(iii) Estimation of Ti(IV) + V(V) + Mo(VI). With an accurate and sufficiently monochromatic apparatus, this can be achieved by working at three wavelengths: 330, 410, and 460 mμ

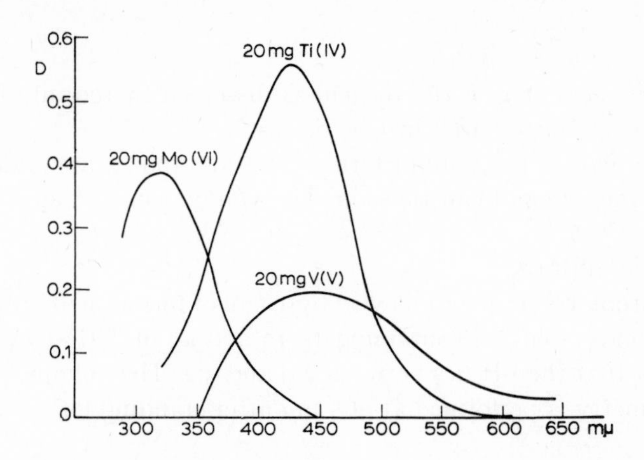

Fig. 72. After A. WEISSLER, *Ind. Eng. Chem., Anal. Ed.*, 17 (1945) 695.

(iv) With thymol. With Ti(IV) in concentrated sulphuric acid, thymol gives an orange coloration. The reaction is better than the preceding one for trace amounts.

Extinction coefficient: $\varepsilon \sim 5{,}000$ at 425 mμ.

The reaction is practically specific: only W(VI) and large amounts of Mo(VI) and Cr(III) alter the coloration, while Nb(V) gives a coloration of low sensitivity. This method should be used when vanadium is present. Unfortunately, many sulphates – Fe(III), Al(III), Ni(II), Co(II), etc. – are sparingly soluble in concentrated sulphuric acid and give rise to turbidity.

cf. J. V. GRIEL AND R. J. ROBINSON, *Anal. Chem.*, 23 (1951) 1871.

(v) With chromotropic acid. With Ti(IV), chromotropic acid gives various complexes according to the pH. The existence of an isosbestic point at 470 mμ enables the estimation to be carried out between pH 2 and 5.

Sensitivity. $\varepsilon \sim 17{,}000$ at 460 mμ.

Interfering ions. Interference is obtained with Fe(III), Nb(V), F^-, phosphates, and large amounts of Mo(VI), W(VI), and V(V)[1]. U(VI) interferes owing to its coloration; by working at 500 mμ, however, it is possible to estimate 1 ppm of Ti(IV), even in the presence 2,000 ppm of U (VI).

Fe(III) should be pre-reduced, *e.g.* by ascorbic acid.

The estimation may also be performed in the presence of EDTA at pH 5.4–

6.0[1]. Under such conditions, 1 ppm of Ti(IV) may be estimated in the presence of 500 ppm of Fe(III), 800 ppm of Th(IV), 300 ppm of Zr(IV), and 500 ppm of Mo(VI).

Nb(V) can be complexed with oxalate.

REAGENTS

Chromotropic acid: 1 g of the reagent is dissolved in 100 ml of water. This solution should be prepared just before use.

Formic acid buffer: to 57 ml of formic acid (98–100%) are added 200 ml of water and then 28 g of caustic soda. The volume is made up to 1 l.

OPERATING PROCEDURE

10 ml of chromotropic acid, followed by 10 ml of formic acid buffer are added to 10 ml of acidic solution containing 10 to 200 μg of Ti(IV). A pH-meter is used to check that the pH lies between 2.9 and 3.2. The volume is made up to 50 ml, colorimetry is performed at 465 mμ after standing for 30 minutes.

(1) L. SOMMER, *Z. Anal. Chem.*, 164 (1958) 299.
A. OKÁČ AND L. SOMMER, *Coll. Czech. Chem. Comm.*, 22 (1957) 433.
L. SOMMER, *Coll. Czech. Chem. Comm.*, 22 (1957) 1793.

In steels
W. KOCH AND H. PLOUM, *Archiv. Eisenhüttenw.*, 24 (1953) 393.
(1) R. ROSOTTE AND E. JAUDON, *Anal. Chim. Acta*, 6 (1952) 149.

(vi) With 'tiron'. (Sodium 1,2-dihydroxybenzene–3,5-disulphonate).

This gives a yellow coloration between pH 4 and 10. Fe(III), and V(V) give interfering colorations. Fe(III) + Ti(IV) may be estimated together by working at two wavelengths, or alternatively, Fe(III) may be reduced to Fe(II) or complexed with EDTA[1].

$\varepsilon \sim 15{,}000$ at 410 mμ.

cf. J. H. YOE AND A. R. ARMSTRONG, *Anal. Chem.*, 19 (1947) 100.
G. V. POTTER AND C. E. ARMSTRONG, *Anal. Chem.*, 30 (1948) 1208.
(1) P. SZARVAS AND B. CSISZAR, *Acta Chim. Hung.*, 7 (1955) 403.

In Al
B. MONLEZUN, *Chim. Anal.*, 42 (1960) 377.

In rocks
T. RIGG AND H. A. WAGENBAUR, *Anal. Chem.*, 33 (1961) 1347.

Tungsten

W = 183.9

(1) Separation of trace amounts

(i) The dry method. After digestion with fused sodium carbonate or caustic soda and treatment with water, the tungsten passes into solution together with Mo(VI), Cr(VI), V(V), P(V), etc. It can be separated completely from certain elements, but Sn(IV), Nb(V), and Ta(V), which distribute themselves between the insoluble matter and the solution, may retain it partially in the residue.

(ii) Precipitation with ammonia. Most of the tungstates are sparingly soluble. The method is suitable for recovering traces of W(VI) after the addition of Fe(III).

cf. M. Ishibashi, T. Shigematsu and Y. Nakagawa, *Bull. Inst. Chem. Res., Kyoto Univ.*, 32 (1954) 199.

(iii) With cupferron. W(VI), V(V) and Mo(VI) are precipitated. The compounds formed can be extracted with *iso*-amyl alcohol.

cf. S. H. Allen and M. B. Hamilton, *Anal. Chim. Acta*, 7 (1952) 483.

(iv) With oxine. Precipitation occurs at pH 4.5–5.0, in the presence of EDTA. V(IV), in particular, does not precipitate.

cf. B. Rehak and M. Malínek, *Z. Anal. Chem.*, 153 (1956) 166.

(v) Extraction with α-benzoin oxime. W(VI) precipitates in an acid medium together with Mo(VI), Nb(V), Ta(V), and U(VI).

The compound formed can be extracted with chloroform.

Reagents

α-Benzoin oxime, 2% in alcohol.

Chloroform.

Operating procedure

(To be used with traces of tungsten). 2 ml of α-benzoin oxime solution are added to the solution of the sample made 1 *N* with respect to HCl. The mixture is placed in a separating funnel, successively extracted with 10, 5, 5, and 5 ml of chloroform, shaking for one and a half minutes each time.

The solvent can then be driven off from a 50 ml flask. 1 ml of 9 *N* (1/4) sulphuric acid is added, and the reagent is destroyed by successive evaporations to white fumes in the presence of nitric acid and then perchloric acid. The colorimetric determination is then carried out.

cf. P. C. TEFFERY, *Analyst*, 81 (1956) 104.

(vi) Standard solutions. Guaranteed purity sodium tungstate, $Na_2WO_4 \cdot 2H_2O$ may be used.

$$W/Na_2WO_4 \cdot 2H_2O = 0.557$$

$$WO_3/Na_2WO_4 \cdot 2H_2O = 0.703$$

(2) Colorimetry

(i) With thiocyanate after reduction. In hydrochloric acid, in the presence of thiocyanate and after reduction, W(VI) gives a yellow complex of W(V). The compound formed is soluble in alcohols, alcohol-chloroform mixtures, ethers, and esters. Various methods of reduction have been proposed: Sn(II), in concentrated hydrochloric acid, tin amalgam, and, more conveniently, a mixture of Sn(II) and Ti(III) in 3 to 4 *M* hydrochloric acid.

Accuracy and sensitivity. As little as 60 μg of tungsten in 10 ml can be estimated to within a few percent.

Extinction coefficient: this varies with the operating procedure, assuming values between 12,000 and 20,000 in water and various solvents.

Interfering ions. Mo(VI) interferes at 10 ppm and above, Cu(II) gives a precipitate above 4 ppm, and V(V) and nitrates interfere.

Fe(III), Mn(II), Ni(II), or Co(II) do not interfere when present in a concentration of 120 ppm.

REAGENTS

Stannous chloride: 45 g of stannous chloride are dissolved in 50 ml of concentrated hydrochloric acid, and 50 ml of water are added.

Titanous chloride: commercial solution containing 15% of $TiCl_3$ are used.

Potassium thiocyanate, 20% in water.

OPERATING PROCEDURE

15 ml of concentrated hydrochloric acid, 5 ml of stannous chloride, 0.2 ml of titanous chloride, and finally 5 ml of thiocyanate are added to 10 ml of solution containing 100 to 600 μg of W(VI). The volume is made up to 50 ml, and colorimetry is performed at 400 mμ after standing for 15 minutes.

cf. G. ECKERT AND E. BAUERSACHS, *Z. Anal. Chem.*, 163 (1958) 161.
C. H. R. GENTRY AND L. G. SHERRINGTON, *Analyst*, 73 (1948) 57.
W. WESTWOOD AND A. MAYER, *Analyst*, 72 (1947) 464.
H. NISHIDA, *Japan Analyst*, 3 (1954) 25.
C. E. CROUTHAMEL AND C. E. JOHNSON, *Anal. Chem.*, 26 (1954) 1284.

In steels
I. GELD AND J. CARROL, *Anal. Chem.*, 21 (1949) 1098.

In rocks
E. B. SANDELL, *Ind. Eng. Chem., Anal. Ed.*, 18 (1946) 163.

In Ti and Zr
D. F. WOOD AND R. T. CLARK, *Analyst*, 83 (1958) 326.

In Mo alloys
G. H. BUSH AND D. G. HIGGS, *Analyst*, 80 (1955) 536.

In the presence of Nb(V)
B. McDUFFIE, W. R. BANDI AND L. M. MELNIK, *Anal. Chem.*, 31 (1959) 1311.

(ii) With dithiol. Toluene-3,4-dithiol (4-methyl-1,2-mercaptobenzene) reacts with W(VI) to give a green compound which can be extracted with esters and hydrocarbons.

The extinction coefficient is about 20,000.

Interfering elements. Mo(VI) gives an analogous reaction, but can, under suitable conditions of acidity, be extracted and estimated by a similar method.

V(V), Ta(V), Ti(IV), and Zr(IV) do not interfere. Sn(IV), which interferes, is not extracted by hydrocarbons.

REAGENTS

Stannous chloride: 20 g of stannous chloride are dissolved in 50 ml of concentrated hydrochloric acid, and 50 ml of water are added.
Titanous chloride: commercial solution containing 15% of $TiCl_3$.
Hydrochloric acid, 1/2.
Dithiol solution: 0.5 g of dithiol are dissolved in 100 ml of amyl acetate.

OPERATING PROCEDURE

(1) *Separation and estimation of molybdenum.* 5 ml of solution containing 10 to 20 µg of Mo(VI) are mixed with 15 ml of 1/2 hydrochloric acid, 3 ml of stannous chloride, and then 10 ml of the dithiol solution. The mixture is shaken for 10 minutes, the aqueous phase is separated, and the amyl acetate is washed twice with 3 ml of 1/2 hydrochloric acid. The washing solutions and the aqueous phase are then combined.

The organic solution containing Mo(VI) is made up to 25 ml with amyl acetate and colorimetry is performed at 685 mμ.

(2) *Estimation of tungsten.* To the aqueous solution containing 10 to 20 μg of W(VI), are added 25 ml of concentrated hydrochloric acid, 3 ml of titanous chloride and 10 ml of the dithiol solution. The mixture is heated for 20 minutes on a water bath at 85°, shaking the flask frequently, cooled, and transferred to a separating funnel. The flask is rinsed three times with 3 ml of amyl acetate. The aqueous phase is separated and rejected, and the organic phase is washed twice with 5 ml of 4/5 hydrochloric acid, and made up to 25 ml with amyl acetate. Colorimetry is performed at 640 mμ.

cf. S. H. Allen and M. B. Hamilton, *Anal. Chim. Acta*, 7 (1952) 483.
P. G. Jeffery, *Analyst*, 81 (1956) 104.

In steels
C. C. Miller, *Analyst*, 69 (1944) 109.
B. Bagshawe and R. J. Truman, *Analyst*, 72 (1947) 189.

W(IV) and Mo(VI) in sea water
M. Ishibashi, T. Shigematsu and Y. Nakagawa, *Bull. Inst. Chem. Res., Kyoto Univ.*, 32 (1954) 199.

In Ti and Zr
D. F. Wood and R. T. Clark, *Analyst*, 83 (1958) 326.

In Ti
H. G. Short, *Analyst*, 76 (1951) 710.

In Ta, Zr and Ti
P. Greensberg, *Anal. Chem.*, 29 (1957) 896.

In steels and Ti
L. A. Machlan and J. L. Hague, *J. Res. Natl. Bur. Std.*, 59 (1957) 415.

In soils–W(VI) and Mo(VI)
A. A. North, *Analyst*, 81 (1956) 660.

Uranium

U = 238.1

(1) Separation of trace amounts

(i) From an alkaline medium. Alkali metal uranates are precipitated, but if the ammonia or the caustic soda are carbonated, they are partially redissolved owing to the formation of carbonate complexes.

V(V) prevents the precipitation of U(VI).

Entraining agents: Fe(III) and Al(III).

(ii) Electrolysis at a mercury cathode. The uranium remains in solution. Phosphates interfere by precipitating U(IV).

cf. F. T. RABBITTS, *Can. Mining J.*, 70 (1949) 84.

(iii) With oxine. The oxinate of U(VI) can be precipitated in the presence of EDTA, and may thus be separated from the rare earth metals, Th(IV), Zr(IV), V(IV), Bi(III), and phosphates. The oxinate may also be extracted with chloroform.

cf. R. N. SEN SARMA AND A. K. MALLIK, *Z. Anal. Chem.*, 148 (1955) 179.
R. A. J. SHELTON, *Analyst*, 82 (1957) 531.

Extraction

R. F. CLAYTON, W. H. HARDWICK, M. MORETON-SMITH AND R. TODD, *Analyst*, 83 (1958) 13.
C. L. RULFS, A. K. DE, J. LAKRITZ AND P. J. ELVING, *Anal. Chem.*, 27 (1955) 1802.

(iv) Extraction with oxine from a carbonate medium. The compound formed by the reaction of uranium and oxine may be extracted in the presence of quaternary ammonium ions $R_4N(UO_2Ox_3)$.

In the presence of EDTA, thorium and the rare earth metals do not interfere.

cf. W. E. CLIFFORD, E. P. BULLWINKEL, L. A. MCCLAINE AND P. NOBLE, JR. *J. Am. Chem. Soc.*, 8 (1958) 2959.
K. MOTOJIMA, H. YOSHIDA AND K. IZAWA, *Anal. Chem.*, 32 (1960) 1083.

(v) Extraction of uranyl nitrate. Suitable solvents consist of ethers, esters, and ketones. Uranyl nitrate is particularly easily extracted in the presence of large concentrations of nitrates, from a nitric acid solution.

Methyl *iso*-butyl ketone ('hexone') is often used as the solvent. The elements extracted by ether are: Au(III), Ce(IV), Th(IV), As(V), Bi(III), Tl(III), Zr(IV),

Hg(II), Cr(VI), and traces of Fe(III), V(V), etc. Zr(IV) can be retained by the inclusion of phosphoric acid.

Limited amounts of F^- and H_3PO_4 do not interfere. SO_4^{2-} is also inactive. Cl^- interferes, since it promotes the extraction of certain elements.

With ethyl acetate.

REAGENTS

Ethyl acetate.

Aluminium solution: 485 g of aluminium nitrate $Al(NO_3)_3 \cdot 9H_2O$ are dissolved in the hot and made up to 500 ml. If trace amounts of U(VI) are involved, this solution should be purified by shaking with 150 ml of ether.

OPERATING PROCEDURE

10 ml of solution are evaporated almost to dryness, and the residue is dissolved in 10 ml of aluminium solution. The solution is shaken with 10 ml of ethyl acetate for 2 minutes and allowed to settle for 5 minutes.

For example

J. A. S. ADAMS AND W. J. MAECK, *Anal. Chem.*, 26 (1954) 1635.
R. J. GUEST AND J. B. ZIMMERMAN, *Anal. Chem.*, 27 (1955) 931.

In zircons

F. CUTTITA AND G. J. DANIELS, *Anal. Chim. Acta*, 20 (1959) 430.

In Zr

P. A. VOZELLA, A. S. POWELL, R. H. GALE AND J. E. KELLY, *Anal. Chem.*, 32 (1960) 1430.

With tributyl phosphate. The partition coefficient is more favourable, but the separation is less selective.

With 4.7 *M* HNO_3 and with a solvent consisting of equal volumes of tributyl phosphate and ether, traces of Fe(III), Cu(II), Ni(II), Bi(III), the rare earth metals, and Cr(III), are extracted, as well as large amounts of Th(IV), Zr(IV), and Ce(IV). These can be removed by washing with nitric acid[1].

cf. D. J. FISHER AND P. F. THOMASON, *Anal. Chem.*, 28 (1956) 1285.
B. E. PAIGE, M. C. ELLIOTT AND J. E. REIN, *Anal. Chem.*, 29 (1957) 1029.
(1) A. R. EBERLE AND M. W. LERNER, *Anal. Chem.*, 29 (1957) 1134.

In monazite

V. T. ATHAVALE, L. M. MAHAJAN, N. R. THAKOOR AND M. S. VARDE, *Anal. Chim. Acta.*, 21 (1959) 353.

In the presence of a quaternary ammonium ion. In the presence of tetrapropylammonium, $R_4N(UO_2(NO_3)_3)$ may be extracted with methyl *iso*butyl ketone (see also Colorimetry).

The following are also extracted partially: Au(III), Ce(IV), Hg(II), In(III), Th(IV), Y(III), and traces of many other elements.

cf. W. J. Maeck, G. L. Booman, M. C. Elliott and J. E. Rein, *Anal. Chem.*, 30 (1958) 1902, and *Anal. Chem.*, 31 (1959) 1130.

With phosphine oxides.

cf. C. A. Horton and J. C. White, *Anal. Chem.*, 30 (1958) 1779.
Analytical Chemistry in Nuclear Reactor Technology, 1957, U.S. Atomic Energy Commission.

With tridecylphosphine oxide. Very minute traces can be extracted at pH 1 provided that the concentration of NO_3^- is at least 0.3 *M*. SO_4^{2-} does not interfere.

The reagent consists of a 0.1 *M* solution of tridecylphosphine oxide in kerosene or cyclohexane.

Th(IV) is not extracted.

cf. C. A. Blake, K. B. Brown and C. F. Coleman, *U.S. Atomic Energy Commission Report*, ORNL (1955) 1964.

(vi) Extraction from a chloride medium. Tributyl phosphate can be used to extract U(VI) from 7 *M* HCl, and this may therefore be separated from Zr(IV) and Th(IV).

cf. A. R. Eberle and M. W. Lerner, *Anal. Chem.*, 29 (1957) 1134.
D. F. Peppard, G. W. Mason and M. V. Gergel, *J. Inorg. Nucl. Chem.*, 3 (1957) 370.

(vii) Amines in a sulphate medium. U(VI) can be extracted with a symmetrical tertiary amine (tri-*n*-octylamine) in carbon tetrachloride.

Interfering ions. NO_3^-, Cl^-, and F^- prevent the extraction. The method is highly selective, since only Mo(VI) and traces of Fe(III) and V(V) are extracted as well.

Reagents

Tri-n-octylamine, 0.1 *M* in CCl_4.

Sodium carbonate, 50 g per litre.

Operating procedure

20 ml of the amine solution are added to 20 ml of uranium solution, (which should be 0.30 ± 0.05 *N* with respect to sulphuric acid,) in a 125 ml flask. The mixture is shaken for 15 minutes, allowed to settle, and then a given aliquot of the organic phase is filtered.

U(VI) can be reclaimed into the aqueous by shaking with sodium carbonate.

cf. *From an* SO_4^{2-} *medium*
F. L. Moore, *Anal. Chem.*, 29 (1957) 1660.
C. Boirie, *Bull. Soc. Chim.*, (1958) 1088.

From an acetate medium

F. L. MOORE, *Anal. Chem.*, 32 (1960) 1075.

(viii) Extraction of the thiocyanate. Uranyl thiocyanate can be extracted with a number of solvents – for example by tributyl phosphate in presence of carbon tetrachloride. The presence of EDTA renders this method of separation selective: See Colorimetry.

(ix) Partition chromatography. A column filled with powdered cellulose is generally used. This method depends upon the fact that uranyl nitrate is soluble in organic solvents.

In this way, U(VI) may be separated from Th(IV) by being repeatedly eluted with ether containing 5% of concentrated nitric acid, and from Th(IV) by elution with the same solvent containing 12.4% of concentrated nitric acid[1].

REAGENTS

Eluting solution: 25 ml of concentrated nitric acid are added to 475 ml of ether.

Cellulose column: 50 g of powdered cellulose are boiled with 475 ml of water and 25 ml of nitric acid. The mixture is filtered and washed with alcohol and then with ether. A tube 20 cm high and 2 cm in diameter is filled to a height of 8 cm with the cellulose in suspension in ether containing 5 ml of concentrated nitric acid per 100 ml of ether.

OPERATING PROCEDURE

10 ml of solution containing U(VI) in 3 *N* (1/4) nitric acid are poured down the column, followed by 100 ml of eluting solution at the rate of 5 ml per minute. 25 ml of water are added to the eluate and the ether is evaporated by directing a current of air onto the surface of the liquid.

cf. J. A. S. ADAMS AND W. J. MAECK, *Anal. Chem.*, 26 (1954) 1635.
(1) N. F. KEMBER, *Analyst*, 77 (1952) 78.
T. PALÁGYI. *Acta Chim. Hung.*, 22 (1960) 239.

With paper

E. C. HUNT, A. A. NORTH AND R. A. WELLS, *Analyst*, 80 (1955) 190.
H. P. RAAEN AND P. F. THOMASON, *Anal. Chem.*, 27 (1955) 936.

Review of a number of extraction procedures

A. R. EBERLE AND M. W. LERNER, *Anal. Chem.*, 29 (1957) 1134.
G. H. MORRISSON AND H. FREISER, *Solvent Extraction in Analytical Chemistry*, Wiley, 1957

(x) With anion-exchangers. As little as 1 μg of U(VI) may be isolated by stirring with an anionic resin, using a sulphuric[1] or acetic[2] acid medium, or else, media containing nitrate[3], chloride[4], or carbonate[5] ions.

With a sulphate medium. Interfering ions. Mo(VI), Cr(VI), Fe(III), and V(V) should first be reduced with SO_2. Most anions interfere.

Preparation of the resin. Amberlite XE-117 or Dowex 1, 40–60 mesh are treated for 20 minutes with 3–4 *N* (1/10) sulphuric acid at the ratio of 3 volumes of acid per 1 volume of resin. The resin is washed with water until methyl red. develops a neutral colour, and is then bottled under water.

OPERATING PROCEDURE

5 ml of resin are placed in a column. The solution, containing 0.5–40 mg of uranium in 1–2 *N* H_2SO_4 and of pH 1.0–1.5 is passed through at the rate of 2 ml/minute. The resin is then washed with 50 ml of hot water. The uranium can be eluted with 50 ml of hot *N* $HClO_4$.

(1) A. ARNFELT, *Acta Chem. Scand.*, 9 (1955) 1484.
S. FISHER AND R. KUNIN, *Anal. Chem.*, 29 (1957) 400.
G. BANERJEE AND A. H. A. HEYN, *Anal. Chem.*, 30 (1958) 1795.
H. J. SEIM, R. J. MORRIS AND D. W. FREW, *Anal. Chem.* 29 (1957) 443.
S. M. KHOPKAR AND A. K. DE, *Anal. Chim. Acta*, 23 (1960) 147.
(2) F. HECHT, J. KORKISCH, R. PATZAK AND A. THIARD, *Mikrochim Acta*, (1956) 1283.
(3) H. M. OCKENDEN AND J. K. FOREMAN, *Analyst*, 82 (1957) 592.
J. KORKISCH, M. R. ZAKY AND F. HECHT, *Mikrochim. Acta*, (1957) 485.
(4) *Separation from Th(IV), Zr(IV) and Ti(IV)*
E. TOMIC, I. M. LADENBAUER AND M. POLLAK, *Z. Anal. Chem.*, 161 (1958) 28.
G. W. C. MILNER AND J. H. NUNN, *Anal. Chim. Acta*, 17 (1957) 494.
L. R. BUNNEY, N. E. BALLOU, J. PASCUAL AND S. FOTI, *Anal. Chem.*, 31 (1959) 324.
(5) *Separation from V(V), P(V), and Mo(VI)*
T. K. S. MURTHY, *Anal. Chim. Acta*, 16 (1957) 25.
S. M. KHOPKAR AND A. K. DE, *Anal. Chim. Acta*, 23 (1960) 147.

(xi) With cation-exchangers in the presence of EDTA. In the presence of ethylenediaminetetraacetic acid at pH 3–5, UO_2^{2+} is one of the few cations which remain free, the majority being bound in the form of anionic complexes – FeY^-, ZnY^{2-} etc. The following interfere: Be^{2+}, Ti(IV), P(V), As(V), and F^-.

REAGENTS

EDTA: ethylenediaminetetraacetic acid, 10% in dilute ammonia, neutralized to litmus.

Exchanger. Column 30 cm × 1 cm². The exchanger should not be in the H^+ form in order to prevent variations of the pH of the solution. It can be converted to the NH_4^+ form by repeated washing with a 20% solution of ammonium chloride, 20% ammonium acetate, and then water.

OPERATING PROCEDURE

20 ml of EDTA are added to the solution and the latter is acidified if a precipitate is formed. After neutralization with ammonia until methyl orange turns yellow, the volume is made up to 150 ml and the solution is passed

through the column at the rate of 2–3 ml per minute. The resin is then washed with 150 ml of water. The UO_2^{2+} may subsequently be eluted, with 150 ml of 6 *N* hydrochloric acid.

cf. S. M. KHOPKAR AND A. K. DE, *Anal. Chim. Acta*, 22 (1960) 153.
H. HOLCOMB AND J. H. YOE, *Anal. Chem.*, 32 (1960) 612.

(xii) Standard solutions. Guaranteed purity uranyl nitrate $UO_2(NO_3)_2 \cdot 6H_2O$ is used.

$$U/UO_2(NO_3)_2 \cdot 6H_2O = 0.474$$
$$U_3O_8/3UO_2(NO_3)_2 \cdot 6H_2O = 0.560.$$

Alternatively, pure U_3O_8 may be dissolved in an oxidizing acidic medium.

$$U/U_3O_8 = 0.848.$$

(2) Colorimetry

Numerous methods are available. Hydrogen peroxide may be used in alkaline solutions. This selective but not very sensitive method may be applied after certain separations, *e.g.* by carbonate etc. The thiocyanate is also used in certain cases. This method is neither very sensitive nor very selective.

The method using dibenzoylmethane is sensitive but very unselective. Absorption measurements can often be carried out directly on the solution after separation by extraction. The same characteristics apply to the methods using arsenazo and PAN.

Absorptiometry after extraction of uranyl tetrapropylammonium trinitrate is specific and very accurate, but also highly insensitive.

Differential absorptiometry can be carried out with pure concentrated solutions.

For the determination of very small traces, fluorimetry in fused salts is used.

(i) With hydrogen peroxide. Yellow complexes are formed in alkaline solutions particularly in the presence of carbonate. The method is not very sensitive – $\varepsilon \sim 1{,}300$ at 370 mμ – but is quite selective. The sample can be digested directly with sodium peroxide, centrifuged and estimated colorimetrically. V(V) does not then interfere. Cr(VI) interferes.

cf. T. R. SCOTT, *Analyst*, 75 (1950) 100.
K. L. CHENG AND P. F. LOTT, *Anal. Chem.*, 28 (1956) 462.

(ii) Thiocyanate. The complex formed by U(VI) and SCN^- can be extracted with a mixture of tributyl phosphate and carbon tetrachloride. The operation is carried out at pH 3.7 ± 0.2 in the presence of EDTA, which complexes a large number of elements.

Sensitivity: $\varepsilon \sim 3{,}850$ at 375 mμ.

Interfering ions. The following do not interfere under the conditions described: 200 mg of Cu(II), Ni(II), Mn(II), Pb(II), Sn(II), Bi(III), Fe(III), Cr(III), Sb (III) and the rare earth metals, 100 mg of Al(III) and Ti(IV), 80 mg of Th(IV) and V(V), 10 mg of W(VI), 5 mg of Co(II), and large amounts of chlorides, nitrates, and sulphates.

Mo(VI), Zr(IV), and Sn(IV) precipitate and entrain a little U(VI). Hg(II) gives a coloration. F^- and large amounts of phosphates also interfere. An alternative procedure consists in the previous extraction of uranyl nitrate, which is then directly estimated in the solvent[1].

REAGENTS

Solvent: 675 ml of carbon tetrachloride are added to 325 ml of tributyl phosphate.

EDTA: 10 g of disodium ethylenediaminetetraacetate are dissolved in 100 ml of water.

Ammonium thiocyanate: 50 g are dissolved in 100 ml of water.

OPERATING PROCEDURE

20 ml of solution containing 50 to 300 μg of U(VI) are mixed with 20 ml of EDTA. The pH is adjusted to 3.7 ± 0.2, and 10 ml of thiocyanate are added followed by 10 ml of solvent, accurately measured. The mixture is shaken for 30 seconds and the organic phase is separated and transferred to a flask containing 0.5 g of anhydrous sodium sulphate. The solution is left to dry for 5 minutes, and colorimetry is then performed at 350 mμ.

cf. J. CLINCH AND M. J. GUY, *Analyst*, 82 (1957) 800.
H. T. TUCKER, *Analyst*, 82 (1957) 529.
(1) O. A. NIETZEL AND M. A. DE SESA, *Anal. Chem.*, 29 (1957) 756.
L. SILVERMAN AND L. MOUDY, *Anal. Chem.*, 28 (1956) 45.

(iii) With dibenzoylmethane. With U(VI), a compound is obtained which can be extracted with numerous solvents.

The method is more sensitive than the thiocyanate technique: $\varepsilon \sim 20{,}000$ at 395 mμ.

Very many ions interfere, and the operation may therefore be performed in the presence of EDTA and tartrate.

Uranium can also be separated by extracting the nitrate with ethyl acetate. Colorimetry is then performed with the aid of dibenzoylmethane in the same solvent. This method is described below.

REAGENTS

Dibenzoylmethane solution: 1 g of dibenzoylmethane, 12.5 g of EDTA, and 10.9 of magnesium nitrate [$Mg(NO_3)_2 \cdot 6H_2O$] are dissolved in 500 ml of water. 500 ml of pyridine are added.

OPERATING PROCEDURE

After the uranyl nitiate has been extracted (see p. 417) 5 ml of the solvent containing 50 to 200 μg of U(VI) are made up to 25 ml with the dibenzoylmethane solution, allowed to stand for 10 minutes and estimated colorimetrically at 410 mμ.

cf. C. A. HORTON AND J. C. WHITE, *Anal. Chem.*, 30 (1958) 1779.
C. A. FRANÇOIS, *Anal. Chem.*, 30 (1958) 50.

(3) Other methods

(i) Arsenazo. 3-(2-arsenophenylazo)-4,5-dihydroxynaphthalene-2,7-disulphonic acid: $\varepsilon \sim 23{,}000$.

The reaction is performed in the presence of EDTA at pH 7.5; Th(IV) interferes.

cf. J. S. FRITZ AND M. JOHNSON-RICHARD, *Anal. Chim. Acta*, 20 (1959) 164.
H. HOLCOMB AND J. H. YOE, *Anal. Chem.*, 32 (1960) 612.

(ii) 'PAN' 1-(pyridyl-2-azo)-2-naphthol. In alkaline solutions PAN and U(VI) give a red precipitate which is soluble in chloroform.

Sensitivity: $\varepsilon \sim 23{,}000$ at 560 mμ.

Interfering ions. A large number of cations which give colorations may be complexed with cyanide and EDTA. P(V) and Sn(IV) interfere. In the presence of F , 0.5 mg of Ti(IV) = 4.3 μg of U(VI). Cl^- and SO_4^{2-} do not interfere.

The method described allows the determination of 20 to 60 μg of U(VI) in the presence of 1 mg of Hg(II), Mo(II), Mn(II), Ni(II), Zn(II), Pb(II), V(V), W(VI), 0.5 mg of Th(IV), Bi(III), Ga(III), In(III) and 0.3 mg of La(III). Larger quantities require greater additions of complexing agents. The excess of EDTA must be small.

REAGENTS

0.1 M EDTA: 37.2 g of the disodium salt of ethylenediaminetetraacetic acid are dissolved in 1 litre of water.

1% potassium cyanide solution.

20% sodium chloride solution.

Ammoniacal buffer: 60 g of ammonium chloride are dissolved in 200 ml of water; 570 ml of concentrated ammonia are added, and the solution is made up to 1 l.

PAN: 100 mg of reagent are dissolved in 100 ml of methanol. This solution should be stored in a brown bottle.

OPERATING PROCEDURE

2 ml of EDTA, 5 ml of the buffer solution and 2 to 3 ml of potassium cyanide

are added to 10 ml of a solution containing 2 to 100 μg of U(VI). The pH is checked and if necessary adjusted to 9.5–10, 10 ml of sodium chloride are added, followed by 2 ml of PAN, and the mixture is allowed to stand for 5 minutes. 10 ml of chloroform are added, the mixture is stirred for 1–2 minutes, and is then decanted. The measurement is carried out at 560 mμ, comparing this with a blank.

cf. S. SHIBATA, *Anal. Chim. Acta*, 22 (1960) 479.

(iii) Tetrapropylammonium trinitrate. $R_4N(UO_2)(NO_3)_3$ at 452 mμ in hexone: $\varepsilon = 30$.

cf. W. J. MAECK, G. L. BOOMAN, M. C. ELLIOTT AND J. E. REIN, *Anal. Chem.*, 31 (1959) 1130.

(iv) Absorption of UO_2^{2+}. At 418 mμ a very accurate, but neither sensitive nor specific colorimetric estimation may be performed. The method is therefore suitable for large amounts of pure uranyl salts.

cf. C. D. SUSANO, O. MENIS AND C. K. TALBOTT, *Anal. Chem.*, 28 (1956) 1072.
A. BACON AND G. W. C. MILNER, *Analyst*, 81 (1956) 456.
C. V. BANKS, K. E. BURKE, J. W. O'LAUGHLIN AND J. A. THOMPSON, *Anal. Chem.*, 29 (1957) 995.

(v) Fluorimetry in fused sodium fluoride. Very minute traces are commonly estimated by this method. The substance to be analysed is fused with a known weight of sodium fluoride (or $NaF + Na_2CO_3$) in a platinum boat.

In ultraviolet light, a yellow-green fluorescence visible to the eye is observed. The sensitivity of the method depends on the fluorimeter used. The method is effective with as little as 10^{-8} to 10^{-9} g to within $\pm$ 5 to 10%, and even with as little as 10^{-10} g.

Many elements inhibit or attenuate the fluorescence.

The fluorescence is specific, except perhaps in the presence of cerium.

cf. H. ERLENMEYER, W. OPPLIGER, K. STIER AND M. BLUMER, *Helv. Chim. Acta*, 33 (1950) 25.
F. H. BURSTALL AND A. F. WILLIAMS, *Handbook for the Determination of Uranium in Minerals and Ores*, H.M. Stationery Office, London, 1950.
G. R. PRICE, R. J. FERRETTI AND S. SCHWARTZ, *Anal. Chem.*, 25 (1953) 322.
C. J. RODDEN, *Anal. Chem.*, 25 (1953) 1598.
F. A. CENTANNI, A. M. ROSS AND M. A. DE SESA, *Anal. Chem.*, 28 (1956) 1651.

Vanadium

V = 50.95

(1) Separation of trace amounts

(i) The dry method. Even traces of V(V) may be separated from sparingly soluble oxides by digestion with fused sodium carbonate, the product being dissolved in water. The following pass into solution with V(V): Mo(VI), W(VI), P(V), As(V), Cr(VI), and Mn(VI).

(ii) Electrolysis at a mercury cathode. The vanadium remains in solution.

(iii) Precipitation of iron (III) vanadate. Traces of V(V) can be separated by precipitation at pH 6–7 in the presence of Fe(III). Cr(III), Mo(V), Ti(IV), Sn (IV), etc. accompany V(V).

cf. K. Sugawara, M. Tanaka and H. Naito, *Bull. Chem. Soc. Japan*, 26 (1953) 417.

(iv) Precipitation and extraction with cupferron. Cupferron precipitates V(V) in very dilute acid; the vanadium may thus be separated from Al(III), U(VI), As(V), P(V), etc.

Trace amounts may if necessary be entrained with Fe(III).

Alternatively, the cupferron may be extracted.

cf. D. Bertrand, *Bull. Soc. Chim.*, 9 (1942) 128.
G. B. Jones, *Anal. Chim. Acta*, 17 (1957) 254.

(v) Extraction of the oxinate. The oxinate can be extracted at pH 4–5 with chloroform. Vanadium may thus be separated from Cr(VI) but not from Fe(III), Mo(VI), or W(VI).

cf. E. B. Sandell, *Ind. Eng. Chem., Anal. Ed.*, 8 (1936) 336.
R. Bock and S. Gorbach, *Mikrochim. Acta*, (1958) 593.

(vi) Extraction of V(V) with tridecylphosphine oxide. The solvent is kerosene. U(VI) is also extracted.

cf. C. A. Blake, K. B. Brown and C. F. Coleman, *U.S. Atomic Energy Commission Report*, ORNL (1955) 1964.

With cation-exchangers. Effective for the separation from H_3PO_4.

J. S. Fritz and J. E. Abbink, *Anal. Chem.*, 34 (1962) 1080.

(vii) With anion-exchangers in a chloride medium. Effective for the separation from Fe(III) and Ti(IV).

cf. K. A. KRAUS, F. NELSON AND G. W. SMITH, *J. Phys. Chem.*, 58 (1954) 11.

(viii) Standard solutions. A solution of sodium or ammonium vanadate is prepared and titrated in an acid medium against ferrous iron, in the presence of sulphonated diphenylamine.

Alternatively, V_2O_5 of guaranteed purity may be dissolved in an alkaline solution.

(2) Colorimetry

Of the two principal methods, that involving hydrogen peroxide does not permit the estimation of minute traces and is subject to interference from Ti (IV). The other, in which the phosphovanadotungstate complex is used, is more accurate and more sensitive if performed at about 400 mμ.

(i) With phosphovanadotungstic acid. When phosphoric, vanadic, and tungstic acids are present together, a yellow to yellow-brown compound is formed. The coloration is stable for 24 hours. Acidity of the solution exerts little influence, but well-defined concentrations of the reagents should be present: W(VI), 0.01 to 0.1 *M*; P(V) 0.5 *M*. The compound formed can be extracted with butanol[1]. Colorimetry can be carried out at 400 mμ.

Sensitivity. $\varepsilon \sim 2{,}000$ at 400 mμ.

Interfering ions. Sn(IV), Ti(IV), Zr(IV), Bi(III), and Sb(III) precipitate, mostly in the form of phosphates. K^+ and NH_4^+ precipitate the complex.

Coloured ions may interfere. The following may be tolerated: 10 ppm of Cr (VI); 0.1 ppm of Co(II) and Cu(II); 0.3 ppm of Ni(II); 0.4 ppm of Mn(II); 1 ppm of U(VI).

The following do not interfere: Mg(II), Ca(II), Sr(II), Ba(II), Zn(II), Cd(II), Al(III), Hg(II), Pb(II), and As(III) at a content of 5 g per litre; Ag(I) at 1 g per litre; Th(IV) at 0.1 g; SiO_3^{2-} at 0.5 g; tartrate, citrate, and oxalate at 1 g; and CN^- at 0.2 g. H_3BO_3, F^-, and Mo(VI) begin to interfere when present in amounts 200 times greater than W(VI). Fe(III) at a content of less than 0.1 g per litre no longer interferes after the solution is boiled.

REAGENTS

Sodium tungstate, 0.5 *M* (16%)

Phosphoric acid, 1/3.

OPERATING PROCEDURE

The solution should contain from 20 to 200 μg of vanadium. It is made 0.5 *N* with respect to sulphuric, nitric, or hydrochloric acid, and 1 ml of 1/3 phosphoric acid and 0.5 ml of tungstate are added. The volume is made up to 10 ml and the solution is boiled, cooled and diluted as necessary.

The colorimetry is performed at 400 mμ, after adding any coloured ions present to the standard sample.

cf. E. R. WRIGHT AND M. G. MELLON, *Ind. Eng. Chem., Anal. Ed.*, 9 (1937) 251.

In rocks

E. B. SANDELL, *Ind. Eng. Chem., Anal. Ed.*, 8 (1936) 336.

In steels

M. D. COOPER AND P. K. WINTER, *Anal. Chem.*, 21 (1949) 605.

P. H. SCHOLES, *Analyst*, 82 (1957) 525.

(1) R. M. SHERWOOD AND F. W. CHAPMAN, JR., *Anal. Chem.*, 27 (1955) 88.

With tungstovanadate

G. W. WALLACE AND M. G. MELLON, *Anal. Chem.*, 32 (1960) 204.

(ii) With hydrogen peroxide. Hydrogen peroxide gives an orange VO_2^{3+} complex with vanadium salts in acid solutions. The coloration is stable for 2 days. Colorimetry is performed at 460 mμ. Only 0.5–3 ml of 3% hydrogen peroxide per 100 ml should be added, since an excess of H_2O_2 attenuates the colour.

The acidity must be fixed – for example, 0.6 to 6 *N* HCl, HNO_3, or H_2SO_4.

The reaction is not sensitive: $\varepsilon \sim 300$.

Interfering ions. Certain other ions also give colorations with hydrogen peroxide: Mo(VI), W(VI), U(VI), and Nb(V) have very low absorptions at 460 mμ. Cr(VI) is reduced by H_2O_2 and interferes. Ti(IV) gives an analogous coloration. Thus, either Ti(IV) and V(V) are estimated together by working at two wavelengths, or the former is complexed by the addition of fluoride. In the latter case, it is necessary to put the same amount of fluoride into the standard samples (HF attacks the colorimetry cell); Fe(III) is complexed[1]. In the absence of Ti(IV) however, Fe(III) can be preferentially complexed by the addition of phosphoric acid; this is also added to the standard samples. P(V) + + V(V) + W(VI) (as they occur in certain steels) give rise to a yellow coloration due to the phosphovanadotungstate complex. The separation of tungstic acid by precipitation entrains vanadium.

REAGENT

Hydrogen peroxide, 3% (10 volume).

OPERATING PROCEDURE

The solution, containing 1 to 5 mg of vanadium, is made up to about 80 ml. Sulphuric acid is added to obtain an acidity of 0.6 to 6 *N*, followed by 1 ml of concentrated phosphoric acid if only Fe(III) is present. 3 ml of hydrogen peroxide are then added, and the volume is made up to 100 ml.

The solution is compared with standards prepared in the same way.

cf. E. R. Wright and M. G. Mellon, *Ind. Eng. Chem., Anal. Ed.*, 9 (1937) 375.
(1) H. C. Davis and A. Bacon, *J. Soc. Chem. Ind.*, 67 (1948) 324.

In steels

M. Chateau and P. Michel, *Compt. Rend. (Paris)*, 220 (1945) 848.
P. H. Scholes, *Analyst*, 82 (1957) 525.

Estimation of Ti(VI) + V(V) and Mo(IV) + Ti(IV) + V(V). (see p. 411)

A. Weissler, *Ind. Eng. Chem., Anal. Ed.*, 17 (1945) 695.

(iii) Colorimetry by means of redox indicators.

Diphenylamine sulphonic acid

F. M. Wrightson, *Anal. Chem.*, 21 (1949) 1543.
S. Hirano, H. Murayama and M. Kitahara, *Japan Analyst*, 5 (1956) 7.

Review of methods

R. Bock and S. Gorsbach, *Mikrochim. Acta*, (1958) 593.

Zinc

Zn = 65.4

(1) Separation of trace amounts

(i) Precipitation of zinc sulphide in a tartaric acid medium. Using Cu(II), Fe (II), or Hg(II) as entraining agents. Hg(II) may later be expelled by calcination.

cf. H. A. Bright, *J. Res. Natl. Bur. Std.*, 12 (1934) 383.

(ii) Extraction of the thiocyanate. The extraction is carried out, from slightly acid solutions with hexone; Fe(III), Co(II), Cu(II), and Cd(II) are partially extracted.

Small amounts of Cu(II) can be complexed with thiourea, and Fe(III) with fluoride.

Zn(II) can be reclaimed into an aqueous ammoniacal buffer.

REAGENTS

Hydrochloric acid, 0.5 *N* (1/20).
Thiourea, 10% in water.
Ammonium thiocyanate, 50% in water.
Ammonium difluoride.
Sodium hypophosphite.
'*Hexone*' (methyl *iso*-butyl ketone).

OPERATING PROCEDURE

100 ml of solution about 0.5 *N* in hydrochloric acid are mixed with 3 g of hypophosphite, 10 ml of thiourea, 30 ml of thiocyanate, and 40 ml of hexone. The mixture is shaken for 2 minutes, allowed to settle, and the aqueous phase is rejected. The hexone phase is then shaken for about 30 seconds with 20 ml of 0.5 *N* hydrochloric acid, 10 ml of thiocyanate, and 2 g of difluoride. If the colour of the organic phase remains red, the latter should be shaken with a further 1 g of difluoride. A new addition of thiocyanate and hydrochloric acid to the hexone should not produce a red coloration after shaking; otherwise more difluoride should be added in small portions.

cf. J. Kinnunen and B. Wennerstrand, *Chemist-Analyst*, 42 (1953) 80.
M. I. Troitskaya, R. G. Pats and A. A. Pozdniyakova, *Chem. Abstr.*, 52 (1958) 1851.

(iii) Extraction in the presence of pyridine, with chloroform. Effective for sep-

arations from Ga(III), In(III), Pb(II), Ti(IV), Bi(III), Sb(III), and U(VI).

cf. J. A. Hunter and C. C. Miller, *Analyst*, 81 (1956) 79.

(iv) Extraction in the presence of tri-iso-*octylamine with hexone.* 2 *N* HCl medium.

cf. L. E. Scroggie and J. A. Dean, *Anal. Chim. Acta*, 21 (1959) 282.

(v) Extraction with dithizone. See Colorimetry.

In the presence of diethanoldithiocarbamate
D. W. Margerum and F. Santacana, *Anal. Chem.*, 32 (1960) 1157.

(vi) Extraction of the diethyldithiocarbamate

cf. G. B. Jones, *Anal. Chim. Acta*, 11 (1954) 88.

(vii) With anion exchangers.

(a) Zn(II) is readily fixed in chloride solutions, whilst Cu(II), Ni(II), Fe(III), Co(II), Mn(II), Al(III), etc. are eluted with 0.5 *N* HCl. Cd(II), Sn(IV), Sb(III), Bi(III), Pb(II), and In(III) accompany Zn(II), at least partially. Zn(II) is then extracted with 0.005 *N* HCl or 3 *N* HNO_3.

Exchanger. 25 to 30 g of Dowex 1 or Amberlite I.R.A. 400 (100–150 mesh) is packed into a column. The column is washed with concentrated hydrochloric acid immediately before use.

Operating procedure

20 ml of a solution in 9 *N* hydrochloric acid, containing at the most 500 mg of the elements to be separated, are poured into the column. About 150 ml of 0.5 *N* hydrochloric acid are then passed through at the rate of 3 to 5 ml per minute.

Zn(II) is then eluted with 100 ml of 3 *N* nitric acid.

cf. D. H. Wilkins, *Anal. Chim. Acta*, 20 (1959) 271.
M. Nishimura and E. B. Sandell, *Anal. Chim. Acta*, 26 (1962) 242.

(b) Separation of Zn(II) from Cd(II). The cadmium is fixed as an iodide complex, and Zn(II) is eluted.

Exchanger. A column 1.5 cm in diameter is packed with about 80 ml of Dowex 1 or Amberlite I.R.A. 400 (100–150 mesh) in the sulphate form, and is brought to equilibrium with 50 ml of 0.5 *N* sulphuric acid solution containing 50 g of potassium iodide per litre.

OPERATING PROCEDURE

About 100 ml of 0.25 *N* H_2SO_4 solution containing not more than 1 g of cadmium is treated with 6 g of potassium iodide, poured into the column and eluted with potassium iodide solution (50 g per litre) until the volume collected is 300 ml. The rate of elution should be 5 ml per minute.

Regeneration of the resin is fairly difficult. It is shaken in a flask with 3 *N* sulphuric acid containing 0.3% hydrogen peroxide until the Cd(II) disappears, then with 2.5 *N* soda. This treatment is then repeated once more.

cf. E. R. BAGGOTT AND R. G. W. WILLCOCKS, *Analyst*, 80 (1955) 53.
J. A. HUNTER AND C. C. MILLER, *Analyst*, 81 (1956) 79.

(viii) Standard solutions. The pure metal may be used. For example, 0.2 g is dissolved in 50 ml of *N* (1/40) sulphuric acid with warming, and the volume is made up to a known value.

Alternatively, $ZnCl_2$ may be dissolved in the presence of a little hydrochloric acid.

$ZnSO_4 \cdot 7H_2O$ may also be used.

$$Zn/ZnCl_2 = 0.480$$

$$Zn/ZnSO \cdot 7H_2O = 0.2274.$$

(2) Colorimetry

(i) With dithizone. This method is very sensitive and quite accurate, but many ions should first be separated or complexed. The red zinc dithizonate may be extracted with chloroform and carbon tetrachloride from an alkaline solution. With an excess of dithizone, it can even be extracted from slightly acidic solutions. The colorimetry may also be carried out on an aqueous solution by the addition of a suitable miscible solvent, for example methyl cellosolve[1]. The colorimetry of the zinc dithizonate is carried out at 530 mμ.

Sensitivity. ε = 94,000 in CCl_4 at 535 mμ.

Interfering ions. The dithizonates of Cu(I), Ag(I), and Hg(II) are first extracted with carbon tetrachloride from a 0.1 *N* acid medium. The zinc dithizonate is then extracted from a medium containing citrate and ammonia. Under these conditions, Pb(II), Cd(II), Bi(III), and Ni(II) are also extracted. In the presence of thiosulphate at pH > 7, small amounts of zinc may be separated from Cu(II), Hg(II), Ag(I), Au(III), Bi(III), and Pb(II), which are complexed.

Separation from Ni(II) and Co(II). In the presence of cyanide and at about pH 5, zinc may be extracted and separated from Ni(II) and Co(II).

OPERATING PROCEDURE

The solution is neutralized with 1/2 ammonia, and 5% potassium cyanide is added drop by drop until the cyanides just redissolve [1 ml = 7 mg of Ni(II) or of Co(II)]. The mixture is acidified with hydrochloric acid to pH 3–4, and the pH is adjusted to 5–5.5 with acetate. The complex is extracted with successive portions of dithizone in CCl_4.

Separation of Pb(II). At about pH 4.75 in the presence of citrate, Zn(II) can be extracted while Pb(II) is left behind.

REAGENTS

Acetate buffer: 136 g of sodium acetate are dissolved in 440 ml of water. 58 ml of concentrated acetic acid are added, and this solution is purified by shaking with a 0.01% solution of dithizone in carbon tetrachloride.
Thiosulphate, 25% in water.
Dithizone: solution at 20 mg per litre of carbon tetrachloride, purified.

OPERATING PROCEDURE

10 ml of slightly acid (not more than 0.01) solution containing 1 to 4 µg of Zn(II) are mixed with 5 ml of acetate buffer and 1 ml of thiosulphate, and then with 5 ml of accurately measured dithizone solution. The mixture is shaken for 2 minutes and colorimetry is performed at 350 mµ (zinc dithizonate) or at 620 mµ (excess of dithizone).

cf. G. KORTÜM AND B. FINCKH, *Die Chemie,* 57 (1944) 73.
L. G. BRICKER, S. WEINBERG AND K. L. PROCTOR, *Ind. Eng. Chem., Anal. Ed.,* 17 (1945) 661.
H. BARNES, *Analyst,* 76 (1951) 220.
H. IRVING, C. F. BELL AND R. J. P. WILLIAMS, *J. Chem. Soc.,* (1952) 356.

In urine without ashing
J. H. R. KAGL AND B. L. YALEE, *Anal. Chem.,* 30 (1958) 1951.

In nonferrous alloys
J. MIGEON, *Chim. Anal.,* 40 (1958) 287.

In plants
R. L. SHIRLEY, D. R. WALDRON, E. D. JONES AND E. J. BENNE, *J. Assoc. Offic. Agr. Chem.,* 31 (1948) 285.

In rocks
E. B. SANDELL, *Ind. Eng. Chem., Anal. Ed.,* 9 (1937) 464.

In soils

G. D. Sherman and J. S. McHargue, *J. Assoc. Offic. Agr. Chem.*, 25 (1942) 510.
E. Shaw and L. A. Dean, *Soil Sci.*, 73 (1952) 341.

In biochemistry

H. Cowling and E. J. Miller. *Ind. Eng. Chem., Anal. Ed.*, 13 (1941) 145.

In aluminium

H. Fischer and G. Leopoldi, *Aluminium*, Berlin, 25 (1943) 356.
L. G. Bricker, S. Weinberg and K. L. Proctor, *Ind. Eng. Chem., Anal. Ed.*, 17 (1945) 661.

In water

A. A. Christie, J. R. W. Kerr, G. Knowles and G. F. Lowden, *Analyst*, 82 (1957) 342.

In cobalt

R. S. Young, *Metallurgia*, 36 (1947) 347.

In cadmium

E. R. Baggott and R. G. W. Willcocks, *Analyst*, 80 (1954) 53.

In foodstuffs

A. C. Francis and A. J. Pilgrim, *Analyst*, 82 (1957) 289.
(1) B. L. Vallee, *Anal. Chem.*, 26 (1954) 914.

Zirconium

Zr = 91.2

(1) Separation of trace amounts

NOTE. In all the methods described, Hf(IV) accompanies Zr(IV). Analytical calculations are thus falsified, since as much as 3% of Hf(IV) may sometimes be present.

(i) The dry method. Digestion with fused sodium carbonate, sodium peroxide, or borax causes the phosphates, fluorides, and sulphates to pass into solution; ZrO_2 remains insoluble provided that Fe(III) is present. In an oxidizing medium, V(V) passes into the solution.

cf. F. S. GRIMALDI AND C. E. WHITE, *Anal. Chem.*, 25 (1953) 1886.

(ii) Precipitation of the phosphate. This is effected in 1 *N* hydrochloric acid, in the presence of hydrogen peroxide to complex titanium, if necessary.

(iii) With arsenic acid and the arsonic acids. Arsenic or, preferably phenylarsonic, *p*-hydroxyphenylarsonic, and *n*-propylarsonic acids precipitate Zr(IV) in $\leqslant 3$ *N* HCl or $\leqslant 1.5$ *N* H_2SO_4 in the form of $ZrO(RAsO_3H)$. For amounts of the order of 50 to 500 μg, the solution should not be too acid – $\leqslant 0.3$ *N* HCl – and should not contain too much sulphate – $\leqslant 0.3$ *M* SO_4^{2-}.

The method is effective for the separation of Zr from Cu(II), Cd(II), V(V), Mo(VI), U(VI), Ce(III), Fe(III), Al(III), Cr(III), Ni(II), Co(II), Zn(II), Mn(II), Mg(II), Ca(II), Ba(II), and Sr(II). Small amounts of Ti(IV) are kept in solution by the addition of hydrogen peroxide. Double precipitation is sufficient to effect separation from Bi(III), Th(IV), and Sb(III).

The following interfere by precipitating: Nb(V), Ta(V), W(VI), and Sn(IV). SiO_2 also interferes.

With *n*-propylphenyl arsonic acid, Ti(IV) and Th(IV) are separated simultaneously. Large amounts of Fe(III) necessitate a double precipitation in all cases, owing to adsorption phenomena.

REAGENTS

p-*Hydroxyphenylarsonic acid.*

Hydrochloric acid, N (1/12).

OPERATING PROCEDURE

The solution containing less than 150 mg of ZrO_2 is made 1.5 *N* (1/8) with respect to hydrochloric acid and adjusted to 150 ml. 1.5 to 2 g of *p*-hydroxyphenylarsonic acid are added, and the mixture is boiled for 1 minute. Completion of the precipitation may be checked by adding one or two drops of reagent to the supernatant solution. The mixture is then filtered and washed with *N* (1/12) HCl.

NOTE. In a medium containing sulphuric acid, which should not exceed 4%, the precipitate obtained is very fine and difficult to filter.

In steels

P. KLINGER AND O. SCHLIESSMANN, *Arch. Eisenhütt*, 7 (1933) 113.

In rocks

I. P. ALIMARIN AND O. A. MEDVEDEVA, *Zavodsk. Lab.*, 11 (1945) 254.

p-Hydroxyphenylarsonic acid

A. CLAASSEN, *Rec. Trav. Chim.*, 61 (1942) 299.

n-Propylarsonic acid

F. W. ARNOLD, JR., AND G. C. CHANDLEE, *J. Am. Chem. Soc.*, 57 (1935) 8 and 591.

In steels

H. H. GEIST AND G. C. CHANDLEE, *Ind. Eng. Chem., Anal. Ed.*, 9 (1937) 169.

Arsenic acid

A. CLAASSEN AND J. VISSER, *Rec. Trav. Chim.*, 62 (1943) 172.
I. SARUDI, *Z. Anal. Chem.*, 131 (1950) 416.

Traces by means of p-dimethylaminoazophenylarsonic acid
Ti(IV) can be complexed by succinic acid.

F. S. GRIMALDI AND C. E. WHITE, *Anal. Chem.*, 25 (1953) 1886.

(iv) Electrolysis at a mercury cathode. Zr(IV) remains in solution.

(v) With cupferron. In dilute acid solutions, Zr(IV) precipitates together with Ti(IV), Fe(III), etc. and may thus be separated from Al(III), etc. In highly dilute acid solutions containing tartrate, cupferron precipitates Zr(IV) at the same time as Ti(IV). The compound formed can be extracted[1].

cf. G. E. F. LUNDELL AND H. B. KNOWLES, *J. Am. Chem. Soc.*, 42 (1920) 1439.
P. KLINGER AND O. SCHLIESSMANN, *Arch. Eisenhütt*, 7 (1933) 113.
(1) J. S. FRITZ, M. J. RICHARD AND A. S. BYSTROFF, *Anal. Chem.*, 29 (1957) 577.

(vi) With mandelic acid. This is probably the most selective method of sep-

aration. Zr(IV) precipitates in 2.5 *N* HCl. Down to 0.1 mg of Zr(IV) may in this way be separated from Ti(IV), Fe(III), V(V), Th(IV), Mo(VI), Cu(II), Sn(IV), Bi(III), Sb(III), W(VI), Al(III), Cd(II), Ce(III), Cr(III), Mn(II), Mg(II) and Ni(II).

When smaller amounts of Zr(IV) are present, acidity of the solution should be reduced, and a period of 18 hours allowed for full precipitation[1].

Interfering ions. Large amounts of sulphates interfere. Nb(V) interferes.

REAGENTS

Mandelic acid, 15% in water.

Washing solution: 2% hydrochloric acid and 5% mandelic acid.

OPERATING PROCEDURE

15 ml of concentrated hydrochloric acid and then 50 ml of mandelic acid are added to 30 ml of solution containing from 50 to 200 mg of zirconium in *N* HCl. The mixture is left for 20 minutes at 85°, filtered and the precipitate is washed in the hot, with the solution indicated.

cf. R. E. OESPER AND J. J. KLINGENBERG, *Anal. Chem.*, 21 (1949) 1509.
(1) E. C. MILLS AND S. E. HERMON, *Analyst*, 78 (1953) 256.
J. H. HILLS AND M. J. MILES, *Anal. Chem.*, 31 (1959) 252.

In steels
G. GAVIOLI AND E. TRALDI, *Met. Ital.*, 42 (1950) 179.

In titanium
M. CODELL, G. NORWITZ AND J. J. MIKULA, *Anal. Chem.*, 27 (1955) 1382.
J. H. HILL AND M. J. MILES, *Anal. Chem.*, 31 (1959) 252.
R. A. PAPUCCI AND J. J. KLINGENBERG, *Anal. Chem.*, 30 (1958) 1062.

Derivatives of mandelic acid
R. BELCHER, A. SYKES AND J. C. TATLOW, *Anal. Chim. Acta*, 10 (1954) 34.
C. L. BRICKER AND G. R. WATERBURY, *Anal. Chem.*, 29 (1957) 558.

In aluminium alloys
J. J. KLINGENBERG AND R. A. PAPUCCI, *Anal. Chem.*, 24 (1952) 1861.

(vii) Extraction with 2-thenoyltrifluoroacetone. The solvent is xylene. This is effective for separations from Al(III), Fe(III), the rare earth metals, Th(IV), and U(VI) in 6 *N* HCl.

cf. F. L. MOORE, *Anal. Chem.*, 28 (1956) 997.

(viii) Extraction with tributyl phosphate. From 10 *M* HNO_3.

cf. E. M. SCADDEN AND N. E. BALLON, *Anal. Chem.*, 25 (1953) 1602.

(ix) Ion-exchangers. In 12 *M* HCl, Zr(IV) is collected by an anionic resin and may thus be separated from *e.g.* Th(IV)[1].

Zirconium may also be collected by an anionic resin from sulphuric acid, and may thus be separated from Al(III), Fe(III), Ti(IV), and Th(IV)[2].

(1) G. B. LARRABEE AND R. P. GRAHAM, *Z. Anal. Chem.*, 156 (1957) 258.
(2) J. KORKISCH AND A. FARAG, *Z. Anal. Chem.*, 166 (1959) 81.

With a cationic resin in HCl, separation from Ti(IV), Fe(III), Al(III), Sn(IV), etc., is achieved. Zr(IV) is collected by the resin.

cf. F. W. E. STRELOW, *Anal. Chem.*, 31 (1959) 1974.

(x) Standard solution. A solution of Zr(IV) is standardized by the gravimetric estimation of ZrO_2.

$$Zr/ZrO_2 = 0.740.$$

(2) Colorimetry

(i) With pyrocatechol violet. A blue coloration is obtained. The pH should be 5.2.

For the same concentration of reagent, the absorption maximum varies from 555 mμ to 625 mμ, according to the content of Zr(IV). The estimation is performed at 590 mμ. Beer's law is not obeyed under these conditions, and a calibration curve is therefore required.

The method is not very accurate, but can be made very selective by working in the presence of EDTA.

Sensitivity: ε = 30,000 — 40,000 at 590 mμ.

Interfering ions. In the presence of EDTA, Sn(IV) interferes by precipitating. Th(IV) should be present in amounts less than 5 mg (red coloration), and Ti (IV) in amounts less than 1 mg. Cr(III) interferes at 1 mg (violet coloration with EDTA). Be(II) does not interfere, but an excess of F^- should be present.

SO_4^{2-} does not interfere.

In the absence of EDTA, the following interfere: Ti(IV), Fe(III), Al(III), Ce(III), and more than 10 ppm of Mo(VI) and Th(IV).

REAGENTS

Disodium ethylenediaminetetraacetate, 40 g per litre.
Sodium acetate, 2 *M* 272 g/l.
Pyrocatechol violet, 40 mg in 100 ml of water.

OPERATING PROCEDURE

The solution should preferably contain 50 to 150 μg of Zr(IV). 3 ml of EDTA are added, and the pH is adjusted to 5.2 (pH meter) by means of sodium acetate. 2 ml of the dye are added, and the volume is made up to 50 ml. To one part of the solution (about half), 200 to 300 mg of sodium fluoride are added, and the mixture is left for 30 minutes. Colorimetry is performed at 590 mμ, in comparison with the solution containing the fluoride as a reference solution.

cf. H. FLASCHKA AND M. Y. FARAH, *Z. Anal. Chem.*, 152 (1956) 401.
J. P. YOUNG AND J. C. WHITE, *Talanta*, 1 (1958) 263.

(ii) With Alizarin S. Of the coloured lakes, formed by various compounds with Zr(IV), the coloration of that formed with Alizarin S remains fairly stable over a period of time. In spite of this, the colorimetric measurements should be made at the end of a pre-determined time, and the concentrations of all reagents used should be defined, as well as the pH.

Accuracy and sensitivity. The method is not very accurate. $\varepsilon \sim 7{,}000$ at 520 mμ.

Interfering ions. All ions of the Al(III) group give analogous lakes, but in 0.02 *N* hydrochloric acid, such interference is only serious if large amounts are present: 4 mg of Al(III), Cu(II), and Fe(III); 40 mg of Th(IV); 18 mg of Zn(II).

Sulphates interfere.

It is generally necessary to separate Zr(IV).

REAGENTS

Hydrochloric acid, 0.10 *N*.

Alizarin S: 50 mg are dissolved in 100 ml of water.

OPERATING PROCEDURE

25 ml of solution, 0.10 *N* with respect to hydrochloric acid and containing 30 to 150 μg of Zr(IV), are mixed with 5 ml of the reagent. The solution is made up to 100 ml with 0.10 *N* hydrochloric acid, left for 15 minutes, and the colorimetric measurement is then performed at 510 mμ.

cf. H. FREUND AND W. F. HOLBROOK, *Anal. Chem.*, 30 (1958) 462.
O. GRÜBELI AND A. JACOB, *Helv. Chim. Acta*, 38 (1955) 1026.

In thorium
L. SILVERMAN AND D. W. HAWLEY, *Anal. Chem.*, 28 (1956) 806.

In rocks
H. DEGENHARDT, *Z. Anal. Chem.*, 153 (1956) 327.

In aluminium
E. C. MILLS AND S. E. HERMON, *Metallurgia*, 51 (1955) 157.

Precision colorimetry
D. L. MANNING AND J. C. WHITE, *Anal. Chem.*, 27 (1955) 1389.

(3) Other methods

(i) With chloranilic acid. Interfering ions: U(IV), Th(IV), Sn(IV), Ti(IV), Fe (III), and SO_4^{2-}.

$\varepsilon = 20{,}000$ at 330 mμ.

In steels

R. B. Hahn and J. L. Johnson, *Anal. Chem.*, 29 (1957) 902.

In Pu

C. E. Bricker and G. R. Waterbury, *Anal. Chem.*, 29 (1957) 558.

(ii) With quercetin. $\varepsilon \sim 30{,}000$ at 440 mμ.

F. S. Grimaldi and C. E. White, *Anal. Chem.*, 25 (1953) 1886.

(iii) Fluorimetry with pentahydroxyflavone. $\varepsilon \sim 1{,}000{,}000$.

R. A. Geiger and E. B. Sandell, *Anal. Chim. Acta*, 16 (1957) 346.

SUBJECT INDEX

Note: Bold face numbers refer to principal entries which may extend for several pages

PRINTED IN THE NETHERLANDS
BY N.V. BOEKDRUKKERIJ F.E. MAC DONALD, NIJMEGEN